D0025406

TREE-RING ANALYSIS

Tree-Ring Analysis

Biological, Methodological and Environmental Aspects

Edited by

R. Wimmer

and

R.E. Vetter

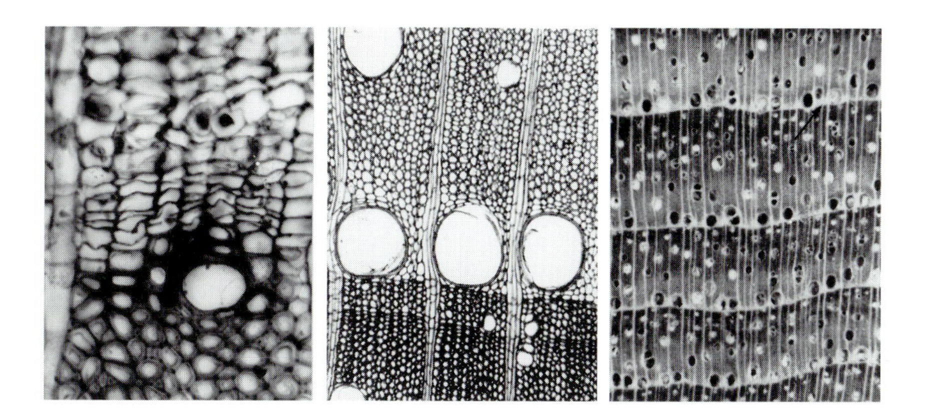

CABI *Publishing*

CABI *Publishing* is a division of CAB *International*

CABI Publishing
CAB International
Wallingford
Oxon OX10 8DE
UK

Tel: +44 (0)1491 832111
Fax: +44 (0)1491 833508
Email: cabi@cabi.org

CABI Publishing
10 E 40th Street
Suite 3203
New York, NY 10016
USA

Tel: +1 212 481 7018
Fax: +1 212 686 7993
Email: cabi-nao@cabi.org

A catalogue record for this book is available from the British Library, London, UK.

Library of Congress Cataloging-in-Publication Data
Tree ring analysis : biological, methodological, and
environmental aspects / edited by R. Wimmer and R.E. Vetter
p. cm.
Includes bibliographical references and index.
ISBN 0-85199-312-5 (alk. paper)
1. Dendrochronology. 2. Tree-rings.
3. Trees—Ecology. 4. Forest ecology. I. Wimmer, R.
(Rupert). II. Vetter, Roland E.
QK477.2.A6T73 1999
582.16—dc21 98-33433
CIP

ISBN 0 85199 312 5

Typeset in 10/12pt Photina by Columns Design Ltd, Reading
Printed and bound in the UK at the University Press, Cambridge

Contents

Contributors

Joseph A. Antos, Department of Biology, University of Victoria, Victoria, British Columbia, V8W 3N5, Canada.

Chris Beadle, CSIRO Forestry and Forest Products, GPO Box 252–12, Hobart, 7001, Australia.

Franco Biondi, Scripps Institution of Oceanography, University of California – San Diego, La Jolla, CA 92093, USA.

Narciso da Silva Cardoso, Technological Institute of Amazonia/UTAM, 69050–020 Manaus-AM, Brazil.

Alessandro Cescatti, Viote del Monte Bondone, I-38040, Trento, Italy.

Margarita M. Chernavskaya, Institute of Geography, Russian Academy of Sciences, Staromonetny str. 29, 109017 Moscow, Russia.

Geoffrey M. Downes, CRC Hardwood Fibre and Paper Science, CSIRO Forestry and Forest Products, GPO Box 252–12, Hobart, 7001, Australia.

Harold C. Fritts, DendroPower, 5703 North Lady Lane, Tucson, AZ 85704, USA.

Tomoyuki Fujii, Forestry and Forest Products Research Institute, Wood Anatomy Laboratory, Tsukuba Norin PO Box 16, Ibaraki 305, Japan.

Takeshi Fujiwara, Forestry and Forest Products Research Institute, Wood Quality Laboratory, Tsukuba Norin PO Box 16, Ibaraki 305, Japan.

Libuse Gandelova, Department of Wood Science, Faculty of Forestry and Wood Technology, University of Agriculture and Forestry, Zemedelska 3, CZ-613 00 Brno, Czech Republic.

William Gensler, Agricultural Electronics Corporation, PO Box 50291, Tucson, AZ 85703–1291, USA.

Michael Grabner, Institute of Botany, University of Agricultural Sciences, Gregor-Mendel-Strasse 33, A-1180 Vienna, Austria.

Henri D. Grissino-Mayer, Department of Physics, Astronomy, and Geosciences, Valdosta State University, Valdosta, GA 31698–0055, USA.

Richard J. Hebda, Department of Biology, University of Victoria, Victoria, V8W 3N5, and Royal British Columbia Museum, Victoria, British Columbia, V8W 9C2, Canada.

Elisabeth Höfs, Universität Köln, Institut für Ur- und Frühgeschichte, Labor für Dendrochronologie, Weyertal 125, D-50923 Köln, Germany.

Petr Horacek, Department of Wood Science, Faculty of Forestry and Wood Technology, University of Agriculture and Forestry, Zemedelska 3, CZ-613 00 Brno, Czech Republic.

Mei Hu, National Institute of Bioscience and Human-Technology, 1–1, Higashi, Tsukuba, Ibaraki 305–8566, Japan.

Mitra Khalessi, Universität Köln, Institut für Ur- und Frühgeschichte, Labor für Dendrochronologie, Weyertal 125, D-50923 Köln, Germany.

Marek Krąpiec, Tree-ring Laboratory, Department of Stratigraphy and Regional Geology, University of Mining and Metallurgy, Al. Mickiewicza 30, 30–059 Cracow, Poland.

Alexander N. Krenke, Institute of Geography, Russian Academy of Sciences, Staromonetny str. 29, 109017 Moscow, Russia.

John S. Kush, School of Forestry, Auburn University, AL 36849, USA.

Kuber Malla, Royal Botanical Garden, Kathmandu, Nepal.

Patrícia Póvoa de Mattos, CNPF-EMBRAPA, Caixa Postal 319, 83411–000 Colombo-PR, Brazil.

Ralph S. Meldahl, School of Forestry, Auburn University, AL 36849, USA.

Daryl Mummery, CSIRO Forestry and Forest Products, GPO Box 252–12, Hobart, 7001, Australia.

Graciela Ines Bolzon de Muniz, Universidade Federal do Paraná, Rua Bom Jesus 650, 80015–010 Curitiba-PR, Brazil.

Dmitri Ovtchinnikov, Institute of Forest, Russian Academy of Sciences, Siberian Branch, Academgorodok, Krasnoyarsk 660036, Russia.

Roberta Parish, Research Branch, BC Ministry of Forests, Victoria, British Columbia, V8W 9C2, Canada.

Won-Kyu Park, Department of Forest Products, College of Agriculture, Chungbuk National University, Cheongju 361–763, Republic of Korea.

Neil Pederson, Marshall Woods Consulting, 14 Roselawn, Highland Mills, NY 10930, and Tree-Ring Laboratory, Lamont-Doherty Earth Observatory, Columbia University, Palisades, NY 10964, USA.

Elena Piutti, Istituto per la Tecnologia del Legno – CNR, Via Biasi, 75, 38010 S. Michele a/A (TN), Italy.

Nathsuda Pumijumnong, Faculty of Environment and Resources Studies, Mahidol University, Nakhonpathom 73170, Thailand.

Andrey V. Pushin, Institute of Geography, Russian Academy of Sciences, Staromonetny str. 29, 109017 Moscow, Russia.

Andrew Tukau Salang, Timber Research and Technical Training Centre, Km 10, Airport Road, Kuching Sarawak, Malaysia.

Yutaka Sashida, School of Pharmacy, Tokyo University of Pharmacy and Life Science, 1432–1, Horinouchi, Hachioji, Tokyo 192–03, Japan.

Burghardt Schmidt, Universität Köln, Institut für Ur- und Frühgeschichte, Labor für Dendrochronologie, Weyertal 125, D-50923 Köln, Germany.

Rudi Arno Seitz, Universidade Federal do Paraná, Rua Bom Jesus 650, 80015–010 Curitiba-PR, Brazil.

Alexander Shashkin, Institute of Forest, Russian Academy of Sciences, Siberian Branch, Krasnoyarsk 660036, Russia.

Paul R. Sheppard, Laboratory of Tree-Ring Research, University of Arizona, Tucson, AZ 85721, USA.

Hiroko Shimomura, School of Pharmacy, Tokyo University of Pharmacy and Life Science, 1432–1, Horinouchi, Hachioji, Tokyo 192–03, Japan.

Jarmila Slezingerova, Department of Wood Science, Faculty of Forestry and Wood Technology, University of Agriculture and Forestry, Zemedelska 3, CZ-613 00 Brno, Czech Republic.

Giorgio Strumia, Institute of Botany, University of Agricultural Sciences, Gregor-Mendel-Strasse 33, A-1180 Vienna, Austria.

Xiao-Jun Tang, Institute of Chinese Material Medica, China Academy of Traditional Chinese Medicine, Beijing, 100700, China.

Hiroko Tokumoto, School of Pharmacy, Tokyo University of Pharmacy and Life Science, 1432–1, Horinouchi, Hachioji, Tokyo 192–03, Japan.

Mario Tomazello, Forestry Department – ESALQ/University of São Paulo, PO Box 09, 13418–900 Piracicaba-SP, Brazil.

J. Morgan Varner III, School of Forestry, Auburn University, AL 36849, USA.

Roland E. Vetter, INPA/CPPF, Caixa Postal 478, 69011–970 Manaus-AM, Brazil.

Tomasz Wazny, Academy of Fine Arts, Faculty of Conservation and Restoration, Wybrzeze Kosciuszkowskie 37, Pl-00379 Warsaw, Poland.

Rupert Wimmer, Institute of Botany, University of Agricultural Sciences, Gregor-Mendel-Strasse 33, A-1180 Vienna, Austria.

Dale Worledge, CSIRO Forestry and Forest Products, GPO Box 252–12, Hobart, 7001, Australia.

Ram R. Yadav, Birbal Sahni Institute of Palaeobotany, 53 University Road, Lucknow 226007, India.

Zhong-Zhen Zhao, Chinese Medicine Programme, Faculty of Science, Hong Kong Baptist University, Kowloon Tong, Hong Kong.

Preface

It is through science that we prove, but through intuition that we discover.

Henri Poincaré (1854–1912)

A tree ring is defined as 'a layer of cells produced in one year in the xylem or phloem'[1] and the basis of all tree-ring studies is a profound understanding of tree biology, particularly understanding the processes of tree-ring growth in the cambium. This book makes an attempt to address such basic aspects of tree life. The analysis of tree rings has been critically important in all kinds of environmental studies including forest decline, ecological prognosis on a large scale and climate trends for the past decades to millennia. In addition, tree-ring analysis provides important knowledge to other fields such as forest management or to the forest product industry. In 1989, during the All-Division 5 Conference in São Paulo, tree-ring analysis was recognized by the International Union of Forestry Research Organizations (IUFRO) as a project group. In 1996, the IUFRO Executive Board accepted a new divisional structure and tree-ring analysis is now a fully established research group within IUFRO Division 5 (Forest Products).

This book is the result of a fruitful All-Division 5 conference that took place in Pullman, Washington, USA, in the year 1997. Nineteen chapters have been compiled and split up into five sections. It starts in section A with basic aspects of tree-ring growth. A mechanistic cambium activity model and field measurements of cambial activity combined with environmental data, as well as new high-resolution techniques to monitor tree growth, are presented. In section B, information is extracted from anatomical features and structures that can be

[1] Kaennel, M. and Schweingruber, F.H. (1995) *Multilingual Glossary of Dendrochronology*. Haupt, Berne, 467 pp.

measured in rings of the xylem and the phloem. Section C is fully dedicated to the challenging field of tropical tree-ring analysis covering studies from periodicity of cambial growth in tropical trees, to various climatic influences. Section D contains tree-ring studies that are predominately historical and the final section E deals with applications in forest monitoring as well as climatic and competition effects on tree growth. The reader will find chapters that present new and challenging thoughts, while others are more conventional, focusing on lesser known species or regions. It is hoped that this book will appeal to a wide audience and also contribute to tree-ring research around the globe through strong linkages and interactions among scientists.

Rupert Wimmer and Roland Vetter,
Vienna and Manaus, April 1999

Acknowledgements

As editors, we wish to thank all of our 53 contributing authors coming from 17 different countries whose names and addresses are given in the list of contributors. We have called upon many colleagues to give us the benefit of their expertise by reviewing the manuscript drafts of each chapter. The chapters were generally reviewed by two specialists in the field, as well as by the editors. We are very grateful to all who assisted us in this task: P. Baas, J. Barnett, J. Bauch, K.M. Bhat, F. Biondi, U. Bräker, P. Cherubini, R. D'Arrigo, M. Dobbertin, D. Dujesiefken, J.L. Dupouey, D. Eckstein, H.C. Fritts, J. Fromm, H.D. Grissino-Mayer, H. Hasenauer, C. Krause, T. Levanic, H. Löppert, C. Sander, F.H. Schweingruber, P.R. Sheppard, K. Smith, H. Spiecker, H. Sterba, J. Tardif, M. Trockenbroth, G. Vieira, E. Wheeler and M. Worbes. The senior editor received support through the Austrian Programme for Advanced Research and Technology (APART) of the Austrian Academy of Sciences. We also thank IUFRO Division 5 for organizing the successful 1997 meeting in Pullman, Washington. Finally, both editors wish to acknowledge the support by the Publisher who patiently helped in completing the manuscripts.

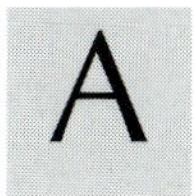

Analysing Radial Growth Processes

A Simulation Model of Conifer Ring Growth and Cell Structure

Harold C. Fritts, Alexander Shashkin and Geoffrey M. Downes

INTRODUCTION

Dendrochronological research throughout the last two decades has made many significant advances in statistical tools and models applied to analysing the characteristics in tree rings and their relationships to climate (Fritts, 1976; Cook and Kairiukstis, 1990). These approaches make assumptions which may or may not fit the systems to which they apply. Perceptive scientists can question whether these models provide the most accurate descriptions of tree growth behaviour over time and space. Long-term planners require assurance that the statistical estimates will be applicable to future conditions that may lie outside the range of past conditions on which model calibrations are based. As a result, there is growing interest in the questions of 'Exactly how do trees record environmental information in the structure of their growth rings in both temperate and tropical environments?' and 'Will these relationships be altered by the anticipated global changes?'

Industrial research on wood quality has focused primarily on measures of stem volume and form as indirect measures of wood quality, because until recently direct measurements of wood properties have not been cost effective (Downes *et al.*, 1997). However, as vertical integration between forest growers and users improves, forest growers will be under increasing pressure to supply wood of known or defined quality. New technology has been developed that makes measurements of wood quality rapid and cost effective (Evans *et al.*, 1995). This allows users of plantation forests to start exploring methods of growing quality wood suited to particular end uses. Several projects are currently under way which investigate relationships between growth and wood property development as a function of changing genotype or silviculture. The

(eds R. Wimmer and R.E. Vetter)

challenge for tree growers in this industry is to understand how wood properties, including cell size and wall thickness, vary as a function of environmental and genetic conditions that influence growth processes in trees (Downes *et al.*, 1997). While the current emphasis is still on volume production, market demands will increasingly drive growers to understand the relationships between growth rate, wood quality and product performance.

Many empirical and process-based models of forest productivity are being developed or are in use (e.g. McMurtrie *et al.*, 1994; Landsberg and Hingston, 1996; Battaglia and Sands, 1997, 1998). Whereas empirical models look for mathematical or statistical relationships between cause and effect, process-based models endeavour to understand the physiological relationships between cause and effect. Consequently the latter tend to be more complicated and, at this time, less accurate. However, empirical models cannot deal adequately with changing environmental and management conditions (Battaglia and Sands, 1998).

A reductionist approach to understanding living systems, although desirable, is not always possible. Investigating cambial dynamics is one of those areas where it is impossible. Process modelling provides an additional approach. TreeRing, version 3, is one such model that describes cambial activity mathematically. It essentially incorporates our current understanding in a way that can predict outputs which can be tested. If the model successfully predicts reality then we can assume our understanding is close to reality until we find those conditions where the model fails. This then gives us the opportunity to refine our knowledge base. There is, to our knowledge, only one other process-based model of wood property development (Deleuze and Houllier, 1998). As the focus in industry shifts from being driven by growth alone, to also being driven by wood properties, this type of model will need to be incorporated into existing commercially used models which at present predict only annual increments of tree growth. In addition, the process of model development itself can improve understanding of how environmental factors influence plant processes important to cell division and cell wall growth. Such clarification is not only important to the forest products industry but it can (i) provide dendrochronologists with a biological basis for constructing more accurate and reliable statistical tools for tree-ring analysis and (ii) lead to new and more powerful dendrochronological applications (Fritts and Shashkin, 1995).

This chapter describes the TreeRing model of xylem growth that operates at daily time steps. We use knowledge of basic physiological processes to characterize the tree growth system controlling cell division, enlargement and maturation at daily time steps. This model is unique in that biomass is converted to cell and wood properties which in turn can be compared to wood properties of trees growing in the forest environment. Thus the actual xylem cell anatomy in past tree rings is used to measure the real tree response to past environmental conditions and in turn to assess the precision of the model estimates. This is accomplished by (i) reading the anatomical measurements from a tree in the simulated environment during each run of the model, (ii) plotting them along with the model estimates, and (iii) calculating a number of statistics that provide an objective comparison.

Our approach is to start with a simple linear model and increase its complexity in a step-wise fashion. At first we described the growth of only a single radial file of cells and assume that the living crown, stem and root mass are constant through time (Vaganov, 1990; Fritts *et al.*, 1991). In a more recent version of the model (Fritts and Shashkin, 1995), the effects of prior environmental conditions were modelled using the dynamics of food storage, growth efficiency and differences in respiration of the various tissues in the tree (Fritts *et al.*, 1997). In the present version the algorithms have been refined. The model is adapted to operating in the southern as well as the northern hemisphere. Outputs from new imaging technology, SilviScan (Evans, 1994), are used with software written by Downes to scan and transform image and X-ray information to files of cell sizes and wall thicknesses for input to the model for verification. Heartwood, sapwood and other measurements derived from the wood samples are used to estimate a number of model parameters.

This version of the model does not yet consider crown and branch growth and changing photosynthetic capacity from one year to the next. These calculations are needed (i) to capture fully the autoregressive and low-frequency behaviour of ring widths through time and other acknowledged changes in growth (Cook and Kairiukstis, 1990) and (ii) to describe the changes in ring structure of the seedling as it grows into a sapling and then a closed canopy or reaches the understorey, codominant, mature and over-mature stages finally ending with death.

THE MODEL

The growth rate of a tree is determined by the amount of available carbon (glucose) produced by photosynthesis and by the rate that the carbon is transformed into biomass. More than 30% of the carbon produced by annual photosynthesis is used for the maintenance of living cells and tissues (maintenance respiration) and the rate of consumption largely depends upon temperature (Linder and Axelsson, 1982; Ryan and Waring, 1992; Ryan *et al.*, 1996). The amount of assimilated carbon used for wood growth depends upon the allocation of the carbon to foliage, roots and stem as well as the growth conditions at the time (Gholtz, 1980; Snowdon and Benson, 1992).

There are four interrelated blocks simulating (1) the microclimate and other input calculations, (2) water balance in the soil, the tree and the stomates, (3) photosynthesis along with photosynthate allocation and utilization by foliage, stem and roots and (4) cambial growth and development (Fig. 1.1). Cambial growth includes separate modules of cell division, enlargement and wall thickening. The model tracks the changing physiological conditions for each day and for each growing tracheid element passing through one of the growth modules until the cell's protoplasm disappears and the cell wall becomes a structural component of the wood (Fig. 1.2).

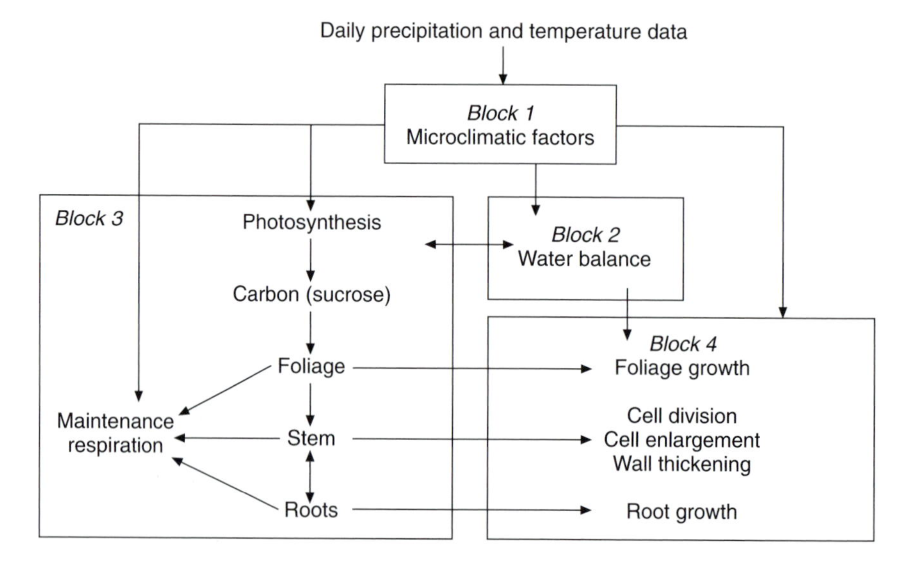

Fig. 1.1. Organizational diagram of the model showing the four basic blocks and the modules they contain.

The rates of processes in these four blocks are influenced by both *regulating* and *limiting* factors where: (i) regulating factors include growth promoters and inhibitors largely in the actively growing crown (Romberger, 1963; Larson, 1969, 1994; Catesson, 1994) and (ii) limiting factors include only external environmental or internal biophysical conditions that limit the rates of growing processes that occur. In TreeRing, these regulating and limiting factors are applied to particular trees growing in dendrochronologically limited sites. The values of both regulating and limiting factors may be different for the different processes and are normalized in that they range from a minimum value (most limiting condition) of 0 to a maximum value (unlimiting condition) of 1. The collective effect of these regulating and limiting factors is expressed as the product of those factors that are most likely to be limiting to a particular process. For example, the effect of limiting factors to growth (g) is described as a function $g_i(s,T,W)$ where T is temperature, s is substrate (sucrose) and w is water stress (a function of stomatal resistance). The regulating factors are expressed as control functions (C_t C_{tm}) related to the growth of new foliage and conditions in the crown. The controlling effects of C_t and C_{tm} begin when the rate of foliage growth declines below two specified critical rates.

Microclimate

The absolute humidity and incoming short-wave radiation are derived from climatological principles (Running *et al.*, 1987; Hungerford *et al.*, 1989; Nikolov

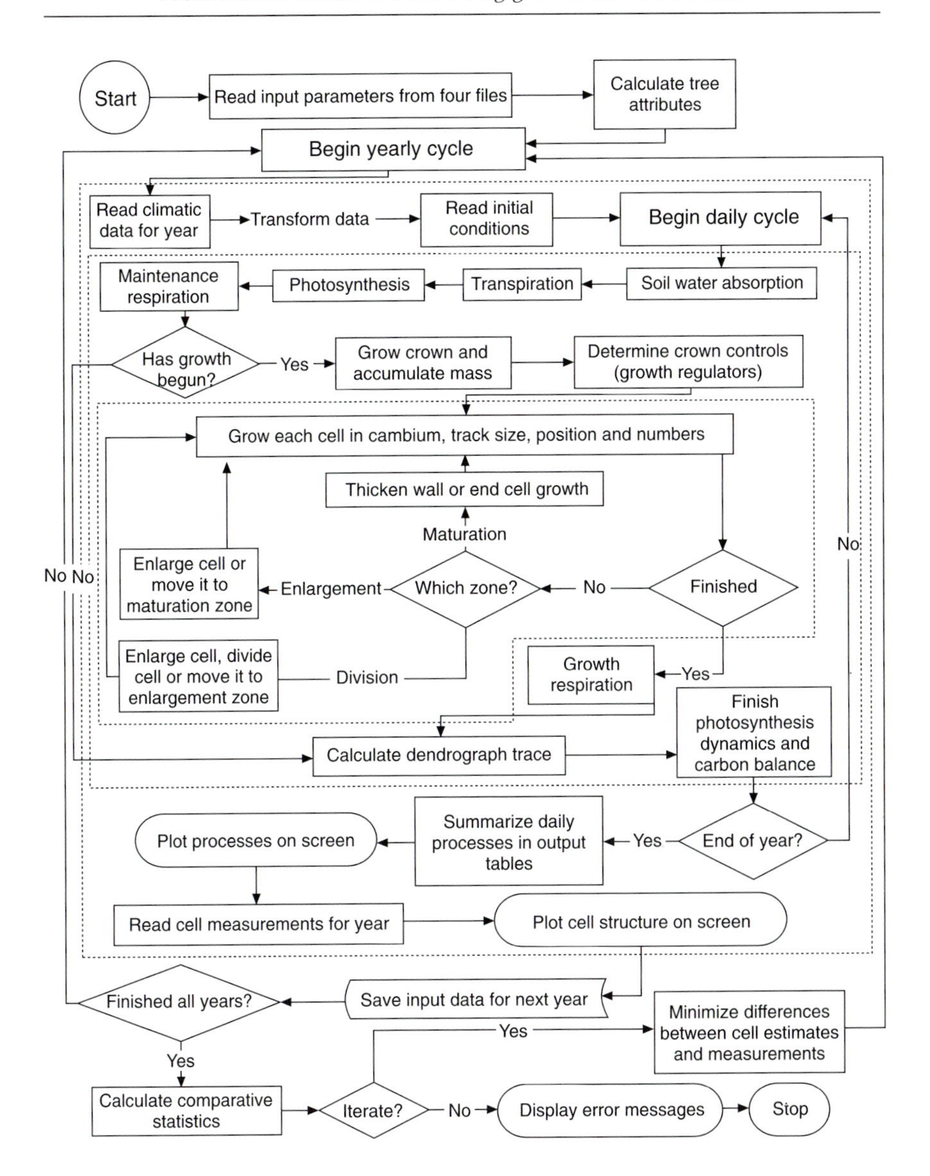

Fig. 1.2. A flow chart outlining the operation of the model. The outer dashed block encloses the annual cycle; the middle block encloses the daily cycle; the inner block encloses the cambial block that grows each cell according to limiting conditions, its size or position in the array and its zone of growth.

and Zeller, 1992), and average daytime temperature is calculated from the maximum and minimum temperatures (Running *et al.*, 1987). Vapour pressure deficit is calculated from the equation relating air temperature to saturation vapour pressure for use in potential evaporation calculations (Murray, 1967;

Running *et al.*, 1987). We use the daily minimum temperature to be equal to dew point as suggested by Running *et al.* (1987) and Hungerford *et al.* (1989). Finally, the estimated dew point for the site is combined with the estimated daily air temperature to calculate the average vapour pressure deficit.

Incoming solar radiation on a horizontal surface is estimated using the algorithm based on meteorological conditions developed by Nikolov and Zeller (1992). Daily potential radiation is a function of latitude, solar declination, sunrise/sunset hour angles and the day of the year (Klein, 1977). The amount of solar radiation received at the earth's surface is less than that at the top of the atmosphere because of scattering and absorption by components of the atmosphere. This attenuation of radiation is expressed as a linear function of undepleted solar radiation at the top of the atmosphere and the average cloudiness (Bristow and Campbell, 1984; Hungerford *et al.*, 1989). The model is applied to the growth of *Pinus ponderosa* at 478 m higher than the weather station at Chiricahua National Monument in southeastern Arizona. The temperatures are corrected for the elevation difference using a lapse rate of $-0.006°C\ m^{-1}$, but no correction was used for elevational differences in precipitation.

In addition to the microclimatic estimations, simple allometric relationships are used to transform readily available information from the study trees into (i) leaf mass, (ii) mass of the living tissues in the stem, (iii) mass of living roots and (iv) the number of cambial initials in the stem. These estimates of living tissues are in turn used to calculate maintenance respiration, growth dynamics of the foliage and roots, photosynthesis, transpiration and absorption of water from the soil by roots. The number of cambial initials determines the initial radial growth mass in the stem. The calculations start with information on the tree height, diameter breast height, the radial size of the dead bark, living phloem and sapwood that is used to calculate the cross-sectional areas for (i) the tree stem at breast height, (ii) bark and dead phloem, (iii) live phloem, (iv) sapwood, (v) heartwood, (vi) an average ring, and (vii) mass and area of leaves (Grier and Waring, 1974; Gholz, 1980; Callaway *et al.*, 1994). Mass of the sapwood and mass of the ring are calculated when the cell size and wall thickness measurements are obtained using density of the wood measured by SilviScan (Evans, 1994). Root mass is calculated from the root/shoot ratio (Kramer and Boyer, 1995; Ryan *et al.*, 1996).

Water Balance

The processes governing water movement through the soil, into the roots, through the plant and into the atmosphere are highly interrelated (Penman and Schofield, 1951; Zahner, 1955, 1968; Noble, 1974; Kramer and Kozlowski, 1979; Gates, 1980; Zimmermann, 1983; Boyer, 1985; Johnson *et al.*, 1991; Ellsworth and Reich, 1992). Transpiration is the dominant factor because evaporation of water produces the water potential gradient in the plant that drives the water movement. It controls the rate of absorption and produces diurnal water deficits in the leaves as well as throughout the entire plant. This influences the water status of guard cells on the leaf surface, and the stomates change in size as they regulate the flux of carbon dioxide into leaves and the rate of photosynthesis in the plant. Water deficits can also affect enlargement of cells in the cambium and cell division can be reduced or even stopped. Stomatal resistance is the most important control of transpiration rate and the tree water balance (Jarvis *et al.*, 1966; Cowan, 1977; Farquhar and Raschke, 1978; Kramer and Kozlowski, 1979; Gates, 1980; Shulz and Hall, 1982; Luo and Strain, 1992; Kramer and Boyer, 1995). Stomatal resistance can be envisioned as an inverse and linear function of water potential if it is less than some threshold value (Slatyer, 1967; Gates, 1980; Running and Coughlan, 1988).

The potential transpiration by leaves is given by

$$tr^{p} = \frac{\delta\rho}{R_{min}} \quad (1.1)$$

where tr^{P} is potential transpiration [kg H_2O m^{-2} s^{-1}], $\delta\rho$ is the water vapour density deficit [kg m^{-3}], R_{min} is the minimum resistance of leaves to water diffusion [m s^{-1}] which is parameter p_4. (All parameters (p_i) and options are included in the model input files.)

The water vapour deficit is determined from equation 1.2

$$\delta\rho = \rho_a \cdot 0.622 \cdot \frac{e_s - e}{P_{atm}} \quad (1.2)$$

where ρ_a is air density [kg m^{-3}], P_{atm} is the atmospheric pressure [mbar], e is the water vapour pressure at temperature T, e_s is the saturated water vapour pressure at the same temperature calculated as follows:

$$e(T) = 6.1078 \cdot e^{\frac{17.269T}{237.3+T}}, \; e_s = e(T_{min}) \quad (1.3)$$

The total potential transpiration from the crown is

$$Tr^{p} = tr^{p} \cdot D \cdot \sigma \cdot M_l \quad (1.4)$$

where D is the day length[s], σ is the coefficient for transforming foliage mass into surface area [m^2 kg^{-1}], M_l is the mass of foliage [kg].

The potential rate of water absorption by the roots is described as

$$W_{max} = q \cdot f(\theta) \cdot m_r \quad (1.5)$$

where W_{max} is the potential rate of water absorption by roots from a unit of soil volume [kg m^{-3} day^{-1}], q is the 'activity' of roots [kg H_2O kg^{-1} day^{-1}], p_{18}, $f(\theta)$ is a normalized function describing dependence of water uptake on soil water content θ [v/v]. This function takes a trapezoidal form (similar to the temperature-dependence curve (equation 1.12) with parameters θ_f – field capacity, p_{13}, θ_{min} – wilting point, p_9, $\theta_1 - \theta_2$ – the range of the optimal soil moisture, p_{10} and p_{11}, θ_{max} – saturation when soil oxygen is absent and water uptake cannot take place, p_{12}.

If the root system has mass M_r, the tree occupies a soil volume, v_s, with a surface area, A_s, p_{15}, and a depth, h, p_{14}, ($v_s = A_s h$) the potential water absorption by roots of the tree will be

$$W_{max} = q \cdot f(\theta) \cdot M_r \cdot v_s = q \cdot f(\theta) \cdot M_r \qquad (1.6)$$

The tree water balance is calculated depending upon the potential rates of absorption – W, loss of water due to transpiration – Tr, and resistance of the leaves – R and is described as $W = Tr$, where $Tr = \min(Tr^p, W_{max})$ and $R = R_{min} \cdot \max(1, Tr^p/W_{max})$.

If photosynthesis is not limited by CO_2, resistance is controlled by the rate of photosynthesis and is increased to value $R^c > R$. In that case

$$W = Tr = \min(Tr^p R_{min}/R^c, W_{min}) \quad R = R_{min} \cdot \max(1, Tr^p/W_{max} R_{min}/R^c) \qquad (1.7)$$

The content of water in the soil of volume $v_s = A_s h$ is calculated each day as

$$\frac{d\Theta}{dt} = A_s \cdot \min(P(t)a_1, a_3) - W - a_2\Theta$$
$$\theta = \frac{\Theta}{v_s} \qquad , \theta_w \le \theta \le \theta_f \qquad (1.8)$$

where Θ is the water content in soil volume v_s [kg], θ is soil moisture [kg m^{-3}], $P(t)$ is the precipitation [mm day^{-1}], $P(t)(1-a_1)$ is the interception of precipitation by the crown, $P(t)a_1$ is the precipitation that goes into the soil. If $P(t)a_1$ is greater than the value a_3, there will be surface runoff equal to $P(t)a_1 - a_3$. Additional loss of water occurs when the soil water content exceeds field capacity, $\theta_f \cdot a_2\Theta$ is the rate of drainage of water from the soil.

Photosynthesis

The control of daily photosynthetic rates by external and internal factors is a complicated relationship. The chemical reactions of photosynthesis depend upon the availability of light and carbon dioxide, while the rates of these reactions are temperature dependent. Leaf temperature is determined by the energy budget of the leaf, which relates radiation, air temperature, wind speed and humidity to the temperature of the leaf (Gates, 1980). When stomates close, carbon dioxide becomes limiting under conditions of water stress increasing the resistance to diffusion of carbon dioxide into the leaf. Different factors may limit photosynthesis at different times during the day. In the morning, increasing light enables photosynthesis to begin, the stomates open, and rising temperatures accompanied by increasing light intensity cause photosynthetic rates to increase. During midday, whenever water loss is greater than water absorption, stomatal apertures may decrease and temporary closure may take place, especially when osmotic concentrations of solutes in the guard cells are low. The degree of stomate opening and the rate of photosynthesis are highly correlated because the aperture controls the CO_2 diffusion rate (Wong *et al.*, 1979; Farquhar and Sharkey, 1982; Campbell *et al.*, 1988; Bowes, 1991; Harley *et al.*, 1992).

It is important for the model to distinguish seasonal variations in the photosynthetic rates in the tree crown caused by leaf development, metabolic activity and changing environmental conditions. In this version of the model, the rate of daily net photosynthesis depends upon light (the average daily incoming radiation for the latitude of the site), average daytime temperature of the air and carbon dioxide inside the leaves which is dependent upon the condition of the stomates. There are several approaches to modelling whole-leaf photosynthesis (e.g. Chartier and Prioul, 1976; Hall, 1979; Gates, 1980; Farquhar and von Caemmerer, 1982; Running and Coughlan, 1988; Buwalda, 1991; Running and Gower, 1991). The rate of photosynthesis is described in the model by the following system of equations

$$p = \frac{C_a - C_i}{R^c} \qquad p = P_{max} f_c(C_i) f_T(T) f_I(I) \tag{1.9}$$

where C_a is the concentration of CO_2 in the air [mM m^{-3}], p_{25}, C_i is the concentration of CO_2 inside the leaf [mM m^{-3}], R^c is the resistance of leaves to diffusion of CO_2 [s m^{-1}], P is the photosynthetic rate [mM CO_2 m^{-2} s^{-1}], f_c, f_T and f_I are normalized functions of dependence of photosynthesis on C_i, temperature T, incoming irradiation I, P_{max} is p_{56}. $R^c = \lambda R$ where λ is the transformation coefficient from water resistance, R, to CO_2 diffusion resistance R^c.

The dependence of photosynthesis on C_i is

$$f_c(C_i) = \begin{cases} 0, & C_i<a \\ (C_i - a)/(b - a), & a<C_i<b \\ 1, & C_i>b \end{cases} \tag{1.10}$$

where a and b are constants p_{26} and p_{27}.

With an average light flux, the function for I ([J m^{-2} s^{-1}] corresponding to photosynthetically active radiation), f_I, is a Michaelis–Menten type curve

$$f_I = I/(I + I^*) \tag{1.11}$$

where I^* is the Michaelis–Menten constant p_2. The temperature dependency of photosynthesis is

$$f_T = \begin{cases} 0, & T<T_{min} \\ (T - T_{min})/(T_1 - T_{min}), & T_{min} \le T \le T_1 \\ 1, & T_1 \le T \le T_2 \\ (T_{max} - T)/(T_{max} - T_2), & T_2 \le T_{max} \end{cases} \tag{1.12}$$

where T_{min}, T_1, T_2, and T_{max} are constants p_{21} to p_{24}.

From equations (1.9) and (1.10) the rate of photosynthesis is

$$p = P_m(T,I)\frac{C_a - a}{b - a + P_m(T,I)R^c}, \quad \text{if } P_m R^c \geq C_a - b$$
$$p = P_m,(T,I) \qquad \text{if } P_m R^c \leq C_a - b \tag{1.13}$$

where P_m is $P_{max} f_T(T) f_I(I)$. For conditions when CO_2 inside the leaf is not limiting the rate of photosynthesis, the leaf resistance increases according to the equation

$$R^c = (C_a - b)/p_m(T, I)/\lambda, \ (R^c \leq R_{max}) \tag{1.14}$$

Daily photosynthesis for the entire crown is

$$P = p \cdot D \cdot \sigma \cdot M_l \tag{1.15}$$

where D is the day length [s], σ is the coefficient for transforming foliage mass into surface area [$m^2 \ kg^{-1}$], M_l is the mass of foliage [kg].

Growth

Leaf Growth

The leaf is an organ of limited growth and limited life span. The increase in dry mass of leaves is a highly complex series of biochemical events, and numerous attempts have been made to model it (Thornley *et al.*, 1981; Dale, 1988). The growth and morphogenesis of the leaf primordium and the unfolding of the associated lamina are intimately associated with both cell enlargement and cell division. While the volume increase of their primordia includes some cell size increase, it largely reflects the increase in the number of cells, and this number ultimately determines the size of leaves even though cell sizes can vary greatly (see Dale, 1988).

The foliage produced during the current year has mass M_{l0} (t), the 1-year-old leaves have mass M_{l1} (t), etc. The mass of meristematic cells is proportional to the potential mass of new foliage M_{l0}^*, $M_{lm} = \eta l \ M_{l0}^*$. The dynamic of foliage growth is

$$\frac{dM_{l0}}{dt} = \mu_l M_{lm} - \lambda_{l0} M_{l0}$$
$$\frac{dM_{li}}{dt} = -\lambda_{li} M_{li} \tag{1.16}$$

where λ_{li} is the rate of leaves lost at age i, μ_l is the rate of new foliage growth by the foliage meristem. The photosynthetic active foliage is $M_l = \Sigma M_{li}$. The rate of leaf growth is

$$\mu_l = \mu_{l0}(1 - M_{l0}/M_{l0}^*)F_l(sl,T,W) \tag{1.16a}$$

where M_{l0}^* is the potential mass of new foliage, F_l is a normalized function relating the growth rate to limiting conditions of substrate concentration, temperature and water balance.

Root Growth

The root system is treated as a carbon sink using substrate during the growing process as it produces new mass (surface area) enabling water uptake from the soil. Water uptake is a function of the mass of living cells in the fine roots while growth of these roots is a function of the root meristematic tissue. The growth of roots is simulated in the model as

$$\frac{dM_r}{dt} = \mu_r M_{rm} - \lambda_r M_r \quad (1.17)$$

where μ_r is the rate of fine root production by the root meristem which varies as a function of limiting factors, $M_{rm} = \eta r\ M_r$ is the mass of root meristem, $\mu_r = \mu_{r0} F_r(s_r, T, W)$ and λ_r is the rate of mortality of roots.

For this current version of the modelled dynamic of root and foliage growth, we suggest that the mass of root, foliage, M_l, and potential mass of new foliage, M_{l0}^*, can be considered constant and equal to M_r.

Stem Growth

The term stem includes all parts of the tree in which the growth is secondary growth in the cambium. This includes branches, the main stem and coarse roots. The living cells in the stem include various cellular types such as ray cells, parenchyma, living phloem cells and the population of cambial initials. The average life span of the initial cell is many years and all cells in the radial file are clones of this one initial cell (Iqbal and Ghouse, 1990). It is assumed that the ratio of cambial initials to all other living cells is constant (the number or mass of cambial initials is constant and proportional to the mass of all living cells in the stem, M_s).

The differentiation of mature tracheid elements is modelled for one radial file which is assumed to be characteristic for all radial files in the tree. Growth in the radial file involves division of the cambial initials and xylem mother cells in the cambial zone, followed by enlargement and wall thickening in zones of enlargement and maturation. As a cell grows, sugar is converted to cell wall material. This consumption of sugar is included as one of the carbon sinks along with the growth and respiration of all other living cells of the tree.

The production of new xylem cells is related in part to the mass of cambial initial cells. The number of cells, n_i, in the radial file includes the initial cell, the xylem mother cells – n_c, the elongating cells – n_e and the maturating cells – n_m.

Each cell in the radial file is characterized by values of several parameters. The controls to these are input parameters Cmb_i in file CAMBINI:

j – is the position of each cell in the radial file. The initial cell is position 1 and those derived from it are numbered by their actual position in the file at the current time

x_j – is the cell size in the radial direction. The tangential cell size is assumed constant and is entered as a parameter of the model
w_j – is the cell wall area in cross-section

Cell size is the most important characteristic in determining the behaviour of each cell. While in the cambium zone, each cell increases in size until it reaches a maximum size and divides (moves through the cell cycle). The resulting daughter cells are half the maximum cell size after each cell division. The innermost cells in the cambial zone lose their ability to divide and enter the enlargement phase where the cell size continues to increase but at a diminishing rate. When the rate of size increase reaches a critical low value, the cell loses its ability to enlarge and enters the zone of maturation where cell wall thickening and wall synthesis continue until the cell dies.

Cell Division

Unless the cambium is dormant, all cambial cells pass through the phases of the cell cycle: G1, S, G2 and M. The size of each cell when it enters a particular phase is D_G1, Ds, D_G2 and Dm (see Cmb_{11}–Cmb_{14}. These parameters refer to those in the input file CMB61.PAR). However, the rate of division varies as a function of the growth rate of cambium cells (Cmb_{10}) while the cell is in phase G1. Growth rates (duration) while the cell passes through all other phases are constant and specified by input parameters. The duration of the full cell cycle will be:

$$\frac{D_{G1} - D_m/2}{V_c} + \frac{D_m - D_{G1}}{Cmb_{10}}$$

However, V_c is not constant but changes with position in the cellular file, varying limiting conditions and controlling factors which change through time. As a result, duration of the cell cycle cannot be expressed as a simple equation.

The rate of growth is a function of cell position, j, and is related to limiting conditions and numbers of other growing and maturing cells as follows:

$$V_c = [Cmb_7 - (10 - j) \cdot Cmb_6 \cdot \frac{x}{x + x^*}] \cdot Cmb_2 \cdot F_c(s_s, T, W) \cdot \left[\frac{1}{Cmb_{37}} + \left(j - \frac{1}{Cmb_{34}}\right) \cdot (C_t)\right] \quad (1.18)$$

where Cmb_7 is the rate of enlargement in position j (Fig. 1.3); Cmb_6 is the slope regulating the increasing growth rate of cambial cells across the cambial zone; x is the sum of the cell size in the division, enlargement and maturation zone; x^* (Cmb_{18}) is the ring width of growing cells [μm] when the slope of the division rate is half the maximum value; Cmb_2 is a scalar of growth rate; Cmb_{37} is a

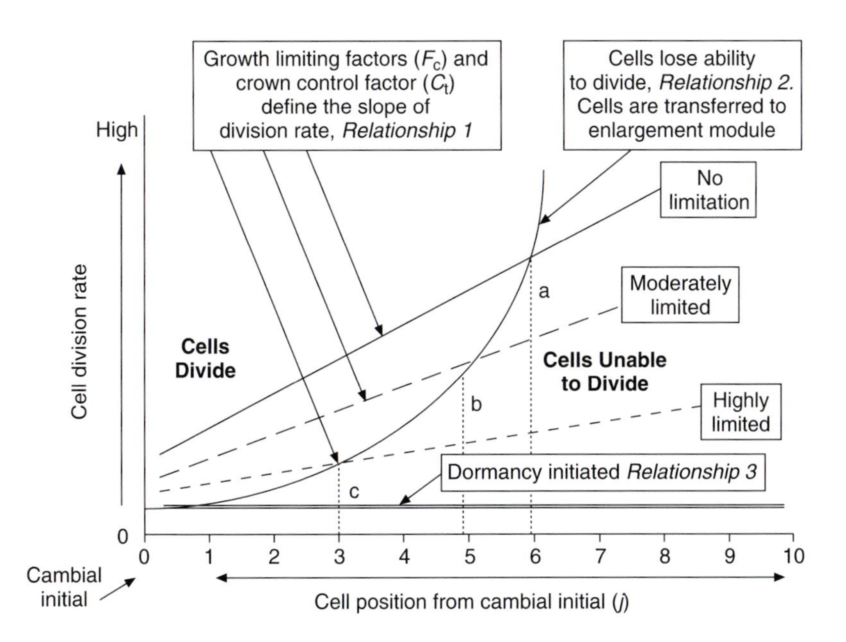

Fig. 1.3. A diagram of the functions that control cell division. Rate of division is a linear function of cell position, *j*, starting with the cambial initial at position $j = 0$. The higher the order of the position, the greater the division rate. However, the change in rate with position is great under conditions that are unlimiting to division. The change in rate diminishes with increasing limitation represented by the change from solid to dashed sloping lines. A second relationship corresponding to the curved line controls the ability of the cell to divide (cells move to the enlargement module when their position exceeds the area inside the curved line). A third relationship, a horizontal line, determines the rate at which the cambium becomes dormant. In this example, under no limitation, six cells are dividing (see dotted line a) and the rate of division is high; under moderate limitation, five cells are dividing (see dotted line b) and the rate of division is moderate; under highly limited conditions, only three cells are dividing (see dotted line c) and the rate of division is low.

scalar for the rate of enlargement; and Cmb_{34} is the sensitivity of enlargement to C_t ranging from 1 to 10.

The relationship between the rate of growth Cmb_7 and x^* of the differentiating cells is used to control the rate of division of the cambial initial and mother cells at the beginning of the growing season through a feedback loop; F_c is a normalized function ranging from 0 to 1; and C_t is the control from the crown.

The cell leaves the zone of cell division and enters into the enlargement stage if $V_c < V_{min}$ (Relationship 2, Fig. 1.3) and the cell is in the G1 phase of the cellular cycle. If the cell is in any other phase, the cell continues to divide until it completes the division process. The function V_{min} (Relationship 2) is determined as

$$V_{min} = Cmb_2 \cdot \exp(Cmb_8 (j + Cmb_9)) = Cmb_2 \cdot \exp(Cmb_8 \cdot Cmb_9) \cdot \exp(Cmb_8 \cdot j) \quad (1.19)$$

where Cmb_2, Cmb_8 and Cmb_9 are coefficients of the equation.

Cell division will stop reversibly (the cambium becomes dormant) for all cells in phase G1 if $V_c < V_{cr}$ (Relationship 3, Fig. 1.3).

Cell Enlargement

The exact mechanisms that control cell enlargement and its variation over time are not known (Catesson, 1994; Fukuda, 1994; Haigler, 1994; Savidge, 1994). However, there are many observations of cell size variations and cell enlargement including experimental data from trees growing under controlled conditions (Zahner, 1968; Larson, 1969, 1994; Wodzicki, 1971; Creber and Chaloner, 1984). In addition, there is considerable information available concerning the molecular control of enlargement and the influence of growth regulators (Fukuda, 1994; Savidge, 1994).

For trees growing on dendrochronologically selected stress sites, temperature, water supply and perhaps substrate availability may be the most important limiting conditions (Zahner, 1968; Larson, 1969, 1994; Wodzicki, 1971). Based upon studies in roots, the rate of enlargement can be divided into accelerating and decelerating phases (Pritchard, 1994). Limiting conditions have the least effect on the first and the greatest effect on the second phase of growth when it is decelerating. Thus, we assume in our model that limiting conditions have the greatest effect on growth rate and potential cell size in the later stages of cell enlargement.

There has been much discussion over the years as to what factors control the transition from large earlywood cells to smaller latewood cells (see Larson, 1969). Environmental factors may hasten or delay this transition, but it does not appear to be directly connected to environmental changes but rather to the physiology of the growth process itself, related to new xylem tissue and hormone production primarily in the crown (see Larson, 1969; Creber and Chaloner, 1984). Since the specific action of these growth regulators is not well known, we simply have modelled the transition from earlywood to latewood as a function of growth and development of foliage in the crown.

As in the cell division module, the enlargement phase of growth is described as a function of limiting factors $F_e(s,T,W)$ and a controlling factor C_t. The relative importances of these factors are not necessarily the same as those in cell division. In addition the rate of enlargement in the model, V_e, decreases with increasing cell size.

This relationship is a linear function (Relationship 1, Fig. 1.4).

The dynamic for enlargement of cells is described as

$$\frac{dx}{dt} = V_e$$
$$\frac{dV_e}{dt} = -k_e \cdot V_e, \quad xt = 0 = x0 \quad \text{and} \quad V_e \quad t = 0 = V_{e0} \tag{1.20a}$$

where x is radial cell size [μm]; V_e is the growth rate [μm day^{-1}]; V_{e0} is the initial growth rate calculated as

$$Cmb_{22} \cdot Cmb_{37} \cdot \left(1 - \frac{x_0}{Cmb_{21}}\right)$$

where Cmb_{22} is the maximum growth rate of the cambial cell in position 10; Cmb_{37} is a scalar mentioned above; x_0 is the initial radial size when the cell entered the zone of enlargement; and Cmb_{21} is the maximum radial size. k_e is:

$$k_e = k_{e\,min}\,[b_{35} + (1 - Cmb_{35}) \cdot C_t] \cdot [y + (1 - y) \cdot F_e] \tag{1.20b}$$

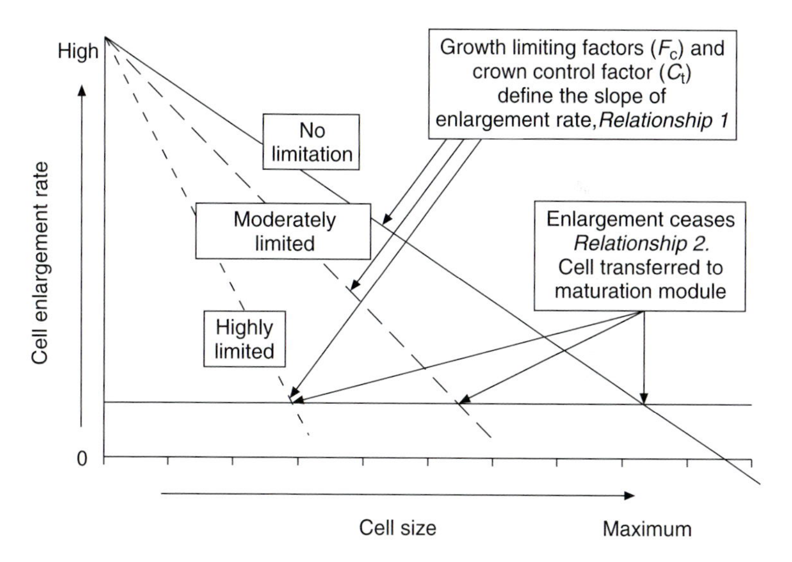

Fig. 1.4. A diagram of the functions that control cell enlargement. The rate of enlargement is an inverse linear function of cell size. A cell enters this zone with maximum enlargement rate and the slope of *Relationship 1* increases with increasingly limited conditions. As conditions become more limiting, the cell remains in the zone for a shorter duration of time and the cell is smaller. If the crown growth rate is rapid, the slope of the line is gradual and large cells are formed. When the enlargement rate reaches a critical rate, *Relationship 2,* the cell loses its capacity for enlargement and moves to the zone of maturation.

where Cmb_{35} ranges from 1 to 10 expressing the sensitivity of the enlargement process to C_t, F_e is a normalized function ranging from 0 to 1, and

$$y = \frac{k_{e\max}}{k_{e\min}},\ k_{e\min} \le k_e \le k_{e\max}$$

where:

$$K_{e\min} = \frac{Cmb_{22} - Cmb_{24}}{Cmb_{21}}$$

$$K_{e\max} = \frac{Cmb_{22} - Cmb_{24}}{Cmb_{23}}$$

where Cmb_{21} and Cmb_{22} are described above, Cmb_{24} is V_{ecr}, the critical rate when the enlarging cell enters the zone of maturation (Relationship 2, Fig. 1.4) and Cmb_{23} is the minimum mature radial cell size. The cell stops enlarging and enters the zone of maturation when $V_e \le V_{ecr}$.

Cell Maturation

The dynamics of cell wall synthesis and wall thickening are similar to those controlling cell enlargement (Fig. 1.5). The rate of cell wall synthesis is

$$\frac{dw_j}{dt} = V_m$$

$$\frac{dV_m}{dt} = -k_m \cdot V_m, \quad \begin{array}{l} w_j(t=0) = 2.0y_o(x_j + x_t - 2y_o) \\ V_m(t=0) = V_{mo} \end{array} \tag{1.20c}$$

where w_j is the cell wall area of the jth cell [μm^2], V_m is the rate of cell wall synthesis [$\mu m^2\ day^{-1}$], y_o is initial cell wall thickness [μm], x_t is the tangential cell size [μm], V_{mo} is the initial rate of wall synthesis, and k_m is a function through which the process is controlled. This function is

$$k_m = k_{m\min}\,[(Cmb_{36} - 1) \cdot (C_{tm} + 1)] \cdot [y + (1 - y) \cdot F_m] \tag{1.20d}$$

where Cmb_{36} (V_{mcr}) is the critical level of the rate of wall thickening (Relationship 2, Fig. 1.6) when the synthesis irreversibly stops; C_{tm} is the crown control

$$y = \frac{k_{m\max}}{k_{m\min}}$$

where:

$$K_{m\min} = \frac{Cmb_{31} - Cmb_{27}}{P_{area} - P_0} \quad \text{and} \quad K_{m\max} = \frac{Cmb_{31} - Cmb_{27}}{P_{\min} - P_0}$$

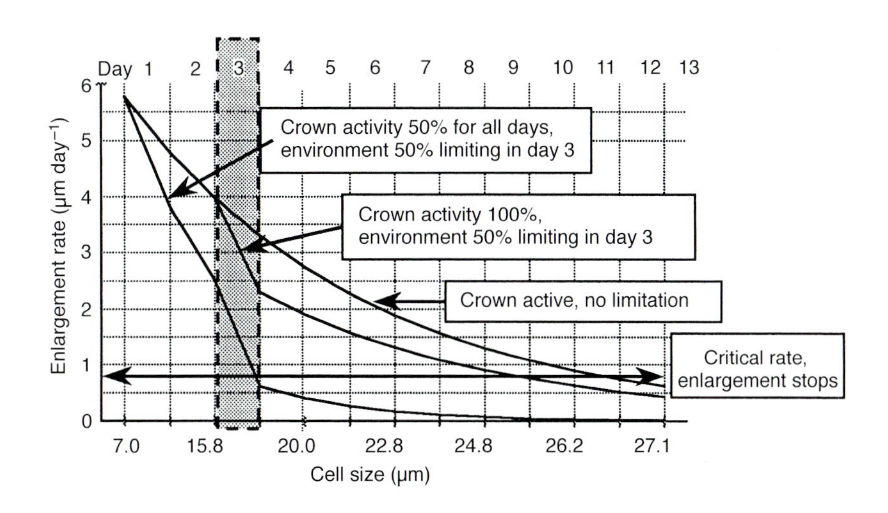

Fig. 1.5. Enlargement rates decrease each day as a function of cell age, crown activity and environmental conditions. Three scenarios are shown. The upper curve shows the gradual decrease in cell enlargement rate with no limitation. The middle curve is the same as the upper curve for days 1 and 2. In day 3 crown activity is unlimiting but environment is 50% limiting and the rate declines rapidly. After day 3 there is no limitation and the rate declines gradually as in the upper curve but at lower rates. The lower curve represents the rate changes when crown activity is 50% limiting for all days and environment becomes 50% limiting only in day 3. Dashed lines surround rate changes for day 3 when environment becomes 50% limiting for only that one day.

P_0 is the initial area of the cell wall:

$$P_0 = 2 \cdot Cmb_{29} \cdot (x + Cmb_{28} - 2 \cdot Cmb_{29})$$

where Cmb_{29} is the initial wall thickness, x is the radial cell size, and Cmb_{28} is the tangential cell size.

$$P_{min} = 2 \cdot Cmb_{33} \cdot (x + Cmb_{28} - 2 \cdot Cmb_{33})$$

where Cmb_{33} is the minimum wall thickness.

An estimate of the potential area of the wall, P_{area}, is

$$P_{area} = 2 \cdot wall_j \cdot (x + Cmb_{28} - 2 \cdot wall_j)$$

However, P_{area} cannot be larger than Cmb_{32}, the maximum wall thickness for the particular species and site measured from wood samples, but must maintain a minimum lumen area, Cmb_{30}. When wall thickening stops, the cell dies and becomes a part of the fully differentiated ring.

If one plots the relationships between cell wall area or wall thicknesses for each cell in a large number of rings expressed as a function of radial cell size (see figure 5 in Fritts *et al.* (1997)) and keeps track of the relative sequence of each

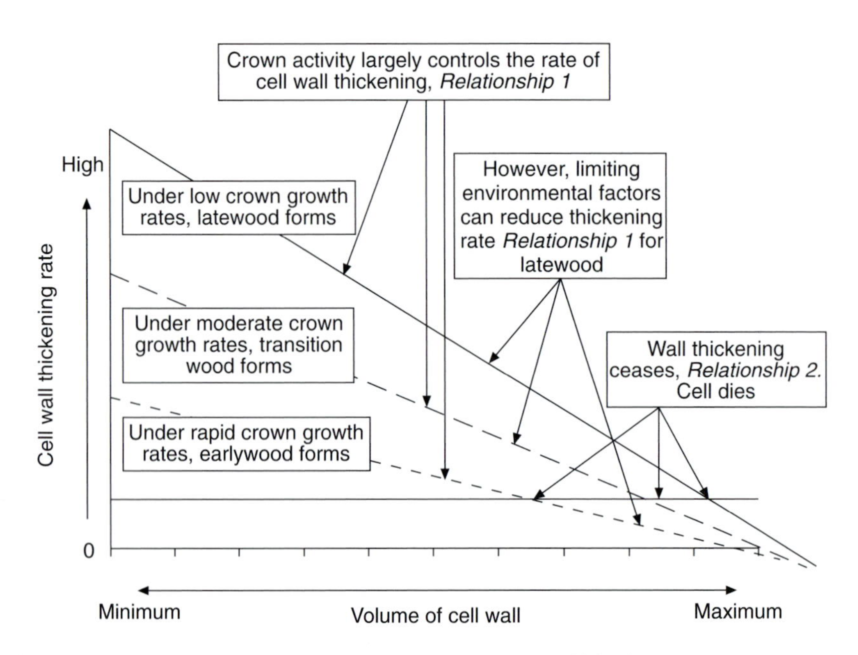

Fig. 1.6. A diagram of the functions that control cell wall thickening. Crown activity largely controls wall thickening, *Relationship 1*. If crown growth is rapid, the rate of wall thickening is low and declines with increasing volume of the cell wall and earlywood is produced. When the rate of crown growth diminishes past a threshold condition, the initial wall thickening rate increases reaching maximum rate under very low crown growth rates. Only under conditions of latewood formation can environmental factors reduce the rate of wall thickening. When the wall thickening rate is lower than a critical rate, thickening stops, the protoplasm disappears and the tracheid cell dies.

cell within its particular ring, two clusters of points are visible. The first cluster of points appears on the lower right of the plot and corresponds to cells from the earlywood portion of the ring. The second cluster of points is in the upper left, corresponding to cells from the latewood in the rings. In the plots of wall thickness as a function of radial cell size, the points for earlywood cluster around a horizontal line, indicating that wall thickness is completely independent of cell size during earlywood formation. The scatter in points around this horizontal line is most likely the error in measurement, random variations from one cell to the next, or possibly some limiting conditions at that time. In the plot of cell wall area, the points fall along a slope indicating that this measurement is dependent on cell size (also see Wimmer, 1991). However, the slope for earlywood is markedly less than the slope of the relationship for latewood cells.

Since wall area is more related to cell size than wall thickness, we use wall area and express it as a linear function of cell size where sl_{min} is the minimum slope of the relationship for earlywood and sl_{max} is the maximum slope expressing the relationship with cell size for the latest part of the latewood. The possible effects of limiting conditions on the wall thickness of the earlywood appears so small that we include no limitation until latewood begins to form. The slope is steeper for latewood, sl_{max}, and is reduced if environmental conditions become limiting. This implies that limiting environmental conditions can reduce the rate of cell wall thickness only during the phase of latewood development. It is well known that limiting conditions such as temperature or moisture are related to the colour, appearance and density of the latewood (Denne and Dodd, 1981).

The Substrate Dynamics in the Tree

Different approaches have been used to deal with the tree carbon balance (Hesketh and Jones, 1975; Gifford and Evans, 1981). Some have used allometric relationships (Monsi and Marata, 1970; Patefield and Austin, 1971; Sheehy *et al.*, 1979); some use a 'nutritional control' that is based on the 'functional equilibrium' between organs (de Wit *et al.*, 1970); and some use transport gradients and resistances (Thornley, 1976) to model allocations within the tree. We assume in the model that substances are translocated rapidly within a day by diffusion processes (Canny, 1973) and that the maintenance respiration is proportional to the weight of the living biomass following Thornley (1970).

The substrate, sucrose, is produced by photosynthesis and is consumed by maintenance respiration and growth of new tissues throughout the tree. Maintenance respiration is dependent upon temperature and, of course, substrate availability (Kramer and Kozlowski, 1979; Amthor, 1989; Buwalda, 1991).

$$Rm_i = \beta_{0i} \cdot \frac{s_i}{s_i + s_i^*} \cdot e^{\beta_{1i}T} \cdot M_i, \quad i = 1, s, r \tag{1.21}$$

where Rm_i is the rate of maintenance respiration [mM CO_2 day^{-1}], β_{0i} is the substrate uptake for maintenance respiration, parameters rmt_i where i varies from 1 to 3 in units of mass per day [mM CO_2 kg^{-1} day^{-1}], s_i^* is a Michaelis–Menten constant, p_{34}, β_{1i} is the temperature constant.

We follow Chung and Barnes (1977) and Williams *et al.* (1987) and assume that the respiration required for growth is proportional to the growth increment of new foliage (leaves), roots and stem

$GR_l = \alpha_{gl} \cdot \mu_l \cdot M_{lm}$ growth respiration (GR) by foliage

$GR_r = \alpha_{gr} \cdot \mu_r \cdot M_{rm}$ by roots,

$$GR_c = \alpha_{gc} \cdot \rho_w \cdot N \cdot \left(\sum_{n_c} V_c + 2 \cdot x_t \cdot d \right) \quad \text{by cambium,}$$
$$GR_e = \alpha_{ge} \cdot \rho_w \cdot N \cdot \sum_{n_e} V_e \quad \text{by enlarging cells,} \qquad (1.22)$$
$$GR_m = \alpha_{gm} \cdot \rho_w \cdot N \cdot \sum_{n_m} V_m \quad \text{by maturing cells,}$$

where α_{gl} and α_{gr} are quantities of substrate assimilation per units of growth per day [mM CO_2 kg^{-1}], α_{gc}, α_{ge} and α_{gm} are the substrate assimilation per units of cell wall mass [mM CO_2 kg^{-1}], ρ_w is the specific gravity of cell wall [kg μm^{-3}]. N is the number of cellular files or cambium initials per tree. If 1 kg of wood tissue is equal to 0.375 kg °C or $3.125 \cdot 10^4$ mM CO_2 the coefficient of efficiency is

$$\frac{3.125 \cdot 10^4}{\alpha_{gi}}$$

where α_{gi} is the value from GR_{ci}, i varies from 1 to 5. The dynamic of substrate content in the leaves is

$$\frac{dS_l}{dt} = P - Rm_l - GR_l - \xi_{ls}(s_l - s_s) \qquad (1.23)$$

in the stem is

$$\frac{dS_s}{dt} = \xi_{ls}(s_l - s_s) - Rm_s - GR_c - GR_e - GR_m - \xi_{sr}(s_s - s_r) \qquad (1.24)$$

and in the root is

$$\frac{dS_r}{dt} = \xi_{sr}(s_s - s_r) - Rm_r - GR_r \qquad (1.25)$$

where S_l is substrate content [mM CO_2], s_l is substrate concentration [mM CO_2 kg^{-1}], ξ_{ls} and ξ_{sr} are coefficients of diffusion for the substrate from leaves to the stem and from the stem to the roots [kg t^{-1}].

The Growth Control Functions

The growth rates in the crown and in the three cambial zones depend upon temperature, substrate concentration and tree water balance. The same functions are used in all four cases but coefficients vary depending upon the relative importance of each condition on the particular process. The relationship is

$$F_i = f_i(T) \frac{s_i}{s_i + s_i^*} \min\left(1, \frac{R_i^*}{R_w}\right), \; i = l, c, e, m, r \qquad (1.26)$$

where s_i is substrate concentration, s_i^* is a Michaelis–Menten constant, gc_1–gc_5, R_w is the resistance determined from water balance, R_i^* are the coefficients p_{46}–p_{50}, $f_i(T)$ is the temperature function (see equation 1.12).

The values C_t describing 'hormonal control' for the cambium and zone of enlargement and C_{tm} for maturation are determined in the model through the dynamic of new foliage growth

$$C_t = MIN\left(\frac{z}{z_t^*}, 1\right) \tag{1.27a}$$

$$C_{tm} = MIN\left(\frac{z}{z_m^*}, 1\right) \tag{1.27b}$$

where z_t^* and z_m^* are p_{28} and p_{29} and z is $\frac{\mu_1}{\mu_{10}}$ (see equation 1.16a). z is an average for an interval of 1 or more days, p_7. If p_7 is positive the average is unweighted, if negative the average is weighted around the central value. Therefore, at any moment t, C_t is the weighted or unweighted average for the period $(t - \Delta 1, t)$. C_t and C_{tm} are compared each day to several critical values. If $C_t \leqslant p_{28}$, latewood begins to form; if $C_t \leqslant p_8$ crown growth stops. If $C_{tm} \leqslant p_{29}$ wall thickening begins. The net effect of these controls are portrayed on the computer screen along with the temperature and precipitation input and the daily rates of selected plant processes (Fig. 1.7).

Other Calculations

Starting with 1 March, growth can begin following the day when the sum of temperature (in °C) over some interval of time (a parameter of the model) becomes greater than an assumed constant threshold value (Lindsay and Newman, 1956; Landsberg, 1974; Valentine, 1983; Cannell and Smith, 1986; Hanninen, 1990) and resistance is lower than an assumed threshold value. These were added as dendrochronological observations made by Christopher Baisan (personal communication) suggesting that at times of severe water stress, growth may not begin until soil moisture is replenished during the summer rainy season. This is consistent with the observations of Shepherd (1964) in *Pinus radiata*.

The calculation of all growth processes in the model begins when time $t > t_{beg}$. Leaf growth ends when the growth rate is less than a specified threshold, and the cambial division stops when leaf growth stops. The enlargement and maturation phases continue until their rates fall below the critical values specified in the parameter file.

A number of standard statistics are calculated to help evaluate the similarities and differences between the simulations and actual measurements of ring width, cell numbers, cell sizes, wall thicknesses and ring indices. The

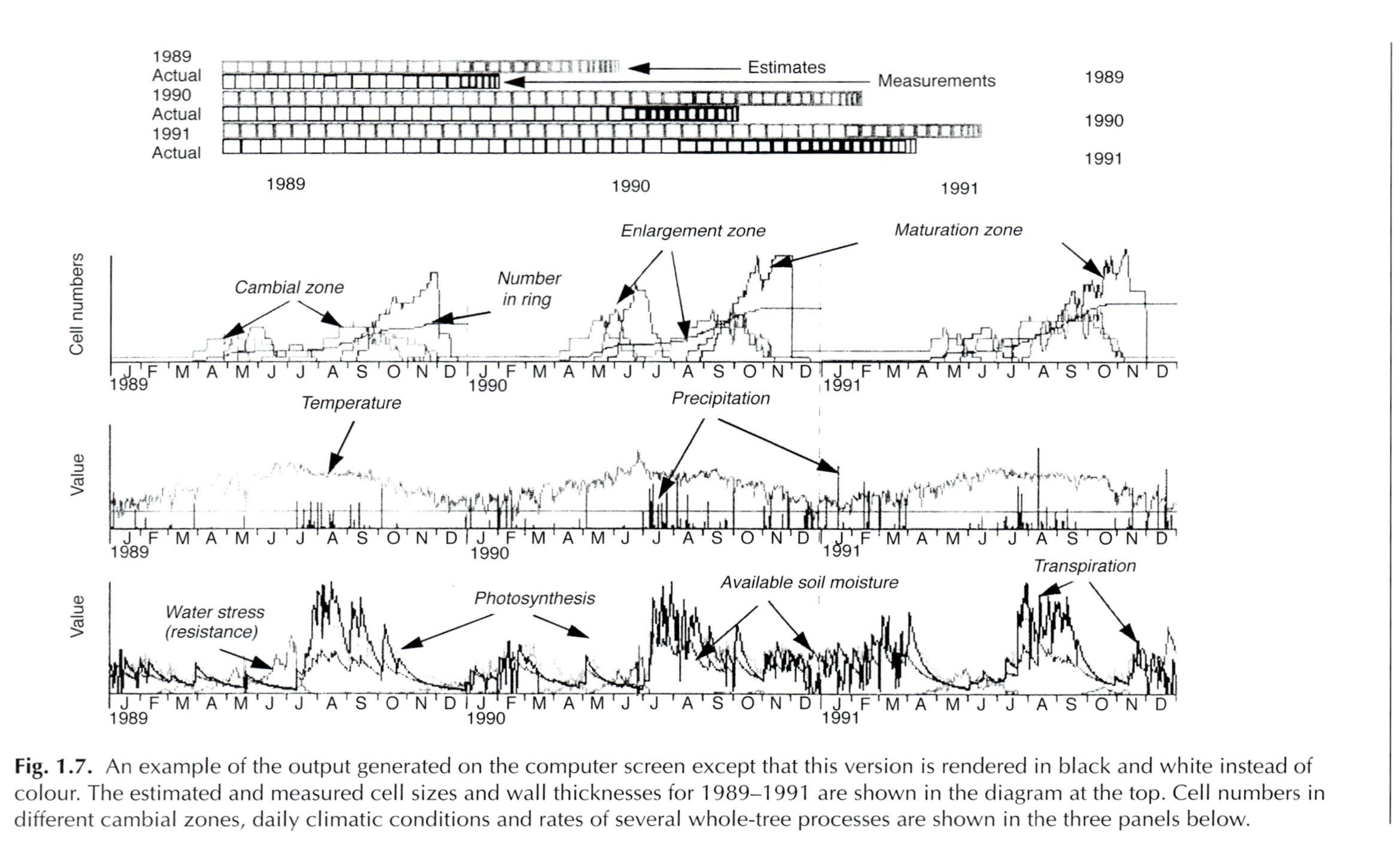

Fig. 1.7. An example of the output generated on the computer screen except that this version is rendered in black and white instead of colour. The estimated and measured cell sizes and wall thicknesses for 1989–1991 are shown in the diagram at the top. Cell numbers in different cambial zones, daily climatic conditions and rates of several whole-tree processes are shown in the three panels below.

means, standard deviations, correlation and regression coefficients, standard error and other statistics are calculated following Ezekiel (1941), but they may be found in most textbooks on basic statistics.

An iteration procedure is designed to optimize any selected parameter or contiguous group of parameters keeping all of the other parameters constant. The residual variance is minimized between the simulated and actual measurements of one or more attributes: (i) ring widths, (ii) cell numbers, (iii) cell sizes, (iv) wall thicknesses for a single core, and (v) ring-width index. Different statistics of the residual can be minimized which allow one to use different optimization strategies. These include (i) absolute value of the mean, (ii) standard deviation + absolute value of the mean, (iii) 1 − correlation squared or (iv) standard deviation + absolute value of the mean residual divided by standard deviation of the measurements. The mean minimizes only the mean residual so that the actual and simulated means can be brought into line; the standard deviation minimizes the variance of the residual; the correlation gives equal weight to all controls but does not consider differences in the mean values; and the last statistic attempts to scale the variance in terms of the variance in the independent variable for that attribute so that the statistics of the residuals weigh each variable equally. The values of these statistics can be used singly or the sum of the statistics for attributes (i) to (iv) or (i) to (v) can be calculated to control the iteration. The statistics of the ring-width index, if run for a different interval of time, will be calculated for a different time period than those of the other attributes and these values can be summed to control the iteration process.

DISCUSSION AND CONCLUSIONS

TreeRing 3 is only a beginning in the immense task of developing a simulation model for cambial activity (Larson, 1994) in both conifer and angiosperm species. However, the statistics provide an objective confirmation that the simulations can explain over 50% of the variance in ring width, cell numbers and cell sizes. Wall thickness estimates are less precise but explain more than 40% of the variance. Results from analysis of variance and correlation statistics of between and within tree-ring-width variance (Fritts, 1976) confirm that measurements from a single core, analogous to a single radial file of model estimates, are not likely to account for a great deal more than 50% of the larger scale stand variance. Therefore, we conclude that the basic form of the model produces reasonable and reliable estimates of ring width, cell number, cell size and wall thickness measurements obtained from the modelled trees in spite of the many model inadequacies. The model is ready to be refined and developed further depending upon the particular application to be made.

Most of the mathematical equations will be changed in future versions and many of the assigned parameters are little more than intelligent guesses based upon a reading of the literature. Many values must be changed when the model is applied to different sites and species. There is a great need for more field investigation and controlled experiments not only to measure the parameters but also to evaluate how well the model captures the dynamics of growth as it progresses over the season and from one year to the next.

The initial impetus behind the model was the need for a more fundamental approach to some important dendrochronological questions. It assumes that the cell information is placed in the correct year in which the ring grew using either dendrochronologically dated materials or plantation-grown trees with known history. The model implies that some variations in climatic factors have affected the growing trees by limiting growth-controlling processes (Fritts, 1976). We developed the model using only daily maximum and minimum temperature and precipitation so that the model could be applied to any area around the world where standard daily climatic records were available and the rings could be precisely dated.

Certainly some processes such as photosynthesis as described by Farquhar and von Caemmerer (1982) can be more precisely described using hourly or minute by minute records of environmental conditions. Very few environmental records with hourly detail are available so that hourly time steps would have to be generated from daily information. It can be easily argued that such detail contributes other errors and, since the model generates cell features that integrate and average the effects of environmental conditions over a number of days, any precision lost is trivial.

A stand or higher level model is certainly possible, but its development will depend upon our capacity to generalize the specific ring and cell features from actual wood measurements. Validation using dendrochronological indices based upon replicated samples is a first step and is already included in version 3.

The focus has been on the development of tree-ring structure. Our attempts to describe photosynthesis, water relations and assimilation are practical and pragmatic means to obtain reasonable daily inputs for the growth models. We solicit help and would welcome the opportunity to incorporate any more sophisticated and better researched models of these processes that can generate daily values for a single tree. Linking TreeRing or elements of it to existing or future models of forest productivity should not only improve the estimates of tree-ring structure, but the host model could benefit by using the estimates of cell structure as rapid and objective validation of the overall model precision.

Applications of the Model

Some general applications of the model are described in the Introduction. A few of the many possible applications are described here. The specific form of each application will, of course, depend on the interests and creativity of the investigators who use it.

Applications of the model to studies of wood and fibre quality are readily apparent. The model could suggest climatic conditions and site characteristics that would yield wood and fibre of a desired quality. The pulp and paper industry in Australia is interested in using the model for future planning in purchasing the appropriate land for growing pulp and paper of a desired quality. Similar interest has been expressed in North America and South Africa.

The rings of trees are a rich source of environmental information on past climate variations, stand conditions and events, as well as biochemical changes occurring in past years. The model output can suggest specific ring and wood features and biochemistry that are related to environmental and climatic conditions and have specific environmental interpretations. The model can serve as an experimental tool for examining a variety of hypotheses using the past record in tree-ring structure.

Geochemical changes in wood can be modelled, studied and compared to field measurements. Current investigations on monthly variations in stable isotopes throughout the year could be assisted by an expanded model with compartments for different fractionations of carbon, hydrogen or oxygen along with coefficients of fractionation for different processes in the carbon cycle.

The ring structure serves as a history of tree productivity and the model can be incorporated in forest productivity models and modelled ring structure compared to measurements from trees growing in the modelled forests. The statistical comparison between modelled and measured wood quality could be a powerful validation tool for such ecological models.

Considerable interest centres around a model for hardwood species. The existing model can serve as a basis by making changes in a step-by-step manner, first constructing simple representations of hardwood anatomy, then expanding them to include more detail.

Last, but not least, the model equations clarify our understanding of the processes affecting cambial activity, cell structure and radial growth. The equations are explicit, can be criticized and replaced with improvements. Thus it is a formal medium allowing concrete discussion and debate for further developments and refinements. The TreeRing model represents a first step in an effort to create a simulation model of cambial growth and the development of wood structure.

SUMMARY

TreeRing, version 3, is a process model of daily cambial activity simulating the ring width, cell number, cell size, wall thickness and ring-width index of growth variations in conifer tree rings. Input data are daily precipitation and temperature, actual cell measurements for trees growing in the simulated area and the annual ring-width index. Input parameters define mathematical constants and initial data for the equations describing the processes governing daily changes in the tree water balance, photosynthesis, carbon allocation, crown growth and cambial activity. Outputs from the whole-tree processes along with limiting environmental conditions of temperature, water stress and available sucrose (carbon) govern the initiation of cambial activity, rates of cell division, cell enlargement, wall thickening and growth cessation. Derivatives of crown growth represent net growth regulator flux. The growth rate, cell size, wall thickness and state (dividing, enlarging, maturing, or dead) of each tracheid is calculated for each day of the year. Data are stored in tabular form and selected information is plotted on the screen as the simulation progresses through time. Actual measurements from a tree in the simulated forest are compared to the simulated annual ring widths, cell numbers, cell sizes, wall thicknesses and ring-width indices (a replicated time series portraying the relative ring-width response of many trees in the forest), and the statistics of comparison are calculated at the end of each simulation run. A variety of initial conditions and input measurements including altitude, tree height, sapwood and heartwood, root depth and spread, soil characteristics, stored sucrose, and needle retention are used to estimate crown surface, proportion of leaves, stems, and roots, living and growing masses, carbon allocation and water balance. An iteration subroutine can be used to cycle through simulations while varying one parameter at a time and minimizing the residual variance to allow fine-tuning of the model. Simulations account for 50% or more of the variance in actual cellular and ring measurements from a single tree. This percentage is near the upper limit calculated from dendrochronological evidence. Limitations and some applications of the current model are described.

REFERENCES

Amthor, J.S. (1989) *Respiration and Crop Productivity*. Springer-Verlag, New York.

Battaglia, M. and Sands, P. (1997) Modelling site productivity of *Eucalyptus globulus* in response to climatic and site factors. *Australian Journal of Plant Physiology* 24, 831–850.

Battaglia, M. and Sands, P. (1998) Process-based forest productivity models and their application in forest management. *Forest Ecology and Management* 102, 13–32.

Bowes, G. (1991) Growth at elevated CO_2: photosynthetic responses mediated through Rubisco. *Plant, Cell and Environment* 14, 795–806.

Boyer, J.S. (1985) Water transport. *Annual Review of Plant Physiology* 36, 473–516.

Bristow, K.L. and Campbell, G.S. (1984) On the relationship between incoming solar radiation and daily maximum and minimum temperature. *Agricultural and Forest Meteorology* 31, 159–166.

Buwalda, J.H. (1991) A mathematical model of carbon acquisition and utilization by kiwi fruit vines. *Ecological Modelling* 57, 43–64.

Callaway, R.M., Delucia, E.H. and Schlesinger, W.H. (1994) Biomass allocation of montane and desert ponderosa pine: an analog for response to climate change. *Ecology* 75, 1474–1481.

Campbell, W.J., Allen, L.H. and Bowes, G. (1988) Effect of CO_2 concentration on Rubisco activity, amount and photosynthesis in soybean leaves. *Plant Physiology* 88, 1310–1316.

Cannell, M.G.R. and Smith, R.I. (1986) Climatic warming, spring bud burst, and frost damage of trees. *Journal of Applied Ecology* 23, 177–191.

Canny, M.J. (1973) *Phloem Translocation.* Cambridge University Press, London.

Catesson, A. (1994) Cambial ultrastructure and biochemistry: changes in relation to vascular tissue differentiation and the seasonal cycle. *International Journal of Plant Sciences* 155, 251–261.

Chartier, P. and Prioul, J.L. (1976) The effect of irradiance, carbon dioxide and oxygen on the net photosynthetic rate of the leaf: a mechanistic model. *Photosynthetica* 10, 20–24.

Chung, H.H. and Barnes, R.L. (1977) Photosynthate allocation in *Pinus taeda*. I. Substrate requirements for synthesis of shoot biomass. *Canadian Journal of Forest Research* 7, 106–111.

Cook, E.R. and Kairiukstis, L.A. (eds) (1990) *Methods of Dendrochronology: Applications in the Environmental Sciences.* Kluwer Academic Publishers, Dordrecht.

Cowan, I.R. (1977) Stomatal behavior and environment. *Advances in Botanical Research* 4, 117–228.

Creber, G.T. and Chaloner, W.G. (1984) Influence of environmental factors on the wood structure of living and fossil trees. *Botanical Review* 50, 357–448.

Dale, J.E. (1988) Control of leaf expansion. *Annual Review of Plant Physiology and Plant Molecular Biology* 39, 267–295.

Deleuze, C. and Houllier, F. (1998) A simple process-based xylem growth model for describing wood microdensitometric profiles. *Journal of Theoretical Biology* 193, 99–113.

Denne, M.P. and Dodd, R.S. (1981) The environmental control of xylem differentiation. In: Barnett, J.R. (ed.) *Xylem Cell Development.* Castle House Publications, Kent, pp. 236–255.

de Wit, C.T., Brouwer, R. and Penning de Vries, F.W.T. (1970) The simulation of photosynthetic systems. In: Setlik, I. (ed.) *Prediction and Measurement of Photosynthetic Productivity. PUDOC,* Wageningen, pp. 47–70.

Downes, G.M., Hudson, I.L., Raymond, C.A., Michell, A.M., Schimleck, L.S., Evans, R. and Dean, G.H. (1997) *Sampling Plantation Eucalypts for Wood and Fibre Properties.* CSIRO Publishing, Melbourne, 132 pp.

Ellsworth, D.S. and Reich, P.B. (1992) Water relations and gas exchange of *Acer saccharum* seedlings in contrasting natural light and water regimes. *Tree Physiology* 10, 1–20.

Evans, R. (1994) Rapid measurement of the transverse dimensions of tracheids in radial wood sections from *Pinus radiata*. *Holzforschung* 48, 168–172.

Evans, R., Downes, G., Menz, D. and Stringer, S. (1995) Rapid measurement of variation in tracheid transverse dimensions in a radiata pine tree. *Appita* 48, 134–138.

Ezekiel, M. (1941) *Methods of Correlation Analysis.* John Wiley & Sons, New York, 531 pp.

Farquhar, G.D. and von Caemmerer, S. (1982) Modelling of photosynthetic response to environmental conditions. In: Lang, O.L., Nobel, P.S., Osmond, C.B. and Ziegler, H. (eds) *Physiological Plant Ecology. Encyclopedia of Plant Physiology (NS)*, Vol. 12B, Springer-Verlag, Berlin, pp. 549–587.

Farquhar, G.D. and Raschke, K. (1978) On the resistance to transpiration of the sites of evaporation within the leaf. *Plant Physiology* 61, 1000–1005.

Farquhar, G.D. and Sharkey, T.D. (1982) Stomatal conductance and photosynthesis. *Annual Review of Plant Physiology* 33, 317–345.

Fritts, H.C. (1976) *Tree Rings and Climate.* Academic Press, London. Reprinted in 1987 in Kairiukstis, L., Bednarz, Z. and Felikstik, E. (eds) *Methods of Dendrochronology* Vols II and III, International Institute for Applied Systems Analysis and the Polish Academy of Science, Warsaw.

Fritts, H.C. and Shashkin, A.V. (1995) Modeling tree-ring structure as related to temperature, precipitation, and day length. In: Lewis, T.E. (ed.) *Tree Rings as Indicators of Ecosystem Health.* CRC Press, Boca Raton, Chapter 2, pp. 17–57.

Fritts, H.C., Vaganov, E.A., Sviderskaya, I.V. and Shashkin, A.V. (1991) Climatic variation and tree-ring structure in conifers: Empirical and mechanistic models of tree-ring width, number of cells, cell size, cell-wall thickness and wood density. *Climate Research* 1, 97–116.

Fritts, H.C., Shashkin, A.V. and Downes, G.M. (1997) *Documentation and Manual for TreeRing 3.* Dendropower, Tucson, 38 pp.

Fukuda, H. (1994) Redifferentiation of single mesophyll cells into tracheary elements. *International Journal of Plant Sciences* 155, 262–271.

Gates, D.M. (1980) *Biophysical Ecology.* Springer-Verlag, New York.

Gholz, H.L. (1980) Structure and productivity of *Juniperus occidentalis* in central Oregon. *American Midland Naturalist* 103, 251–261.

Gifford, R.M. and Evans, L.T. (1981) Photosynthesis, carbon partitioning, and yield. *Annual Review of Plant Physiology* 32, 485–509.

Grier, C.C. and Waring, R.H. (1974) Conifer foliage mass related to sapwood area. *Forest Science* 20, 205–206.

Haigler, C.H. (1994) Commentary from signal transduction to biophysics: tracheary element differentiation as a model system. *International Journal of Plant Sciences* 155, 248–250.

Hall, A. (1979) A model of leaf photosynthesis and respiration for predicting carbon dioxide assimilation in different environments. *Oecologia* 143, 299–316.

Hanninen, H. (1990) Modeling dormancy release in trees from cool and temperate regions. In: Dixon, R.K., Meldahl, R.S., Ruark, G.A. and Warren, W.G. (eds) *Process Modeling of Forest Growth Responses to Environmental Stress.* Timber Press, Portland, pp. 159–165.

Harley, P.C., Thomas, R.B., Reynolds, J.F. and Strain, B.R. (1992) Modelling of cotton growth in elevated CO_2. *Plant, Cell and Environment* 15, 271–282.

Hesketh, J.D. and Jones, J.W. (1975) Some comments on computer simulators for plant growth. *Ecological Modelling* 2, 235–247.

Hungerford, R.D., Nemani, R.R., Running, S.W. and Coughlan, J.C. (1989) MTCLIM: A Mountain Microclimate Simulation Model. *Research Papers Int-414.* US Department of Agriculture, Forest Service, Intermountain Research Station, Ogden, UT.

Iqbal, M. and Ghouse, A.K.M. (1990) Cambial concept and organization. In: Iqbal, M. (ed.) *The Vascular Cambium.* Research Studies Press, Taunton, Somerset, pp. 1–36.

Jarvis, P.G., Rose, C.W. and Bogg, J.E. (1966) An experimental and theoretical comparison of viscose and diffusive resistance to gas flow through stomatous leaves. *Agricultural Meteorology* 4, 103–117.

Johnson, I.R., Melkonian, J.J., Thornley, J.H.M. and Rina, S.J. (1991) A model of water flow through plants incorporating shoot/root 'message' control of stomatal conductance. *Plant, Cell and Environment* 14, 531–544.

Klein, S.A. (1977) Calculation of monthly average insolation on tilted surfaces. *Solar Energy* 19, 325–329.

Kramer, P.J. and Boyer, J.S. (1995) *Water Relations of Plants and Soils.* Academic Press, San Diego, 495 pp.

Kramer, P.J. and Kozlowski, T.T. (1979) *Physiology of Woody Plants.* Academic Press, New York.

Landsberg, J.J. (1974) Apple fruit bud development and growth: analysis and an empirical model. *Annals of Botany* 38, 1013–1023.

Landsberg, J.J. and Hingston, F.J. (1996) Evaluating a simple radiation/dry matter conversion model using data from *Eucalyptus globulus* plantations in Western Australia. *Tree Physiology* 16, 801–808.

Larson, P.R. (1969) *Wood Formation and the Concept of Wood Quality.* Bulletin No. 74, Yale University, School of Forestry.

Larson, P.R. (1994) *The Vascular Cambium: Development and Structure.* Springer-Verlag, Berlin, 725 pp.

Linder, S. and Axelsson, B. (1982) Changes in carbon uptake and allocation patterns as a result of irrigation and fertilization of young *Pinus sylvestris* stand. In: Waring, R.H. (ed.) *Carbon Uptake and Allocation in Sub-Alpine Ecosystems as a Key Management.* Oregon State University, Corvallis, pp. 38–44.

Lindsay, A.A. and Newman, J.E. (1956) Uses of official weather data in spring time–temperature analysis of an Indiana phenological record. *Ecology* 37, 812–823.

Luo, Y.H. and Strain, B.R. (1992) Leaf water status in velvet leaf under long-term interaction of water stress, atmospheric humidity and carbon dioxide. *Journal of Plant Physiology* 139, 600–604.

McMurtrie, R.E., Gholz, H.L., Linder, S. and Gower, S.T. (1994) Climatic factors controlling productivity of pine stands, a model-based analysis. *Ecological Bulletins* 43, 173–188.

Monsi, M. and Marata, Y. (1970) Development of photosynthetic systems as influenced by distribution of dry matter. In: Setlik, I. (ed.) *Prediction and Measurement of Photosynthetic Productivity.* PUDOC, Wageningen, pp. 115–130.

Murray, F.W. (1967) On the computation of saturation vapor pressure. *Journal of Applied Meteorology* 6, 203–204.

Nikolov, N.T. and Zeller, K.F. (1992) A solar radiation algorithm for ecosystem dynamic models. *Ecological Modelling* 61, 149–168.

Noble, P.S. (1974) *Introduction to Biophysical Plant Physiology.* Freeman, San Francisco.

Patefield, W.M. and Austin, R.B. (1971) A model for the simulation of the growth of *Betula vulgaris. Annals of Botany* 35, 1227–1250.

Penman, H.L. and Schofield, R.K. (1951) Some physical aspects of assimilation and transpiration. *Symposium of the Society of Experimental Biology* 5, 115–129.

Pritchard, J. (1994) The control of cell expansion in roots. *New Phytologist* 127, 3–26.

Romberger, J.A. (1963) *Meristems, Growth, and Development in Woody Plants, an Analytical Review of Anatomical, Physiological, and Morphogenic Aspects.* Tech. Bull. No. 1293, US Department of Agriculture, Forest Service.

Running, S.W. and Coughlan, J.C. (1988) A general model of forest ecosystem processes for regional applications, I. Hydrological balance, canopy, gas exchange and primary production processes. *Ecological Modelling* 42, 125–154.

Running, S.W. and Gower, S.T. (1991) FOREST-BGC, a general model of forest ecosystem processes for regional applications, II. Dynamic carbon allocation and nitrogen budgets. *Tree Physiology* 9, 147–160.

Running, S.W., Ramakrishna, R.N. and Hungerford, R.D. (1987) Extrapolation of synoptic meteorological data in mountainous terrain and its use for simulating forest evapotranspiration and photosynthesis. *Canadian Journal of Forest Research* 17, 472–483.

Ryan, M.G. and Waring, R.H. (1992) Maintenance respiration and stand development in a subalpine lodgepole pine forest. *Ecology* 73, 2100–2108.

Ryan, M.G., Hubbard, R.M., Pongracic, D., Raison, R.J. and McMurtrie, R.E. (1996) Foliage, find-root, woody-tissue and stand respiration in *Pinus radiata* in relation to nitrogen status. *Tree Physiology* 16, 333–343.

Savidge, R.A. (1994) The tracheid-differentiation factor of conifer needles. *International Journal of Plant Sciences* 155, 272–290.

Sheehy, J.E., Cobby, J.M. and Ryle, G.J.A. (1979) The growth of perennial ryegrass: a model. *Annals of Botany* 43, 335–354.

Shepherd, K.R. (1964) Some observations on the effect of drought on the growth of *Pinus radiata* D. Don. *Australian Forest* 28, 7–13.

Shulz, E.D. and Hall, A.E. (1982) Stomatal responses, water loss and CO_2 assimilation rates of plants in contrasting environments. In: Pirson, A. and Zimmermann, M.H. (eds) *Encyclopedia of Plant Physiology* (NS) Vol. 12B, Springer-Verlag, Berlin, pp. 181–230.

Slatyer, R.D. (1967) *Plant–Water Relationships.* Academic Press, London.

Snowdon, P. and Benson, M.L. (1992) Effects of combinations of irrigation and fertilization on the growth and above-ground biomass production of *Pinus radiata. Forestry and Ecology Management* 52, 87–116.

Thornley, J.H.M. (1970) Respiration, growth and maintenance in plants. *Nature* 227, 304–305.

Thornley, J.H.M. (1976) Mathematical models in plant physiology: a quantitative approach to problems in plant and crop physiology. *Experimental Botany* 8, 152–171.

Thornley, J.H.M., Hurd, G.G. and Pooley, A. (1981) A model of growth of the fifth leaf of tomato. *Annals of Botany* 48, 327–340.

Vaganov, E.W. (1990) The tracheidogram method in tree-ring analysis and its application. In: Cook, E.R. and Kairiukstis, L.A. (eds) *Methods of Dendrochronology: Applications in the Environmental Sciences.* Kluwer Academic Publishers, Dordrecht, pp. 63–76.

Valentine, H.T. (1983) Bud break and leaf growth functions for modelling herbivary in some gypsy moth hosts. *Forest Science* 29, 607–617.

Williams, K., Percival, F., Merino, J. and Mooney, H.A. (1987) Estimation of tissue construction costs from heat of combustion and organic nitrogen content. *Plant, Cell & Environment* 10, 725–734.

Wimmer, R. (1991) Beziehungen zwischen Holzstruktur und Holzeigenschaften an Kiefer (*Pinus sylvestris* L.) im Nahbereich eines Fluoremittenten. Dissertationen der Universität für Bodenkultur in Wien, VWGÖ Wien, 37, 227 pp.

Wodzicki, T.J. (1971) Mechanism of xylem differentiation in *Pinus sylvestris* L. *Journal of Experimental Botany* 22, 670–687.

Wong, S.C., Cowan, I.R. and Farquhar, G.D. (1979) Stomatal conductance correlates with photosynthetic capacity. *Nature* 282, 424–426.

Zahner, R. (1955) Soil water depletion by pine and hardwood stands during a dry season. *Forest Science* 17, 466–469.

Zahner, R. (1968) Water deficits and growth of trees. In: Kozlowski, T.T. (ed.) *Water Deficits and Plant Growth, II.* Academic Press, New York, pp. 191–254.

Zimmermann, M.H. (1983) *Xylem Structure and the Ascent of Sap.* Springer-Verlag, New York, 143 pp.

2

Effects of Environment on the Xylogenesis of Norway Spruce (*Picea abies* [L.] Karst.)

Petr Horacek, Jarmila Slezingerova and Libuse Gandelova

INTRODUCTION

Xylogenesis, the process of wood formation, involves meristematic activity of the cambium and formation of woody elements with various radial diameters and different thicknesses of cell walls. The cambium is a lateral vascular tissue the derivatives of which form xylem in the centripetal direction. The term cambium is generally understood to mean the cambial zone (Larson, 1994). In the case of conifers, the cambial zone is represented by a single cambial initial and a variable number of xylem and phloem mother cells (Bannan, 1962; Larson, 1994). Xylogenesis begins with cell division of the cambial initial, followed by cell enlargement of primary cell wall and thickening of secondary cell wall, and ends with the loss of protoplasm (autolysis) of the newly formed cell called a tracheary element or wood fibre (Wilson *et al.*, 1966). The anatomic structure of wood in conifers is relatively simple, spruce wood contains 94.5–96.5% of tracheids, 4.4–5.5% of ray parenchyma, 0–5.8% of longitudinal parenchyma and 0.2–0.3% of epithelial cells of resin canals (Wagenführ and Scheiber, 1974).

A differentiating tracheid reaches its final radial diameter (and primary cell wall) during the phase of radial enlargement after it has emerged from the cambial zone, and secondary cell wall is formed during a subsequent maturation phase. The phases sequence is called cytodifferentiation by Roberts (1976).

Although many details of the physiological and biochemical processes of xylogenesis have been described (Little and Savidge, 1987; Catesson, 1989; Lachaud, 1989; Savidge, 1996), information on the timing, controls and interrelationships of the overall process is incomplete (Roberts *et al.*, 1988). Except for some preliminary results (Wodzicki, 1971), very little is known about either the rate or the duration of radial enlargement and maturation of tracheid differentiation.

Roberts (1983) summarized the roles of physical factors on xylogenesis *in vitro* and postulated that physical factors may alter cytodifferentiation by regulating hormone activity or availability, by influencing the carbon source required for growth, or by modifying some metabolic pathway associated with the different growth processes. These physical conditions include temperature, water, light, atmospheric and soil gases, mechanical stress and acidity (Roberts *et al.*, 1988). In most cases the observed modifications in the cambial growth processes result from an environmental stress. Denne and Dodd (1981) emphasize the effects of climatic conditions on wood fibre diameter and wall thickness. Variations in temperature and water availability can have direct effects on the rates or duration of cell expansion and wall thickening, and indirect effects on carbon availability and growth regulator levels (Denne and Dodd, 1981). Philipson *et al.* (1971) laid the emphasis on the climatic factors as well.

This chapter describes the effects of external environmental factors on duration and rate of wood formation in Norway spruce (*Picea abies* [L.] Karst.). It includes analysis of xylem cell development during seven growing seasons, and the associated environmental factors that appear to limit the processes of cell enlargement and wall thickening.

MATERIAL AND METHODS

The experimental area is in the Drahany Upland, Czech Republic (16°43′ N, 49°29′ E longitude, 625 m a.s.l., NW–SE orientation, sloping ground up to 5%). The average annual temperature is 6.6°C and the average total precipitation is 683 mm. The tree layer consists of a fully canopied spruce monoculture aged 80–90 years (census as of 1989). The soil profile has been formed on colluvial deposits of diverse layers with inclusions of granodiorite gravel. The air temperature in the stand at a level of 2 m above ground was measured daily at 07.00, 14.00 and 21.00 h and average diurnal temperature was calculated as $T_{average} = (T_7 + T_{14} + 2T_{21})/4$. The total soil water supply was measured on the part of soil horizon with dense rootage 0–40 cm at 7-day intervals using the resistivity method on the basis of a calibration curve determined in the locality. Calculated hydrolimits for the soil water supplies within the physiological soil layer (0–40 cm) were 72 mm (wilting point), 90 mm (point of decreased availability) and 140 mm (full field moisture capacity).

The study required repeated examination of the cambial region throughout the season. Preliminary examination of the effect of wounding (Wodzicki and Zajaczkowski, 1970; Horacek, 1995) indicated it is best to sample from the same tree at points at least 2 cm apart around the circumference of the stem at 14-day intervals. In this way six codominant and dominant trees were sampled simultaneously from April to the beginning of November during the years

1983–1989. Samples were extracted at approximately 1.3 m using a modified hammer-driven cylindrical punch 10 mm in diameter (Wodzicki and Zajaczkowski, 1970). The end of the punch was tapered inward and sharp so that the wood core could be removed without crushing the thin-walled cells. The location of the sample from the second and third replicate tree was shifted clockwise one-third of the circumference from the sample orientation of the previous sample tree.

Each sample was placed in a fixative compound of FAA (formaldehyde–acetic acid–ethanol) for 24 h. After rinsing, the samples were sectioned using a sliding microtome (Jung). Thickness of the sections varied between 20 and 40 μm. The sections were stained with safranine and light green SF, and embedded in Canada balsam. Measurements were made along five parallel lines on each sample and included (i) number of cells in the zone of radial enlargement; (ii) number of fully enlarged cells in the zone of wall thickening (maturation); (iii) number of fully mature xylem after autolysis has occurred; (iv) average widths of these layers. It was also impossible to differentiate between the cambial initials and mother cells so these details could not be assessed. Sanio (1873) criteria for identifying cells outside the cambial zone were employed and the tracheids were not observed until they had appeared in the differentiation zones outside the cambial zone. In samples collected at the end of the growing season (November) were recorded (v) radial diameter of tracheids measured at the end of the growing season and (vi) thickness of tangential cell wall of tracheids; the combined thickness of two opposite tangential cell walls was used for calculation. The aim of diameter and wall thickness measurements at the end of the season was to measure the fully formed lignified and mature tracheids. This was impossible with differentiating tracheids during the season.

The total radial growth was calculated as the sum of each layer's width. To check precision of the measurements, the radial growth was measured simultaneously in the same sample trees using tape dendrometers as well. The results obtained from these two types of measurement (microscopic and dendrometric) indicated identical behaviour of radial growth. The measured values were interpolated by splines, and radial growth rate (μm day^{-1}) by first derivation according to time was calculated.

The number of tracheids in the zones of radial enlargement (*G*), maturation (*D*) and fully mature tracheids (*T*), and radial growth rate (*RGR*) obtained at each tree were averaged for a given sampling period.

The measured (each tree) and averaged ('average' tree) numbers of tracheids in *G*, *D* and *T* were analysed according to Wodzicki (1971). From the total radial number of cells in the zones the increment of cells in the zones can be obtained at sampling intervals, and the duration of the radial enlargement and maturation phases can be established for each consecutive tracheid formed during the season. Average number of days necessary for complete emergence of one cell into the given zone in the *i*th period between two sampling terms (VG_i, VD_i and VT_i) was calculated:

$$VG_i = 14/[(G_i + D_i + T_i) - (G_{i-1} + D_{i-1} + T_{i-1})] \quad (2.1)$$

$$VD_i = 14/[(D_i + T_i) - (D_{i-1} + T_{i-1})] \quad (2.2)$$

$$VT_i = 14/(T_i - T_{i-1}) \quad (2.3)$$

where G_i, D_i and T_i are radial number of cells in given zones (G, D and T) after the ith 14-day period of measurement. Index i refers to the sampling intervals and takes values from 0 to 15.

The period of time spent by the jth cell in the given zone (tG_j and tD_j) was calculated:

$$tG_j = (twG_j + 2tpG_j + twD_j)/2 \quad (2.4)$$

$$tD_j = (twD_j + 2tpD_j + twT_j)/2 \quad (2.5)$$

where twG_j, twD_j and twT_j are the full times required for emergence of the jth cell into the given zone and tpG_j and tpD_j are the times at which the whole jth cell entered the given zone.

In determining twG_j (the times twD_j and twT_j are determined in an analogous way) we are dealing with the jth cell as emerging into G only in such ith period for which $(G_{i-1} + D_{i-1} + T_{i-1}) \leq j - 1 < (G_i + D_i + T_i)$. Two cases are possible:

1. The whole jth cell has emerged into G in the ith period. When the condition $j \leq (G_i + D_i + T_i)$ is fulfilled, then $twG_j = VG_i$.
2. A part of the jth cell (r) has emerged into G in the ith period, and the remaining part in the $(i + 1)$ period.

When the condition $j > (G_i + D_i + T_i)$ is fulfilled, then $twG_j = rVG_i + (1 - r)VG_{i+1}$ where $r = (G_i + D_i + T_i + 1) - j$.

The time tpG_j (the time tpD_j is determined in an analogous way) at which the whole jth cell entered the given zone was calculated as $tpG_j = t_1 + t_2 + t_3$, where t_1 is the time in which the given cell emerged completely into G, t_2 is the time at which the given cell begins to emerge from G and t_3 is the full time during which the cell was presented in G.

If $(G_{i-1} + D_{i-1} + T_{i-1}) < j \leq (G_i + D_i + T_i)$ holds true, then $t_1 = (G_i + D_i + T_i - j)VG_i$; and similarly if $(D_{i-1} + T_{i-1}) \leq j - 1 < (D_i + T_i)$, then $t_2 = (j - D_{i-1} + T_{i-1} - j)VD_i$. The time t_3 must fulfil the condition $i' > i + 1$, where i is the time at which the given cell emerged completely into G and i' is the time at which the given cell began to emerge from G, and then $t_3 = (i' - i - 1)14$. If the given cell emerged into G and began to emerge from G within the same ith period, then $tpG_j = t_1 + t_2 - 14$.

The errors in the estimate of time of growth phases following the method described are in the limits of 2.0–2.5 days at the 95% level of confidence. For more details refer to Wodzicki (1971).

The analysed data represent continuous growth process, thus the numbers of cells in G, D and T at the sampling intervals were interpolated by means of cubic splines (Hunt, 1982). Considering the persistence of the growth, the effects of environmental variations become subdued in the growth response, which is more intensive the longer the time interval between stimulus and response. To suppress the variations of measured environmental factors 14-day weighted moving averages were used. The coincidence of environmental changes with changes in *G* zone was used to infer what environmental factors controlled the growth processes, and this was analysed using regression method. The lag of structural response to the filtered diurnal temperatures and soil water supplies was analysed by the method of least-square fit to data.

RESULTS

The mean total radial number of cells found at the end of cambial production (November) varied in the years 1983–1989 from 28 to 56 (Table 2.2). The mean radial number of cells found in each of three zones (*G, D, T*) and measured *RGR* are presented in Table 2.1 for various times during the seasons and plotted as a function of time in Fig. 2.1. The example of the year 1989 is used. The average coefficient of variation for individual measurements is 32% and the maximum error in given measurement attained ± 3 cells. It is interesting to note that the onset of maturation occurred, as a rule, at 27 May ± 7 days in the years 1983–1989.

For further analysis of association between radial growth and environmental factors only the cells in the radial enlargement zone were used. The number of cells in this zone properly simulates changes in the dynamics of growth and the rate of radial growth is in close relationship to the number of cells in *G* ($R^2 = 0.96$, Fig. 2.2) except for the beginning of the growing season, when radial growth did not occur but differentiation was still in progress. At the onset of the growing period, after reactivation of the cambium, we always found one cell in the enlargement zone. We suppose that these cells might overwinter in an undifferentiated state as reported in five ring-porous species (Imagawa and Ishida, 1972). The presence of the cell at the beginning of the growing season was accompanied by similar presence of one cell at the end of the season (Fig. 2.5). This finding supports Wodzicki (1971) who found, by the time cambial production ceased, a rapidly decreasing rate of radial enlargement completion. It should be noted that partially differentiated cells (outside the cambial zone) are discussed here, which are not the same as undifferentiated cells remaining within the cambial zone (Barnett, 1992).

Temperature is most likely to be a limiting factor at the beginning of the growing season. The critical temperature for growth to begin was 5 ± 1°C in the years 1983–1989. The general rise and fall of temperature values during May and June are positively correlated ($R = 0.87$–0.90, $P<0.05$). A marked lag in the growth response was also detected. The average lag of response in the enlargement zone at daily level to the course of diurnal temperatures is 16 ± 2 days at 0.05 probability level. Such a long lag could be attributed to the time required for synthesis and transport of the building and regulating substances

Table 2.1. Radial number of cells in each of three zones of differentiating tracheids and radial growth rate at various times during the year 1989.

Date of sample collection (1989)	19 Apr	3 May	17 May	31 May	14 Jun	28 Jun	12 Jul	26 Jul	9 Aug	23 Aug	6 Sep	20 Sep	4 Oct	18 Oct	1 Nov
Radially enlarging tracheids *G*	0.7	1.1	2.4	8.4	12.6	12.7	12.5	12.2	9.0	6.3	3.7	1.2	1.0	0.9	0.8
Maturing tracheids *D*	0	0	0	0	3.3	8.0	10.5	13.2	14.2	14.7	14.5	13.8	10.3	6.6	0.9
Fully mature tracheids *T*	0	0	0	0	0	2.7	10.7	16.7	23.8	28.4	32.2	35.9	40.0	44.3	50.4
Radial growth rate (μm day^{-1})	0	0	2.0	9.0	11.5	13.5	11.5	13.0	10.0	4.0	2.5	0.5	0	0	0

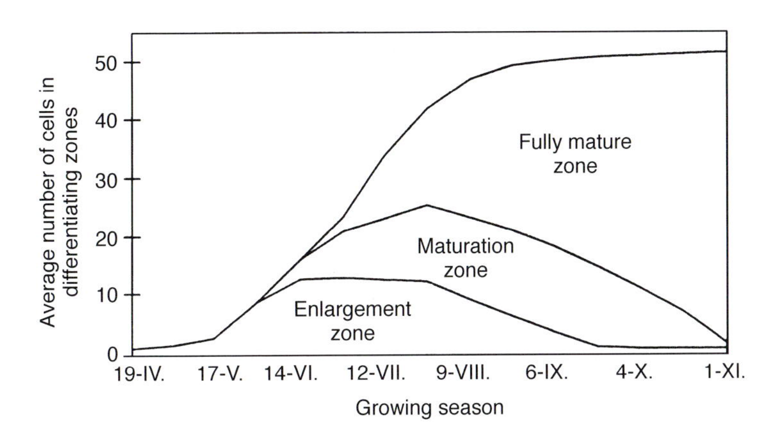

Fig. 2.1. The average number of cells in zones of enlargement, maturation and fully mature tracheids at 14-day intervals during the year 1989.

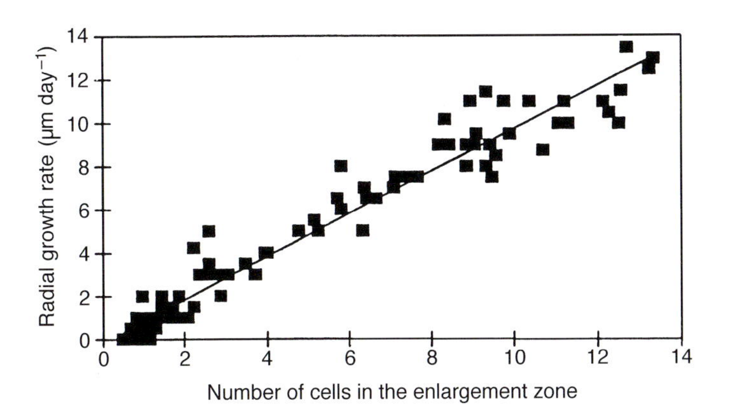

Fig. 2.2. The dependence of radial growth rate (μm day^{-1}) on the number of cells in the enlargement zone (*G*) at 14-day intervals during the years 1983–1989.

that are required for growth to begin and cell division to generate a cell entering the zone of enlargement. The association between number of cells in the enlargement zone and temperature with removed lag is illustrated in Fig. 2.3. Due to the lag, all of the values involved were delayed by the same time constant – 16 days.

Plots of total soil water suggest that water becomes limiting at values below 100 mm (Fig. 2.4). Radial growth is much reduced and ceases at soil water levels ranging from 85 to 90 mm, which is a point of decreased availability (90 mm) or approaching a permanent wilting point (72 mm). The general rise and fall of soil water supply values during June to October are positively

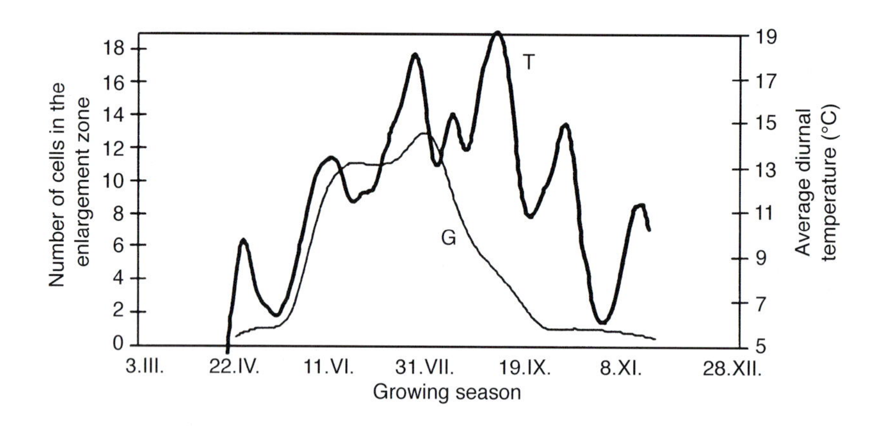

Fig. 2.3. The association of temperature with removed lag (*T*) with the number of cells in the enlargement zone (*G*) in the year 1989 (Horacek, 1994).

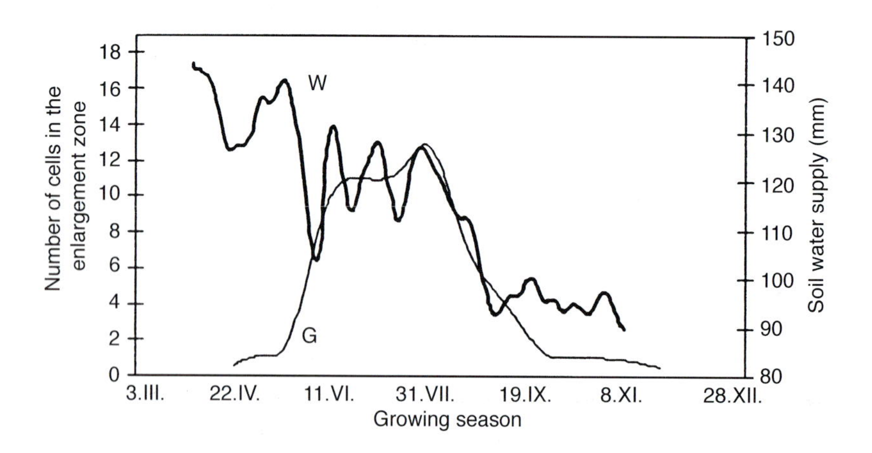

Fig. 2.4. The association of total soil water supply (*W*) with the number of cells in the enlargement zone (*G*) in the year 1989 (Horacek, 1994).

correlated ($R = 0.87–0.94$, $P<0.05$). No lag in time of growth behind changes in moisture were detected at 0.05 probability level in the years 1985, 1986 and 1989. Short lags (4 ± 1 day, $P<0.05$) observed in 1983, 1984, 1987 and 1988 are within the range of error of measurement. The strong relationship between growth and soil water implies that internal water stress occurred within the tree as the rate of water absorption by the roots diminished below the rate of transpiration from the crown. While the water in the measured column of soil was still above the permanent wilting point, the water content in areas with the greatest concentrations of roots may well have been below that point.

Figure 2.5 is a plot of photoperiod (hours) corresponding to 50°N latitude along with the number of cells in the enlargement zone for all seasons corresponding to each day of the growing season. A possible association of radial growth with day length is apparent. However, no causal link between length of photoperiod or a limiting effect of light could be established as soil water or temperature were frequently low during the same time periods. Despite that, it is possible to state that maximum rate of radial growth is attained about the summer solstice (21 June) and cambial production has stopped by the time of the autumnal equinox (22–23 September).

Since temperatures are usually optimum for growth during late June and July, soil moisture may become the most limiting factor during middle and late season. However, this soil moisture effect may be due to its effect on the rate of photosynthesis and subsequent production of carbon reserves available for growth, or upon the activity in crown growth and the production of growth regulators influencing the amount of radial growth. Optimum growth rates during this time are associated with intervals when neither moisture, temperature or photoperiod are limiting to processes affecting the rate of growth. If one assumes that no other factor such as crown activity or growth regulator production have become limiting, a dependency of the radial growth rates on both average daily temperature and total water supply is evident (Fig. 2.6). The area in this figure is based on gridding points representing the numbers of cells in the enlargement zone and corresponding to the average diurnal air temperatures with the 16-day lag removed and soil water supply without the lag. The Kriging gridding method was used which is based on a geostatistical technique to calculate the autocorrelation between data points and produce a minimum variance unbiased estimate. Duplicate data points were averaged.

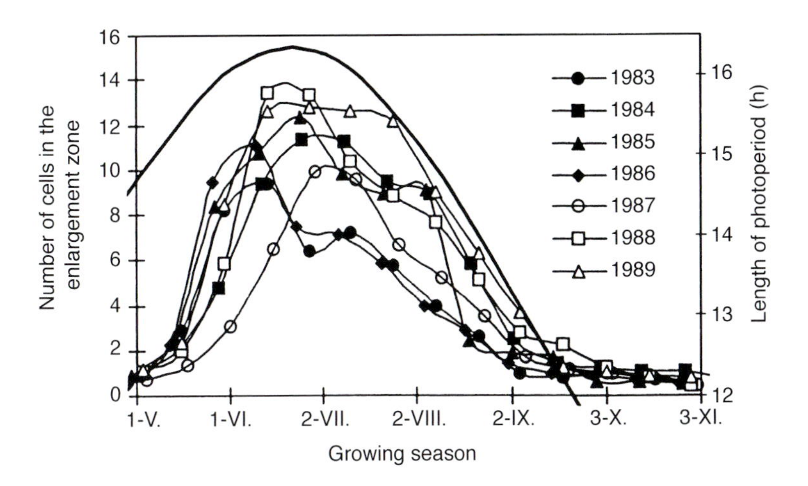

Fig. 2.5. The number of cells in the enlargement zone in years 1983–1989 and the length of photoperiod (thick line).

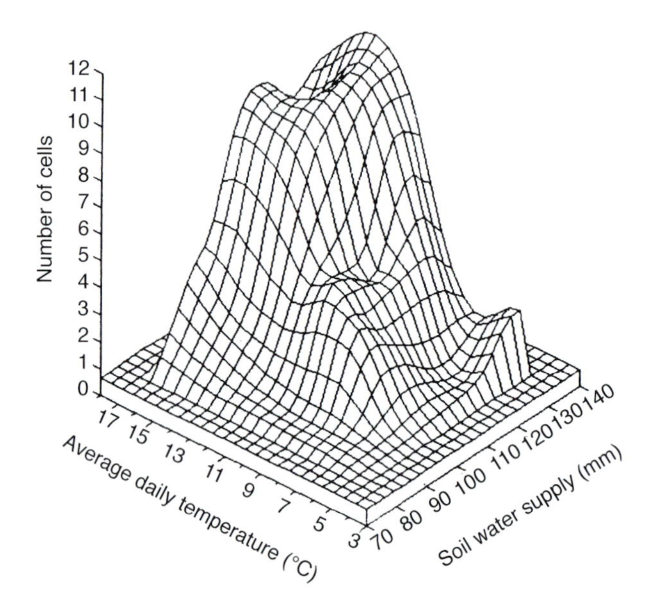

Fig. 2.6. Interaction between average diurnal temperature and soil water supply in their effects on the number of radially enlarging cells over all seasons (Horacek, 1994).

It is evident from this plot that enlargement begins when temperature rises above about 5°C. It appears from this plot that the optimum average diurnal temperature for cell enlargement is 13°C. At higher or lower temperatures the rate of radial growth decreases. The critical soil water content for cell enlargement is approximately 80 mm (which is 8 mm above the permanent wilting point). Between 80 and 140 mm (the maximum capillary water capacity) the rate of cell enlargement is directly dependent on soil water content.

Irregularities in the generalized temperature and moisture relationships may be attributed to other factors such as available carbon, crown condition, growth regulators, time, day length or other factors not considered in this investigation. Under conditions of temperature and water supply abundance, insufficient illumination from low light intensity or short day length carbon levels may have become limiting to either cell enlargement or to other processes affecting the ability of cells to enlarge, perhaps related to changes in osmotic potential.

The duration within the zones of enlargement and maturation, and the total time spent by a given cell in both zones in all years investigated is shown in Table 2.2. The large thin-walled earlywood cells generally spend more time enlarging and less time maturing, while the smaller thick-walled latewood cells spend less time enlarging and more time in the zone of maturation. A plot of the duration of each cell in the zones of enlargement and maturation, and the

Table 2.2. The average duration of each differentiated cell in the zones of enlargement (*G*) and maturation (*D*), the total average radial number of tracheids at the end of cambial production and the average time spent in the zones of enlargement (*G*) and maturation (*D*) during the years 1983–1989.

Year	Duration of *G* (days)	Duration of *D* (days)	Total average number of cells	Time spent in *G* + *D* (days)
1983	14–33	19–60	44	36–78
1984	9–29	20–48	56	38–62
1985	9–27	22–53	46	50–67
1986	13–28	18–60	36	42–79
1987	19–45	18–58	28	46–77
1988	10–28	13–41	50	30–62
1989	15–35	14–51	52	37–77

course of their appearance in the season 1989 is summarized in Fig. 2.7. As a rule, in the years investigated the duration of radial enlargement decreased slowly towards the last-formed tracheid. At the very end of the season, cells which grew for only 9 to 10 days were found. The period of maturation for the first few tracheids decreased in general. For successive tracheids the duration of maturation was increasingly longer until it reached a maximum and decreased slightly at the end of the season in all years.

In the process of enlargement and maturation, cells attain different sizes and exhibit variable wall thickness. If we assume that final radial diameter and wall thickness are achieved at termination of appropriate phase (*G* and *D*), then we are able to lay out the measurements of final radial diameter and wall thickness on a time scale for each tree. The mean tangential diameter was found to be almost uniform for all tracheids along the radial file, it was 36±2 μm for all years. The final radial diameters and tangential cell wall thickness of tracheids from different trees during the growing season 1989 can be seen in Figs 2.8 and 2.9. Each point on the figures refers to one cell at the time when (i) its elongation is finished and radial diameter is achieved (Fig. 2.8), and (ii) its maturation is terminated and cell wall thickness is fixed (Fig. 2.9). Both dimensions were found to be irreversible. The tracheid size decreases throughout the season while the wall thickness increases to a maximum in October and then decreases. The differences in the rates of change later in the year suggest that at this time the processes are more or less independent. Figure 2.10 is a plot of interrelation of radial diameter and tangential cell wall thickness of fully mature tracheids measured on radial files during the years 1983–1989. All tracheids in which the cell wall thickness multiplied by two was greater than the width of the lumen were considered as latewood (Mork, 1928; Denne, 1988). The increase of cell wall thickness is related to cell diameter in latewood, but in the case of earlywood the relationship is missing. This is in accordance

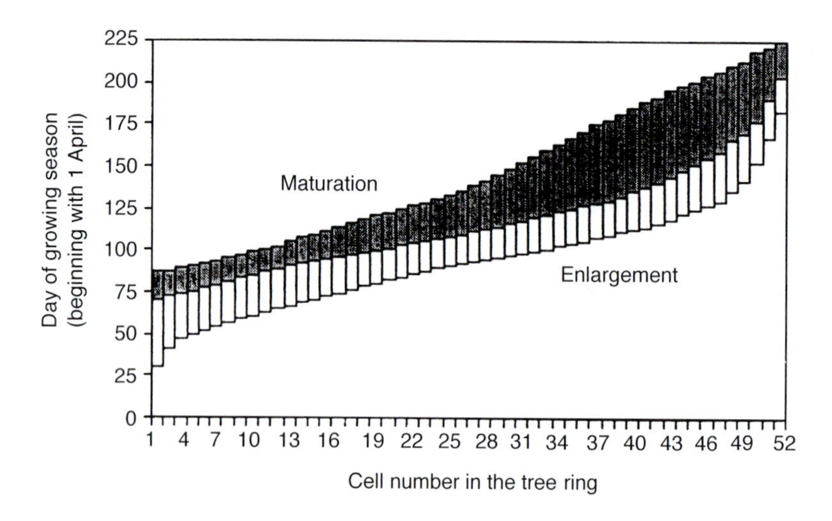

Fig. 2.7. The average duration of each differentiated cell in the zones of enlargement and maturation formed during the season 1989.

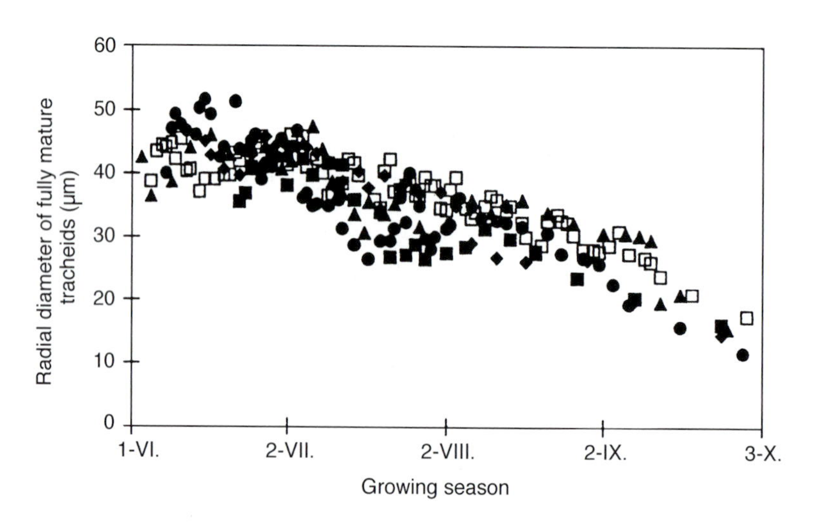

Fig. 2.8. The final radial diameter of tracheids formed during the season 1989. The different symbols represent measurements from different trees.

with Zimmermann and Brown (1971) who concluded that relation of both variables is not causal.

The differences among the six trees are less for cell size than for wall thickness although the measurements of the latter are less precise. From the figures the similar course among different trees during the season is apparent

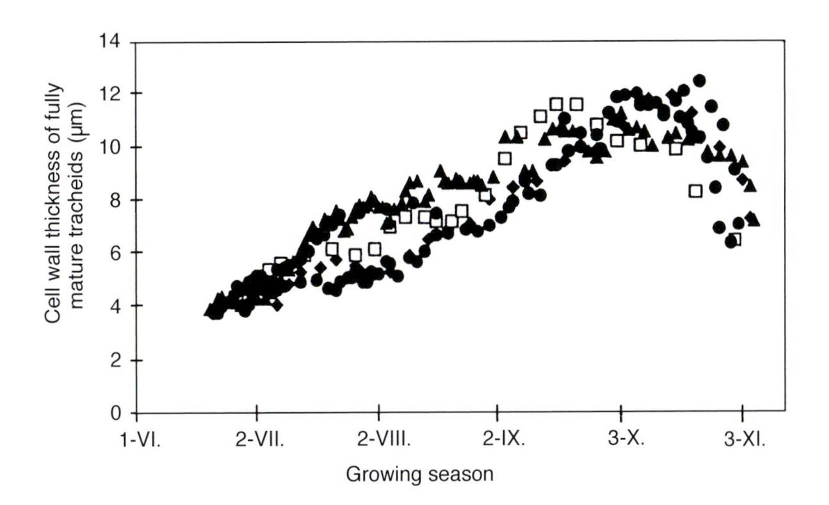

Fig. 2.9. The final tangential cell wall thickness of tracheids formed during the season 1989. The different symbols represent measurements from different trees.

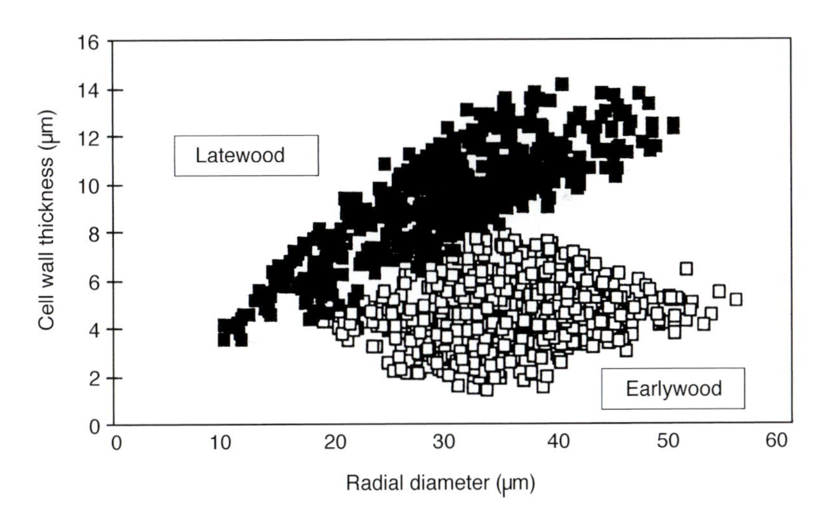

Fig. 2.10. The dependence of tangential cell wall thickness on radial diameter of fully mature tracheids formed in the years 1983–1989. The definition of latewood was based on Mork (1928).

although the radial growth rate and the numbers of cells in *G* and *D* varied remarkably. The total amount of tracheids produced during the season and observed when differentiation ceased were noted: in 1989 there were 24, 27, 44, 50, 64 and 94 tracheids for the six trees with average number of 52 (Table 2.2). The radial diameter varied from 15 to 50 µm and the tangential cell wall thickness from 3 to 13 μm during the growing season for all years. It appears that irrespective of how many cells were produced up to a given date of the year, the anatomical dimensions of tracheids depends on the exact time location during the growing season.

There is a general positive association between radial diameter and wall thickness, and duration in the corresponding growth zone. The final radial diameter is proportional to its duration in the enlargement zone (Fig. 2.11). The wall thickness of a fully differentiated tracheid is closely related to the duration of its maturation phase (Fig. 2.12). The correlations are significant at both variables ($R = 0.60$ at radial diameter and $R = 0.82$ at wall thickness (both at $P<0.05$)) provided the primary and secondary wall is continuously formed at an approximately steady rate during *G* and *D*. The differences of real processes from steady state could explain the moderate dependence. From the foregoing discussion it is obvious that the longer tracheids stay in the differentiating zones, the higher the values of radial diameter and wall thickness attained. In this way the wood quality could be affected.

After its last division, each cell undergoes enlargement and maturation at a rate dependent upon limiting conditions of temperature and water availability while residing in each of these growth phases (Figs 2.13 and 2.14). The same method as for Fig. 2.6 was used. The optimal temperature for the longest duration of enlargement of the growing tracheids (Fig. 2.13) appears to be 5°C in these plots while the optimal soil water supply is 140 mm (maximum capillary water capacity). Interaction is also apparent. The effect of temperature is greater the higher the soil water supply. Duration of maturation appears to be related to the limitations of both temperature and soil water supply (Fig. 2.14). The optimal temperature for cell wall formation is 20°C and becomes more

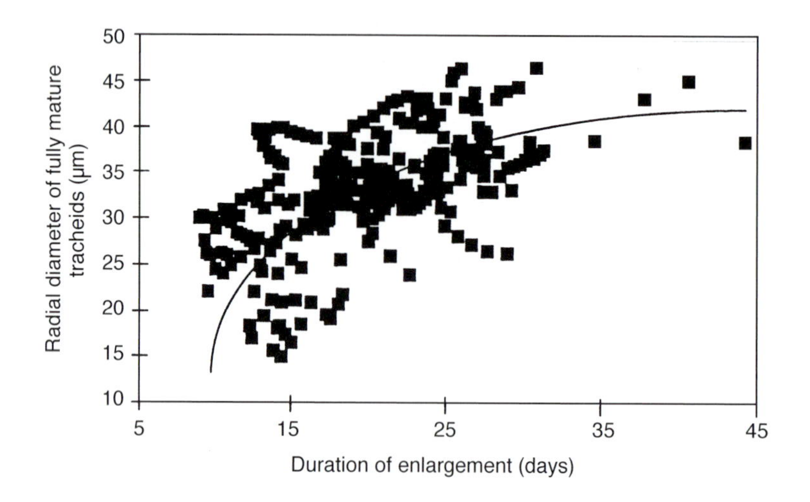

Fig. 2.11. The dependence of final radial cell size on duration in the enlargement zone in the years 1983–1989.

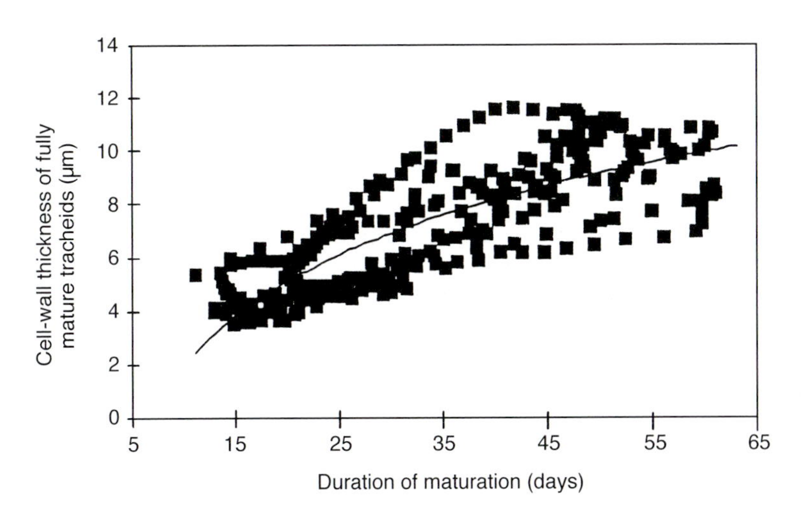

Fig. 2.12. The dependence of final tangential cell wall thickness on duration in the maturation zone 1983–1989.

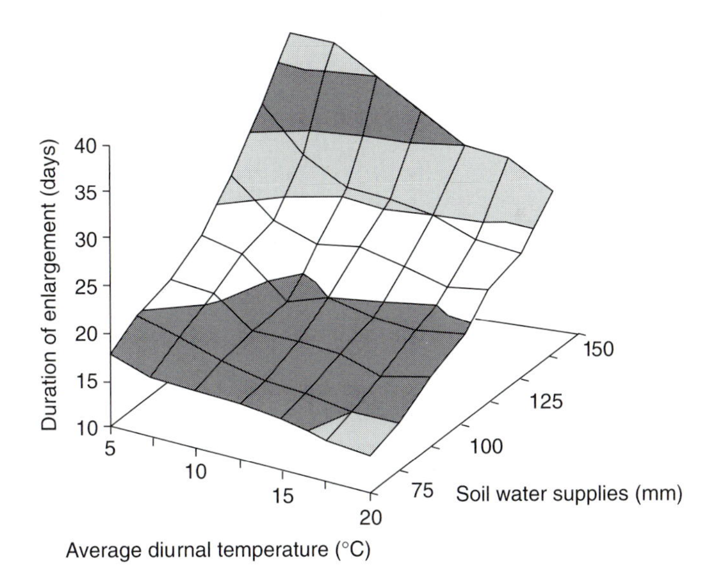

Fig. 2.13. The joint dependence of cell duration in the enlargement zone on average diurnal temperature and soil water supplies.

limiting with decreasing temperature. Soil water supply appears to be optimum at 100 mm and appears more limiting with both higher and lower values. Temperature and soil water interact in that the effect of temperature is greater

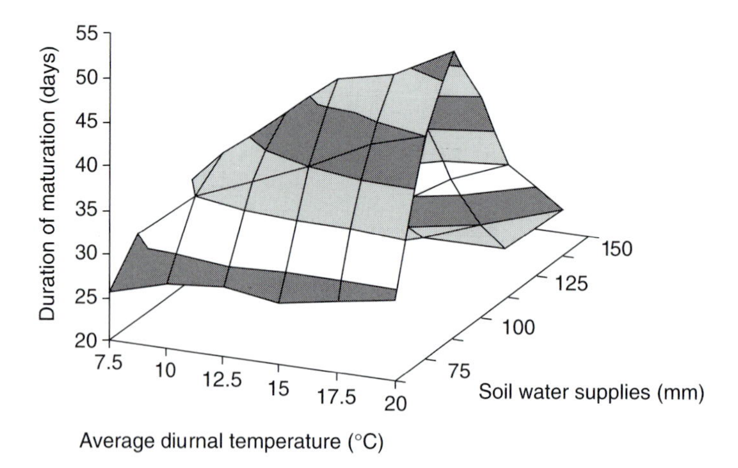

Fig. 2.14. The joint dependence of cell duration in the maturation zone on average diurnal temperature and soil water supplies.

at optimum moisture values than at either the maximum or minimum soil water extremes.

The results presented above show that water deficits have a marked effect on radial enlargement. While temperature and moisture do affect the chemical and physical rates of the growth processes, the strong influence shown here is also due to indirect temperature effects on photosynthesis, respiration, transpiration and related processes. The interaction between temperature and water shown in Fig. 2.13 is probably due to temperature effects that accelerated transpiration and the development of internal water stress, stomatal closure and decreased photosynthesis within the tree. The inverse relationship between soil water greater than 100 mm and duration in the maturation zone (Fig. 2.14) may in part be a result of decreasing aeration in the soil, but it is more likely attributable to (i) decreasing crown activity, (ii) growth regulator changes, (iii) decreasing light levels, (iv) root growth and (v) other factors that were not considered in this particular study. The exact cause of this interrelation between temperature and water on xylem maturation remains unclear.

DISCUSSION

Recognition of the mechanisms that control the cell differentiation provides clues to understanding wood production. Most studies dealing with the effect of environmental factors focus attention only on variations in the tree-ring width and density (Fritts, 1976; Schweingruber, 1983; Heikkinen, 1985; Eckstein *et al.*, 1989). However, within the tree-ring width and density all of the differentiation phases as discussed in this chapter are combined. The environmental effect thus becomes obscured, despite the fact that even with insignificant dependencies the causal processes are concerned. Among the studies that deal with xylogenesis in relation to the environment there are few which observe differentiated phases of cell growth (Wareing, 1958; Waisel and Fahn, 1965; Skene, 1969; Wodzicki, 1971; Fritts *et al.*, 1990, 1991; Vaganov *et al.*, 1990; Antonova and Stasova, 1993).

In the tracheids of conifers in Central Europe the whole process of differentiation takes 1–2 months approximately (Wodzicki, 1971). The rate of division in the xylem mother cells, that is, the interval between two successful divisions, varies from 4–6 days (Zimmermann and Brown, 1971) to 10–28 days (Wilson, 1964). Duration of the phase of radial elongation depends on the time location during the growing season and varies from 2 to 4 weeks (Kozlowski, 1971; Wodzicki, 1971). The length of maturation tends to get longer and average duration varies from approximately 4 to 5 weeks with a maximum of 8 weeks (Wodzicki, 1971). Our results stand in the range indicated by the literature.

The results presented show that temperature and water deficits have a marked effect on wood formation. The cell radial enlargement is caused by cumulation of vacuoles into one large central vacuole whose uptake of water becomes intensive (Kozlowski, 1971; Kozlowski *et al.*, 1991). As a result, osmotic pressure inside the vacuole tends to increase with a passive enhancement of the cell wall plasticity, which provides continual deposition of new polysaccharide chains (Salisbury and Ross, 1991). The course of elongation is therefore influenced, as well as by temperature, by the amount of water available to the plant. Instead of precipitation we monitored the soil water supply, which expresses better the water regime at the ecotope. The presented results do not agree with other authors (Wodzicki, 1971; Larson, 1994) who did not find a direct effect of precipitation on the rate of tracheids production or enlargement. Usually, temperature in temperate regions is suggested as the most important limiting factor at the beginning of the growing season. Temperature controls the onset of cambial reactivation and the level of activity during the first half of the growing season (Priestley, 1930; Wareing, 1958; Bannan, 1962; Kozlowski, 1971; Fritts, 1976) and the cambial activity usually has a positive correlation with temperature (Ladefoged, 1952; Glock, 1955). A period of low temperatures during the growing season may bring cessation of radial growth although the short day length or water deficit is thought to be a limiting factor (Waisel and Fahn, 1965). The relationship between growth and temperature is not surprising because temperature affects all phases of cytodifferentiation. Temperature affects growth indirectly, via photosynthesis and regulation of respiration and transpiration. Although the final thickness of cell wall of tracheids in conifers is known to be related to temperature (Larson, 1967), in our study temperature together with soil water supply were among the most important factors.

The dependence of wall thickness of fully mature tracheids on duration in the maturation zone is supported by earlier findings (Skene, 1969; Wodzicki, 1971). Zimmermann and Brown (1971) suggest that cell wall thickening is a function of the rate and duration of cellulose deposition during the process of tracheid differentiation and maturation. On the other hand, our suggestion of close dependence of radial diameter on duration of radial elongation stays unconfirmed.

CONCLUSIONS

The number of cells emerged from the cambial zone features positive correlation with both temperature and soil water on the assumption that none of the factors involved is limiting. Temperature represents the most important critical factor at the beginning of the growing season, when sufficient water reserves are available. With soil water dropping to the point of decreased availability, radial growth is increasingly exposed to stress due to water deficit. Cambial activity may be influenced by both these factors at any time of the growing season, the higher intensity of one of them always being decisive.

Different trees showed great variability in the number of differentiated cells, but much less variability was expressed for anatomical parameters of cells in time and duration of cell phases. Once divided, a cell outside the cambial zone proceeds in its growth according to environmental conditions which modify the duration of the process. There was found to be a close relationship between the duration of both cell phases on the courses of temperature or water regimes.

Radial diameter and cell wall thickness are both a function of their time of elongation and secondary thickening. Changes in radial diameter and tangential cell wall thickness are accompanied by proportional changes in the duration of cell cycle phases at different dates during the season. The final thickness of cell wall is closely related to the duration of maturation over all seasons which appears to be related to both the temperature and soil water supply courses. The final radial diameter of tracheids is related to the duration of elongation which appears to be connected mainly to the soil water supplies. During the latter part of the season the variation of radial diameter of tracheids does not coincide with changes in thickness of their secondary wall. It is believed that both processes are more or less independent.

SUMMARY

The effects of mean daily air temperature, soil water supplies and day length on cell enlargement, cell wall thickening (maturation), radial diameter and tangential cell wall thickness in Norway spruce (*Picea abies* [L.] Karst.) were investigated. The radial diameter varies from 15 to 50 μm and the tangential cell wall thickness from 3 to 13 μm. Generally, the thin-walled earlywood cells remain longer in the enlargement zone than in the maturation zone. Later in the season the relative proportion of time in these two zones is reversed. Early in the season cell enlargement increases with rising temperatures. Later, declining soil water supplies become the dominant limiting factor although both variables can have direct effects at any time. The relationship of these two factors with cell wall thickening (maturation) is more complex. Temperature is directly related to wall thickening but the effect is most marked when soil water is 100 mm. Above and below this soil water value the apparent wall thickening declines and the influence of temperature is less important. As soil water supply increases from 75 to 100 mm water stress in the tree declines and cells remain in the maturation zone for a longer period of time. Radial diameter and cell wall thickness are both a function of their time in the enlargement and maturation zones.

ACKNOWLEDGEMENTS

This work was supported by grants from Grant Agency of the Czech Republic No. GACR 501/93/0804, and Ministry of Education, Youth and Sport of the Czech Republic No. PG97104.

REFERENCES

Antonova, G.F. and Stasova, V.V. (1993) Effects of environmental factors on wood formation in Scots pine stem. *Trees* 7, 214–219.

Bannan, M.W. (1962) The vascular cambium and tree-ring development. In: Kozlowski, T.T. (ed.) *Tree Growth*. Ronald Press, New York, pp. 3–21.

Barnett, J.R. (1992) Reactivation of the cambium in *Aesculus hippocastanum* L. *Annals of Botany* 68, 159–165.

Catesson, A.M. (1989) Cambial cytology and biochemistry. In: Iqbal, M. (ed.) *Radial Growth of Plants. Vol.1. Vascular Cambium.* Research Studies Press, Taunton, Somerset, pp. 63–112.

Denne, M.P. (1988) Definition of latewood according to Mork (1928). *IAWA Bulletin* 10, 59–62.

Denne, M.P. and Dodd, R.S. (1981) The environmental control of xylem differentiation. In: Barnett, J.R. (ed.) *Xylem Cell Development*. Castle House Publications, Kent, pp. 236–255.

Eckstein, D., Krause, C. and Bauch, J. (1989) Dendrochronological investigation of spruce trees (*Picea abies* /L./ Karst.) of different damage and canopy classes. *Holzforschung* 43, 411–417.

Fritts, H.C. (1976) *Tree Rings and Climate*. Academic Press, London.

Fritts, H.C., Vaganov, E.A., Sviderskaya, I.V. and Shashkin, A.V. (1990) Modeling tree-ring climatic relationships. In: Bartholin, T.S., Berglund, B.E., Eckstein, D. and Schweingruber, F.H. (eds) *Tree Rings and Environment*. Proceedings of the International Dendrochronological Symposium, Ystad, South Sweden, 3–9 September 1990. Lunqua Report, Lund University, pp. 104–108.

Fritts, H.C., Vaganov, E.A., Sviderskaya, I.V. and Shashkin, A.V. (1991) Climatic variation and tree-ring structure in conifers: empirical and mechanistic models of tree-ring width, number of cells, cell size, cell-wall thickness and wood density. *Climate Research* 1, 97–116.

Glock, W.S. (1955) Tree growth. II. Growth rings and climate. *Botanical Review* 21, 73–188.

Heikkinen, O. (1985) Relationships between tree growth and climate in the subalpine Cascade Range of Washington, U.S.A. *Annales Botanici Fennici* 22, 1–14.

Horacek, P. (1994) Norway spruce (*Picea abies* L. /Karst.) cambial activity according to ecological conditions. In: Spiecker, H. and Kahle, P. (eds) *Modelling of Tree-Ring Development, Cell Structure and Environment.* Proceedings of the workshop, Freiburg, 5–9 September 1994, Institut für Waldwachstum, Universitat Freiburg, pp. 39–49.

Horacek, P. (1995) Modelling tree-ring climatic relationships. *Lesnictví-Forestry* 41, 188–193.

Hunt, R. (1982) *Plant Growth Curves. The Functional Approach to Plant Growth Analysis.* Edward Arnold, London, 248 pp.

Imagawa, H. and Ishida, S. (1972) Study on the wood formation in trees. Report III. Occurrence of the overwintering cells in cambial cells in cambial zone in several ring-porous trees. *Research Bulletin Hokkaido University Forest* 29, 207–221 (in Japanese).

Kozlowski, T.T. (1971) *Growth and Development of Trees. Vol. II. Cambial Growth, Root Growth and Reproductive Growth.* Academic Press, New York, 514 pp.

Kozlowski, T.T., Kramer, P.J. and Pallardy, S.G. (1991) *The Physiological Ecology of Woody Plants.* Academic Press, London, 657 pp.

Lachaud, S. (1989) Participation of auxin and abscisic acid in the regulation of seasonal variations in cambial activity and xylogenesis. *Trees* 3, 125–137.

Ladefoged, K. (1952) The periodicity of wood formation. *Kgl. Danske Videnskab. Biol. Skrifter* 7, 1–98.

Larson, P.R. (1967) Effects of temperature on the growth and wood formation of ten *Pinus resinosa* sources. *Silvae Genetica* 16, 58–65.

Larson, P.R. (1994) *The Vascular Cambium. Development and Structure.* Springer-Verlag, Berlin, 725 pp.

Little, C.H.A. and Savidge, R.A. (1987) The role of plant growth regulators in forest tree cambial growth. *Plant Growth Regulation* 6, 137–169.

Mork, E. (1928) Die Qualitat des Fichtenholzes unter besonderer Rücksichtnahme auf Schleif- und Papierholz. *Papier-Fabrikant* 26, 741–747.

Philipson, W.R., Ward, J.M. and Butterfield, B.G. (1971) *The Vascular Cambium. Its Development and Activity.* Chapman & Hall Ltd, London, 182 pp.

Priestley, J.H. (1930) Studies in the physiology of cambial activity. III. The seasonal activity of the cambium. *New Phytologist* 29, 316–354.

Roberts, L.W. (1976) *Cytodifferentiation in Plants. Xylogenesis as a Model System.* Cambridge University Press, Cambridge, 160 pp.

Roberts, L.W. (1983) The influence of physical factors on xylem differentiation. In: Dodds J.H. (ed.) *Tissue Culture of Trees.* Croom Helm, London, pp. 88–102.

Roberts, L.W., Gahan, P.B. and Aloni, R. (1988) *Vascular Differentiation and Plant Growth Regulators.* Springer-Verlag, Berlin, 154 pp.

Salisbury, F.B. and Ross, C.W. (1991) *Plant Physiology.* Wadsworth Publishing Company, Belmont, CA, 350 pp.

Sanio, C. (1873) Anatomie der gemeinen Kiefer (*Pinus sylvestris* L.). *Jahrbücher der Wissenschaftlichen Botanik* 8, 401–420.

Savidge, R.A. (1996) Xylogenesis, genetic and environmental regulation – A review. *IAWA Journal* 17, 269–310.

Schweingruber, F.H. (1983) *Der Jahrring. Standort, Methodik, Zeit und Klima in der Dendrochronologie*. Haupt, Bern, 234 pp.

Skene, D.S. (1969) The period of time taken by cambial derivatives to grow and differentiate into tracheids in *Pinus radiata* D. Don. *Annals of Botany* 33, 253–262.

Vaganov, E.A., Sviderskaya, I.V. and Kondratieva, E.N. (1990) Weather conditions and tree ring structure: simulation model of the tracheidograms. *Lesovedenie* 2, 37–45.

Wagenführ, R. and Scheiber, C. (1974) *Holzatlas*. VEB Fachbuchhandlung, Leipzig.

Waisel, Y. and Fahn, A. (1965) The effect of environment on wood formation and cambial activity in *Robinia pseudoacacia* L. *New Phytologist* 64, 436–442.

Wareing, P.F. (1958) The physiology of cambial activity. *Journal of the Institute of Wood Science* 1, 34–42.

Wilson, B.F. (1964) A model for cell production by the cambium of conifers. In: Zimmermann, M.H. (ed.) *The Formation of Wood in Forest Trees*. Academic Press, New York, pp. 19–36.

Wilson, B.F., Wodzicki, T.J. and Zahner, R. (1966) Differentiation of cambium derivatives: proposed terminology. *Forest Science* 12, 438–440.

Wodzicki, T.J. (1971) Mechanism of xylem differentiation in *Pinus silvestris* L. *Journal of Experimental Botany* 22, 670–687.

Wodzicki, T.J. and Zajaczkowski, S. (1970) Methodical problems in studies on seasonal production of cambial xylem derivatives. *Acta Societatis Botanicorum Poloniae* 39, 509–520.

Zimmermann, M.H. and Brown, C.L. (1971) *Trees – Structure and Function*. Springer-Verlag, New York, 336 pp.

3

Measuring and Interpreting Diurnal Activity in the Main Stem of Trees

William Gensler

INTRODUCTION

Water status and metabolic activities are important factors during tree-ring growth in the stem. An instrument was used (PHYTOGRAM™, Agricultural Electronics Corporation, USA) to measure the water status and metabolic activity level of trees. The basic method is to implant a measuring electrode in the tissue under investigation. Electrochemical testing procedures are then applied to this electrode to determine hydration and proton concentration changes in the region around the electrode surface. In order to relate the PHYTOGRAM to a physical variable related to water status, point dendrometers are employed to simultaneously determine radius changes.

The PHYTOGRAM differs from previously developed instruments using invasive electrodes to determine plant parameters (Davis *et al.*, 1979; MacDougall *et al.*, 1988; Lekas *et al.*, 1990; Smith and Ostrofsky, 1993). The methods employed in the cited papers focus on the bulk volume between two similar implanted electrodes. A measurement is made of the resistance and capacitance of this volume of tissue and their changes under different environments. These methods have their origin in the free space concepts (Fensom, 1966). The PHYTOGRAM focuses on the chemical constituents at the electrode/liquid interface and the extent of this interface. The concern of the PHYTOGRAM is interfacial electrochemistry.

This chapter will emphasize methods and data interpretation as opposed to exposition of new results. Furthermore, the discussion will be limited to the diurnal patterns as opposed to transients which arise out of distinct environmental stimuli.

The objectives of this chapter are: (i) to describe the physical basis and electrochemical mechanisms underlying the assays of proton and hydration concentration by the PHYTOGRAM and the radius change as measured with a point dendrometer, and (ii) to give a physiological interpretation of typical diurnal patterns of these variables.

The objectives are limited to diurnal activity. This activity is highly repetitive and reproducible. Activity associated with transient phenomenon such as growth or the response to environmental stimuli such as rainfall and/or irrigation is beyond the present discussion.

METHODS

Overall Electrochemical Circuit

Figure 3.1 illustrates the electrochemical circuit consisting of the measuring electrode, the path through the tree and root zone and the return electrode. This path can be conceptually as well as quantitatively represented by the equivalent circuit shown in Fig. 3.1. The inside-the-plant interface of the electrode with the

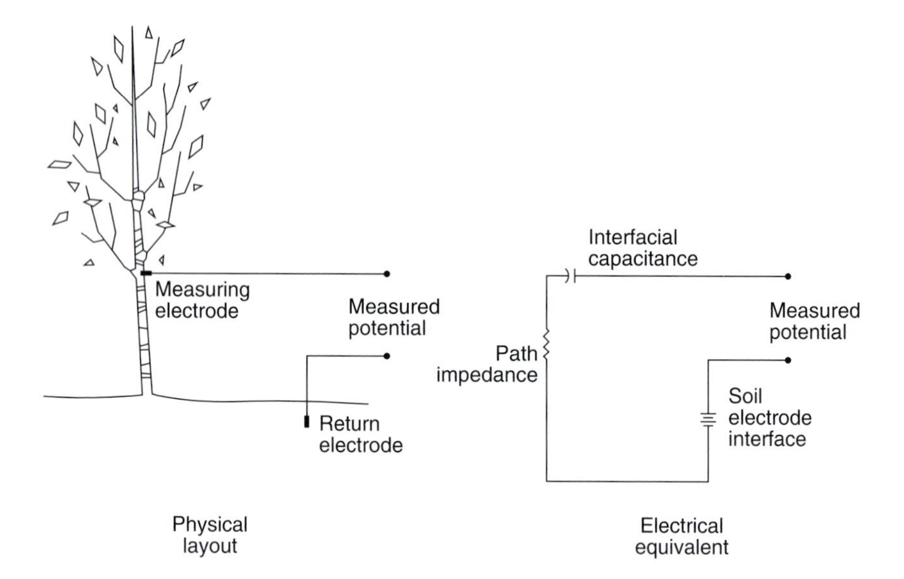

Fig. 3.1. Physical layout and equivalent electrical circuit. A measuring electrode is implanted in the tree. A return electrode is placed in the root zone. Electrical potential and capacitance is measured between the two wires from the electrodes. The equivalent circuit is: the electrode metal/liquid tissue interface acts as a capacitor; the return electrode acts as a battery; the path between the two electrodes acts as a resistor.

plant tissue acts as an electrical capacitor. The path characteristics are equivalent to a series resistor. The return electrode is represented by a battery. A measurement of potential between the two output wires from the two electrodes under conditions of zero current would consist of two series potentials: the potential across the capacitor and the potential across the battery.

The circuit in Fig. 3.1 also serves to point out the conditions of proper operation. Changes in the proton concentration in the tissue change the magnitude of the potential across the capacitor. Changes in hydration cause changes in the size of the capacitor. The current through the circuit is very low (of the order of 10 pA) and the value of the series path resistor is also small. This results in an insignificant potential drop across the series resistor. Furthermore, the potential across the return electrode is assumed to remain constant. Under these two conditions, any change in the capacitor potential drop manifests itself as a change in the measured potential.

Cell Surface–Liquid Bridge–Electrode Surface

Figure 3.2 illustrates the electrode resident in the tissue. The relative scale of this diagram is close to the relative physical scale. The electrode is 150 μm in diameter and the approximate cell size is a cube 20 μm on a side. Electron microscope analysis indicates an annulus or ring-like channel exists between the surface of the electrode and the wall of opposite functioning cells (Goldstein, 1982; Rugenstein, 1982). Assuming the thickness of this annulus is 50 μm and that each cell occupies an area 20 by 20 μm, there are 2000 cells facing a 1 μm length of the electrode surface.

The two surfaces of this annulus are bridged by a liquid. Cellular activity involves the transfer of chemical reactants to and from the liquid on one side of the bridge. The presence and concentration of these reactants is then manifest at the surface of the electrode on the other side of the liquid bridge. Assays of these reactants permit a measure of the magnitude and timing of this cellular transfer activity. This is the essential mechanism whereby the PHYTOGRAM can be used as an indicator of cellular activity.

Figure 3.2 also serves to emphasize the apoplastic nature of the technique. The electrode is outside the cell. A recent review of the apoplast has called for a more precise definition of the term 'apoplast' (Canny, 1995). In the present context, the term 'apoplast' encompasses the region contiguous to the *wetted* cell surface. Figure 3.2 shows in a highly discrete and schematic form the aereoplast. The gas-filled continuum or intercellular space connected to the outer surface of the tree by a direct gas pathway is defined in this exposition as the aereoplast (Gensler, 1994). This distinction follows the development of Raven in dividing transport tissue into three categories: symplast, apoplast and gas-filled continuum (Raven, 1977).

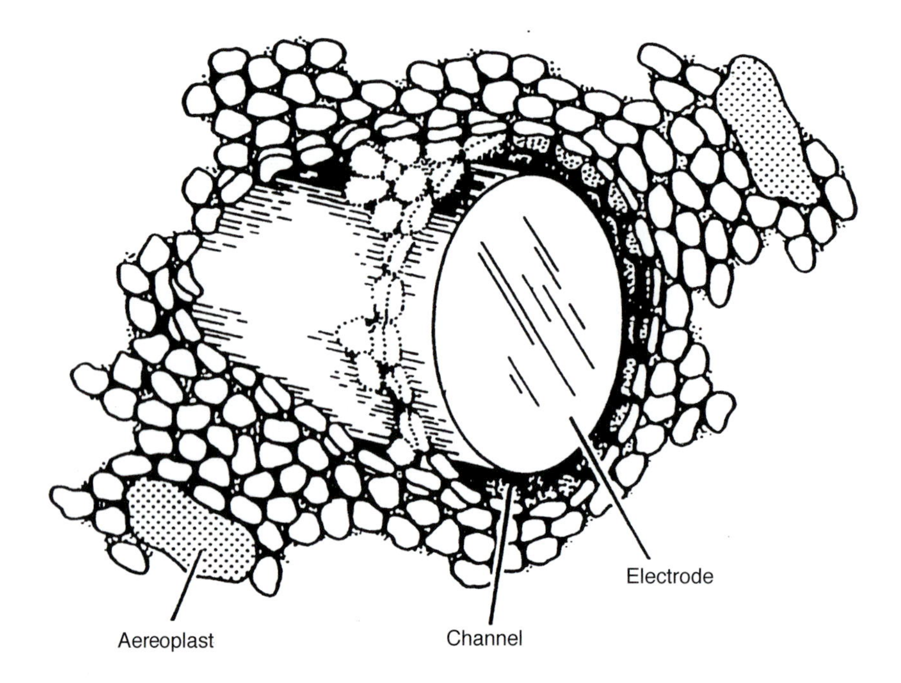

Fig. 3.2. Approximate scale representation of the electrode resident in the tissue. An annulus exists between the electrode surface and layer of adjacent cells. Water within this annulus provides a liquid bridge between the cell surface and the electrode surface. For conceptual clarity, the aereoplast is depicted as a discrete volume. It is actually the cumulative volume of diffuse gas-filled intercellular space formed schizogenously or schizo-lysigenously.

Anatomically, the electrode has damaged the tissue in its entry. The healing response consists of an exponential rise in potential to the equilibrium value (Gensler, 1978, 1979). There is no immediately observable scar tissue if the electrode is inserted at a time corresponding to the upper shoulder of the 'S' curve of growth (Goldstein, 1982; Rugenstein, 1982). The possibility of encapsulation with extended residence will be considered below.

Oxygen and Protons in the Liquid Bridge

Figure 3.3 illustrates the magnified liquid bridge in schematic form. This figure is not drawn to scale. The metal surface is partially covered with a layer of water oriented according to its dipole moment. In addition, ionized oxygen molecules are adsorbed at the surface (Bockris and Reddy, 1970). For example, 27% of platinum is covered by a monolayer of oxygen; 22% of palladium is covered by a monolayer of oxygen. This oxygen layer forms one charge layer of the

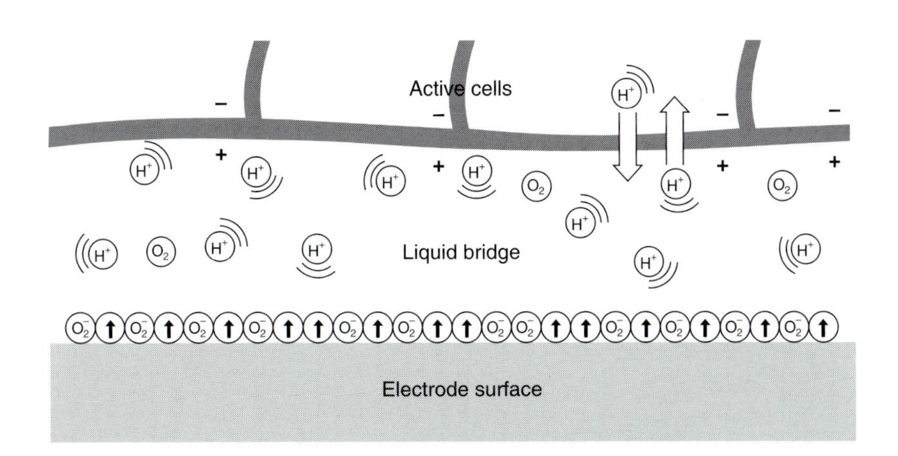

Fig. 3.3. Schematic representation of the liquid bridge and surfaces on each side of the bridge interface. Oxygen molecules are adsorbed and ionized on the electrode surface. Protons move about within the liquid bridge. Protons also move back and forth across the electrically polarized plasmalemma. This representation is NOT to scale. (Interfacial Structure Reference: Bockris and Reddy, 1970, Vol. 2, p. 1163.)

capacitor represented in the measuring electrode shown in Fig. 3.1. While the dominant characteristic of the surface structure is capacitive in nature, there is some charge transfer between metal and liquid. Hoare (1965) has suggested the following set of reactions between oxygen and the noble metal, platinum.

$$(O_2) + e \rightarrow (O_2^-)_{ads} \qquad (3.1)$$
$$(O_2^-)_{ads} + H^+ \rightarrow (HO_2)_{ads} \qquad (3.2)$$
$$(HO_2)_{ads} + e \rightarrow (HO_2^{\ -})_{ads} \qquad (3.3)$$
$$(HO_2^-)_{ads} + H^+ \rightarrow (H_2O_2)_{ads} \qquad (3.4)$$
$$(H_2O_2)_{ads} + Pt - O \rightarrow Pt + O_2 + H_2O \qquad (3.5)$$

Oxygen is adsorbed as a negatively charged oxygen molecule. Two of these reactions involve a proton. This indicates that the potential which accrues from this set of reactions is pH sensitive. In other words, it is not always the same oxygen molecules that are adsorbed at the surface. This layer of oxygen together with the pH sensitivity gives rise to the changes in measured potential due to changes in both oxygen and proton concentration in the liquid bridge.

Calibration curves for this oxygen and proton sensitivity are given in Figs 3.4 and 3.5, respectively. The form of the oxygen sensitivity curve follows from the mechanism of adsorption wherein at low concentrations of oxygen there are many surface sites available to occupy. Then, as the sites fill up, adsorption becomes more difficult and the sensitivity curve begins to flatten out. The form of the adsorption versus concentration shown in Fig. 3.4 is similar to the form of oxygen adsorption on myoglobin molecules (Matthews and Van Holde, 1990). The site occupation theory invoked in organic chemistry to explain this

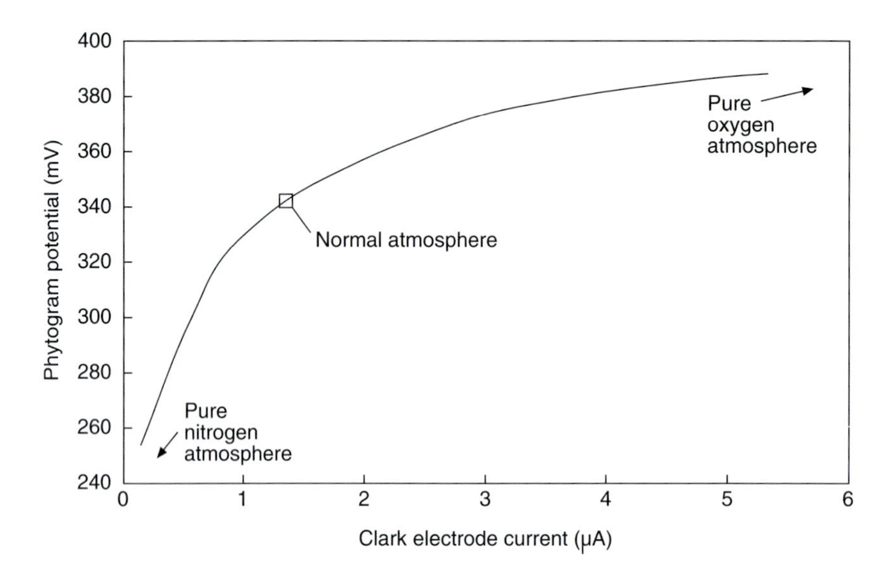

Fig. 3.4. Oxygen calibration curve. A set of electrodes was immersed in a sealed container containing distilled water and a phosphate buffer (Sigma #3288, pH 7.2 at 25°C). The solution was first sparged with nitrogen to the near-zero oxygen level. Then oxygen was slowly ported back into the solution. Oxygen concentration in the solution was measured with a conventional Clark electrode at a bias of −734 mV and no convection (YSI, Model 5739). The potential was measured with respect to a non-polarizable reference electrode (Orion 90–02 double junction).

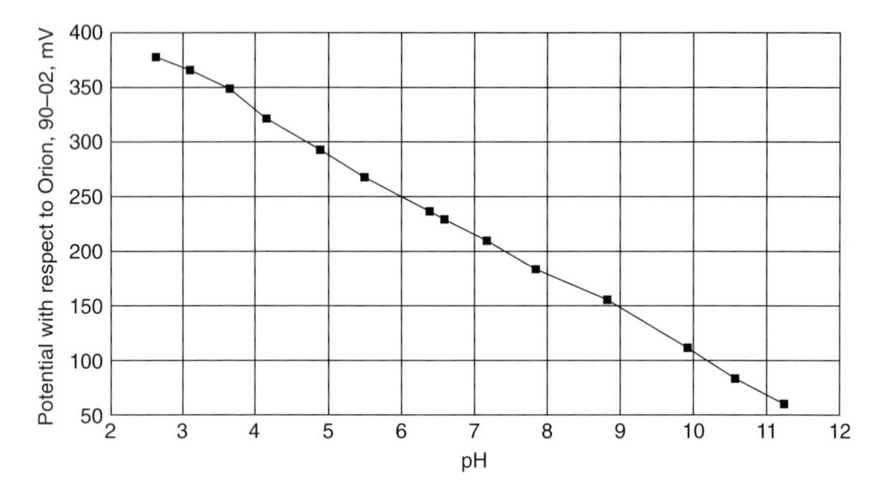

Fig. 3.5. Proton calibration curve. An electrode was immersed in a distilled water solution containing a conventional pH electrode (Beckman #39845) and a silver chloride reference electrode (Orion 90–02 double junction). Sodium hydroxide and hydrochloric acid were sequentially added with a micropipette to adjust the pH back and forth between 11 and 2.5.

form may be applicable here. Proton variations give rise to a linear potential sensitivity of 36 mV per log unit change in concentration. As the proton concentration rises, the potential rises.

This dual sensitivity gives rise to the question of separating out the influence of an oxygen change from a proton change. Background literature is consistent in reporting that the tissue is well aerated due to presence of lenticels in the main stem and regions of hydrophobic cell surfaces (Hook *et al.*, 1972). Raven (1977) considers the 'gas-filled continuum of intercellular spaces' as a transport system on the level of the liquid apoplast and the symplast.

Secondly, the presence of a variable wetted area indicates that the aereoplast is contiguous with the liquid apoplast at the electrode surface. Since the diffusion of oxygen in gas is 10,000 times faster than in liquid water and oxygen demand during the slow diurnal variations is probably stable, the oxygen concentration in the aereoplast at this liquid–gas interface is probably at equilibrium.

Thirdly, the sensitivity curve at atmospheric pressure indicates that a decrease in oxygen concentration in the liquid phase of as high as 20% would result in a change in potential of only 12 mV. The entire variation of potential as the oxygen goes from the atmospheric equilibrium level to near zero is only 90 mV.

Fourthly, when the oxygen is stripped off the electrode surface and further depleted in the liquid apoplast pathway by continuous application of external reducing potentials, the form of the recovery of the oxygen concentration after sudden removal of the external applied potential is exponential with a time constant of approximately 5 min (Gensler, unpublished). This suggests first-order kinetics can be applied to the movement of oxygen from the remote source in the aereoplast to the electrode surface (Nobel, 1991). Since this discussion is limited to diurnal variation in which the time period for oxygen usage during active proton transfer is about 300 min, the oxygen gradient between the aereoplast and the cell surface is very probably in equilibrium during this entire period.

These factors suggest that a valid assumption for non-transient changes in potential such as occur during the normal diurnal cycle are accomplished at no significant change in oxygen concentration. The changes in potential can then be attributed to changes in proton concentration alone with little loss of accuracy.

Since the apoplast contains other chemical constituents, the question further arises as to the influence of these constituents on the measured potential. The interfacial potential can arise from both adsorbed species (the word 'species' is used in the electrochemical sense) or an exchange of electrons across the interface. Adsorption of organic compounds is possible on noble metal surfaces (Bard and Faulkner, 1980). Oxygen adsorption is present and at significant magnitudes. If organic adsorption does occur, it would have to occur at a magnitude which supersedes the influence of oxygen in setting the interfacial potential. Such reactants have not been reported in the literature.

The presence of reactants in the apoplast which could exchange electrons with the metal was directly checked by cyclic voltammetry *in vivo* and the results compared with cyclic voltammetry *in vitro* (Silva-Diaz *et al.*, 1983). The patterns matched. This strongly suggested that reactants other than oxygen and hydrogen were not exchanging electrons at the electrode surface.

Interfacial electrode potential at a noble metal surface is set by what Hoare (1965) refers to as a 'potential determining reaction'. In order for ions such as K^+, Ca^{2+} or Cl^- present in the apoplast to set electrode potential, they would have to acquire electrons from or release electrons to the metal at the surface with an exchange current density that supersedes the potential determining reactions suggested in equations 3.1 through 3.5. Such reactions have not been reported in the literature.

Physiological Significance of the Proton Concentration in the Liquid Bridge

The previous sections have been concerned with electrochemical aspects of the electrode and the liquid bridge. Now some physiological aspects will be considered. Transfer of protons across the cell membrane is a well-known mechanism in cellular metabolism (Fensom, 1959; Hze, 1985; Nobel, 1991; Flowers and Yeo, 1992). Energetically, movement of the proton out of the cell requires energy to surmount the electrical potential gradient across the membrane shown in Fig. 3.3. Movement of the proton back into the cell releases this energy. There are many reasons for proton transfer. In the present context, proton movement is interpreted as the initial step in a water transfer mechanism similar to that of stomatal closure (Nobel, 1991). As the proton moves into the cell, potassium ions move out. Water moves out because of the increased water potential inside the cell with respect to outside of the cell. The movement of water out of the cells and into the apoplast increases the wetted area of the electrode surface and the capacitance rises. This is the underlying mechanism which relates bidirectional proton changes to bidirectional capacitance changes.

Relation between Changes in the Apoplast Proton Concentration and Changes in the Cellular Proton Concentration

Changes in the apoplast pH cannot be linearly related to changes in the cell pH because of the difference in the volume of the two regions, the possibility of masking of proton influence in the Donnan phase of the apoplast (Nobel, 1991) and possible generation of protons or absorption of protons in intracellular reactions.

A theoretical analysis of the first aspect wherein protons move from the apoplast into the cell is possible. Assume that the ratio between the volume of

the cells (considered as a whole) and the volume of the apoplast is given by R_v, the ratio of the proton concentration of the cell to the proton concentration of the apoplast is given by R_c, the number of protons in the apoplast volume before the transfer is N_a and the number of protons in the apoplast volume after the transfer is KN_a. The cellular pH before and after the transfer of protons from the apoplast into the cells is given by:
Before:

$$pH_c = -\log (N_c/V_c) = -\log (N_a/V_a) - \log (R_c) \quad (3.6)$$

After:

$$pH_c = -\log (N_a/V_a) - \log (R_c) - \log \{1 + [(1 - K)/(R_c \cdot R_v)]\} \quad (3.7)$$

The apoplastic pH before and after the transfer of protons from the apoplast into the cells is given by:
Before:

$$pH_a = -\log (N_a/V_a) \quad (3.8)$$

After:

$$pH_a = -\log (N_a/V_a) - \log (K) \quad (3.9)$$

Table 3.1 gives the values of these pH levels for a volume ratio, R_v of 20 (Grignon and Sentenac, 1991) and a peak to peak diurnal electrode potential change of

Table 3.1. pH changes in the apoplast and the cells before and after a transfer of protons from the apoplast to the cells as a function of a difference in initial pH of the apoplast.

	Before	After	Before	After	Before	After
pH of the cells	7	6.98	7	6.82	7	6.22
pH of the apoplast	7	9	6	8	5	7
Fractional concentration, K	0.01		0.01		0.01	
Ratio of the volume of the cells to the volume of the apoplast, R_v	20		20		20	
Ratio of the proton concentration of the cells to the proton concentration of the apoplast, R_c	0.1		0.01		1.0	

K, R_v and R_c are dimensionless.
Example of the values in this table: If the ratio of the cell volume to the apoplast volume is 20 and the ratio of the proton concentration in the cells to the proton concentration in the apoplast is 0.1, a transfer of protons from the apoplast to the cells occurs such that the apoplast begins the transfer at pH 6 and ends the transfer at pH 8. The cells begin the transfer at pH 7 and end the transfer at pH 6.82.

72 mV or two log units at three different values of R_c. If the apoplast and cellular pH values both begin at pH 7, a change in apoplast pH from 7 to 9 corresponds to change of cellular pH from 7 to 6.98. The volume difference completely dominates the influence of the transfer. By contrast, when the initial pH values of the apoplast and cells are 5 and 7, respectively, a change in apoplast pH of two log units corresponds to a change in the cellular pH of 0.72 log units. Only when the initial pH value of the apoplast is significantly below the initial value of the cellular pH does the transfer have a significant influence on the cellular pH. These relations would exist in the absence of other cellular and apoplastic proton absorbing or releasing mechanisms. Since the absolute value of potential relates to the absolute level of pH in the apoplast, this analysis indicates that both the absolute value of potential as well as the difference in potential over the diurnal cycle must be considered in determining the influence within the cell of this proton transfer.

The Donnan phase in the apoplast is primarily negatively charged (Nobel, 1991). As such it can mask the measurement of proton concentration by immobilizing protons moving out of the cell into the apoplast. This mechanism would decrease the number of protons available for the reactions described in equations 3.1 through 3.5.

Raven and Smith (1976) discuss the regulation of cellular pH in terms of a biochemical mechanism and a biophysical mechanism. The biochemical mechanism involves cellular reactions that influence intracellular pH and not transfer across the cell membrane. The biophysical mechanism involves transfer of protons across the cell membrane.

The two mechanisms are interrelated. For this reason, it is difficult to predict the change in cellular pH based on measurement of changes in apoplast pH alone.

Electrochemical Circuit Aspects of the Measurement of Electrode Potential

In the electrochemical circuit shown in Fig. 3.1, the measured potential includes the voltage across the soil–return electrode interface. If changes in the measured potential are observed, the assumption must be made that these changes arise from changes across the tissue–measuring electrode interface in the presence of a constant soil–return electrode interfacial potential. The assumption that the return electrode interface is constant can be checked by examining the measured potential from two or more measuring electrodes which utilize the same return electrode. A common electrochemical circuit path exists for the two electrodes. Part of this common path is the soil–return electrode interface. If the potential across the interface changes, this change would appear in the measured potential pattern of both electrodes. Such commonality of change is easily discerned and can be taken into account in the data analysis. For example, it occurs after a severe rainstorm when a flush of water penetrates down into the soil and passes over the soil–return electrode interface. This chapter focuses on diurnal patterns during which the constancy of the soil–return electrode interfacial potential is readily determined.

In the longer term, an alternate method of stabilizing the soil–return electrode interface is to use a reference electrode as a return electrode such that the reference electrode interface is connected to the soil via a salt bridge. The constituents of the salt bridge can be designed to minimize changes in the potential drop between the soil and the conducting electrolyte within the reference electrode.

The moles of oxygen drawn from the interface by the measuring circuit can be calculated based on an input impedance of the measuring electrode circuit of 10,000 MΩ, a measuring period of 0.1 s and an average potential of 500 mV. If four electrons are exchanged per molecule of oxygen, the number of moles extracted from the interface in each measurement of potential is 0.13 fmol.

The Extent of the Liquid Bridge: Wetted Area, Electrode Capacitance, Extracellular Hydration

Liquid does not completely occupy the annulus between the wall of intact cells and the surface of the electrode. Air is present. In other words, part of the electrode surface is wetted by liquid in the apoplast, and part of the electrode surface is exposed to gas in the aereoplast. As the internal hydration of the plant changes, the area of the electrode surface covered with liquid changes. By a measurement of the electrode wetted surface area, the degree of internal hydration can be measured. Electrical capacitance is directly proportional to this wetted area because only the wetted surfaces of the electrode support the two charge layers. The essential mechanism whereby the internal hydration level can be assayed with the electrode is by a measurement of the electrode capacitance.

A calibration curve indicating the relation between wetted area and electrode capacitance is given in Fig. 3.6. This calibration curve was determined by progressively immersing an electrode into and out of water with a specific conductivity of less than 32 μΩ at pH 7.0. The slope of the curve is 0.143 μF mm^{-1} of immersed electrode length. This corresponds to 0.0456 μF mm^{-2} of wetted electrode surface area.

The slope of this hydration calibration curve can be used to determine the approximate separation distance between the two charge layers forming the tissue capacitor shown in Fig. 3.1. The formula for capacitance is given by

$$\text{Capacitance} = \frac{\text{(area of charge layers) (dielectric constant) (permittivity)}}{\text{(distance between charge layers)}} \tag{3.10}$$

Using a dielectric constant of 80, the distance between the two charge layers is 0.2 nm. This would be the approximate distance from the centre of the adsorbed

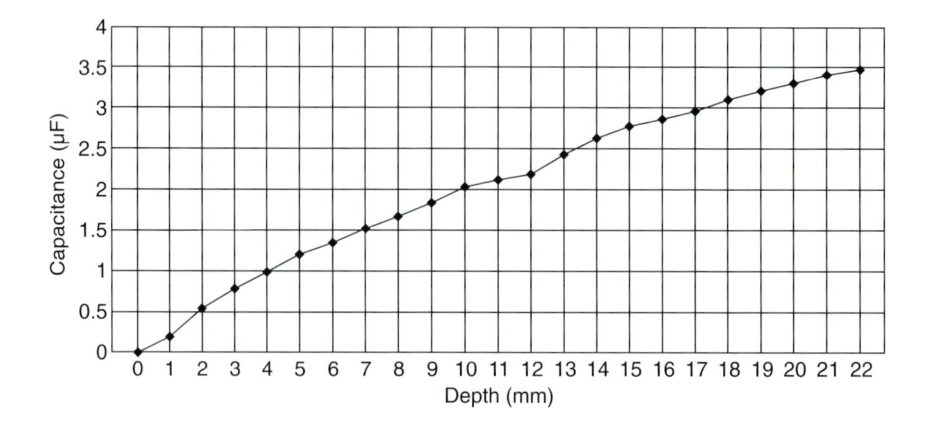

Fig. 3.6. Hydration calibration curve. A set of measuring electrodes were initially suspended above water with a specific conductivity less than 32 µΩ and a return brass electrode. The electrode was then progressively lowered into and out of the water and the capacitance measured at 1 mm steps. This procedure was performed at three levels of potential. All six conditions yielded the same slope. The run shown is for movement out of the water at a potential of 224 mV. The procedure was repeated in tap water with no change of result.

oxygen molecules to the metal surface schematically depicted in Fig. 3.3. This is in agreement with distances for charge separation of other adsorbed species (Bard and Faulkner, 1980).

Physiological Significance of the Variation in the Extent of the Electrode Wetted Surface Area

The apoplast occupies a volume of the aerial organs of 5% or less (Grignon and Sentenac, 1991). If the diurnal shrinkage in radius is 100 µm and the assumed distensible bark radial distance is 2000 µm, the aereoplast occupies 20 µm of equivalent radial length. This implies that the water moving out of the cell must then move out of the apoplast since the volume of the available aereoplast is too small to absorb this water transfer. This implies that the water at the surface of the electrode is in motion during the non-equilibrium daytime period. Presumably this water moves towards the cambium during the day and from the cambium at night.

Electrode Implant Technique

The method of implant of electrodes in mature trees is to excise a square portion of bark radially down to the woody secondary xylem. This is easily discerned by the abrupt change in density as one removes the tissue. The electrode is then inserted circumferentially into undisturbed tissue among the perimeter of the trunk at the base of the square hole in the region slightly more radial than the woody secondary xylem. In this manner, the electrode slides along soft tissue which is approximately 100 µm more radial than the woody/soft tissue interface. This implant location is said to be within the cambial zone. Electrode length into the undisturbed tissue is variable, but a distance of 100–200 cm is easily achieved.

The size of the excised hole varies depending on the age of the tree. For older trees with thicker bark, the hole must be proportionately larger. As an example, for trees with bark approximately 2 cm thick, a square hole about 12 by 12 mm in area is sufficient. The hole and electrode are sealed with asphalt sealant.

Schematic Representation of the Diurnal Potential, Hydration and Radius Pattern

The consistency of the general form of the diurnal pattern of potential, hydration and radius permits representation in schematic form and definition of the salient characteristics of the pattern as shown in Fig. 3.7. The diurnal potential pattern can be visualized as a pulse which moves negatively away from an equilibrium level during postdawn, reaches a minimum level and remains at that level for a period of time before recovering back to the equilibrium level during the late afternoon and evening. The hydration pattern has these same essential characteristics differing only in the direction of the pulse. During the daytime the hydration level increases. The radius pattern has the same basic form as the potential pattern, that is, the radius decreases during the daytime.

The schematic representation in Fig. 3.7 leads to the following definitions applicable to potential, hydration and radius changes from the predawn equilibrium level:

T_0 = Time of onset of the change away from equilibrium
T_1 = Time at which the midday short-term equilibrium is reached
T_2 = Time at which the return to the equilibrium level begins
T_f = Time at which the equilibrium level is reached
M = Magnitude of the excursion away from equilibrium

Point Dendrometers

The electrochemical measurements are combined with a physical measurement of trunk radius changes using point dendrometers. The implant is usually located within about 15 cm of the point dendrometer sensing head. This proximity permits the assumption that the physiological activity of the region of the implant gives rise to kinetically correlated changes in the radius as

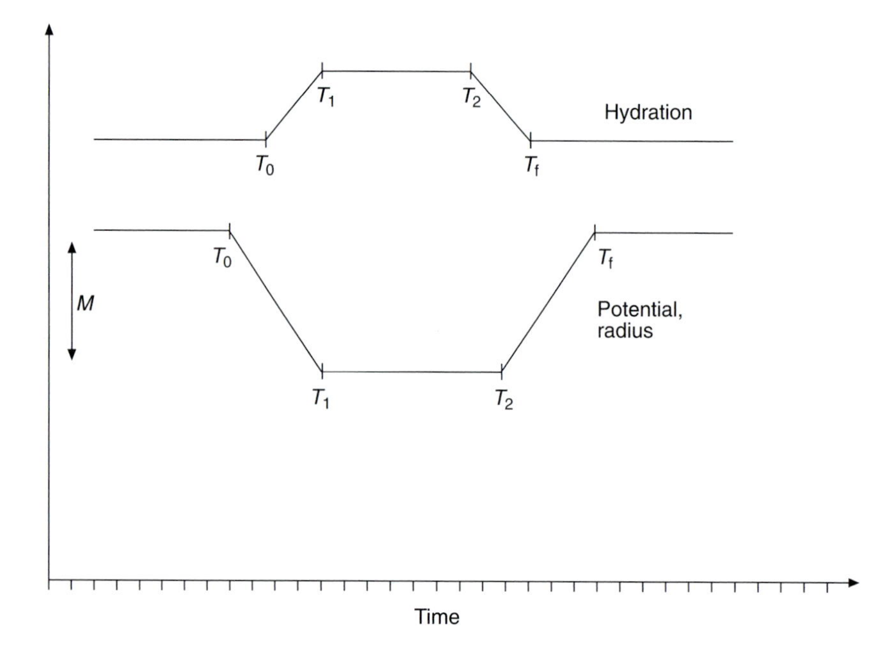

Fig. 3.7. Schematic representation of the diurnal potential, radius and hydration patterns.

measured by the dendrometer. Dendrometer resolution is 4 μm and the linear correlation coefficient relating radius change to electrical output is 0.995 or better over the range of operation.

The point dendrometer sensing head is at the outer surface of the bark. As such it measures the summation of changes in all tissue along the radial vector from the bark surface to the centre of the trunk. Mathematically,

$$\text{Radius} = \text{SUMMATION}\ (C_j(t)) \tag{3.11}$$

where $C_j(t)$ is the radial length between the middle lamellae of cell j in metres at time t. The summation is over the radius vector from radius zero to the main stem surface. The point dendrometer output is

$$\text{Output} = \text{SUMMATION}\ (C_j(t)) - \text{SUMMATION}\ (C_j(t_b)) \tag{3.12}$$

where t_b is a base time. The summation can be broken down into subsets such as C_{stele}, C_{cortex} and $C_{periderm}$. These grouping are apropos for regions in which middle lamellae are no longer discernible or do not elicit changes. For a visual description of the change see Zimmermann (1983).

Experimental Sites

Two sites were monitored with combined PHYTOGRAM/dendrometers. In southeast Arizona, 16 electrodes with a diameter of 150 μm were implanted in the cambial zone of the main stem of four *Pinus engelmannii* in Rhyolite Canyon (lat: 32°00′15″; long: 109°21′10″; alt: 1554 m) in the Chiricahua National Monument. Tree age varied between 70 and 90 years old as determined from increment cores evaluated at the Tree-Ring Research Laboratory of the University of Arizona.

Electrodes were implanted by excising a wedge-shaped portion of the bark of the main stem approximately 1.5 by 1.5 cm exposing the cambial region in May of 1995. Electrodes were inserted horizontally and circumferentially into the cambial region such that the electrode surface extended 1–2 cm into undisturbed tissue. The remainder of the electrode and the exposed area was then covered with asphalt sealant to eliminate any contact with air or water. Automatic point dendrometers were installed at two heights on the trees such that the sensing head of the dendrometers was located about 15 cm diagonally away from the electrode implant.

In Lewisham, Tasmania, near Hobart, nine electrodes with a diameter of 150 μm were implanted in the cambial zone of the main stem of two *Eucalyptus globulus* (lat: 32°00′15″; long: 109°21′10″; alt: 15 m). Tree diameter at the implant location was approximately 20 cm. The implant technique was the same as described above for the Arizona site.

RESULTS

Characteristics of Diurnal Potential, Radius and Hydration Patterns

Cyclic diurnal patterns appeared with great regularity at both sites in all the electrodes. In keeping with the emphasis in this chapter on method and interpretation, examples of the patterns will be given which illustrate the basic characteristics and their interpretation in terms of physical and electrochemical properties.

Figure 3.8a,b shows the hourly pattern of the potential, radius and hydration on 4 January 1997 at the Lewisham site. Figure 3.9a,b shows the summer pattern on 21 July 1995 at the Rhyolite site. This pattern is characteristic of the time period from about late March until late October. Figure 3.10a,b shows the winter pattern exhibited on 17 November 1995 at the Rhyolite site. This is the pattern from late October until late March. Table 3.2 quantifies the characteristics in terms of the four principal time breaks and magnitude defined in Fig. 3.7.

In summary, throughout the year: (i) the diurnal potential pattern is characterized by a drop in the morning after dawn, a midday period of constant potential and then a rise back to equilibrium in the evening and night-time

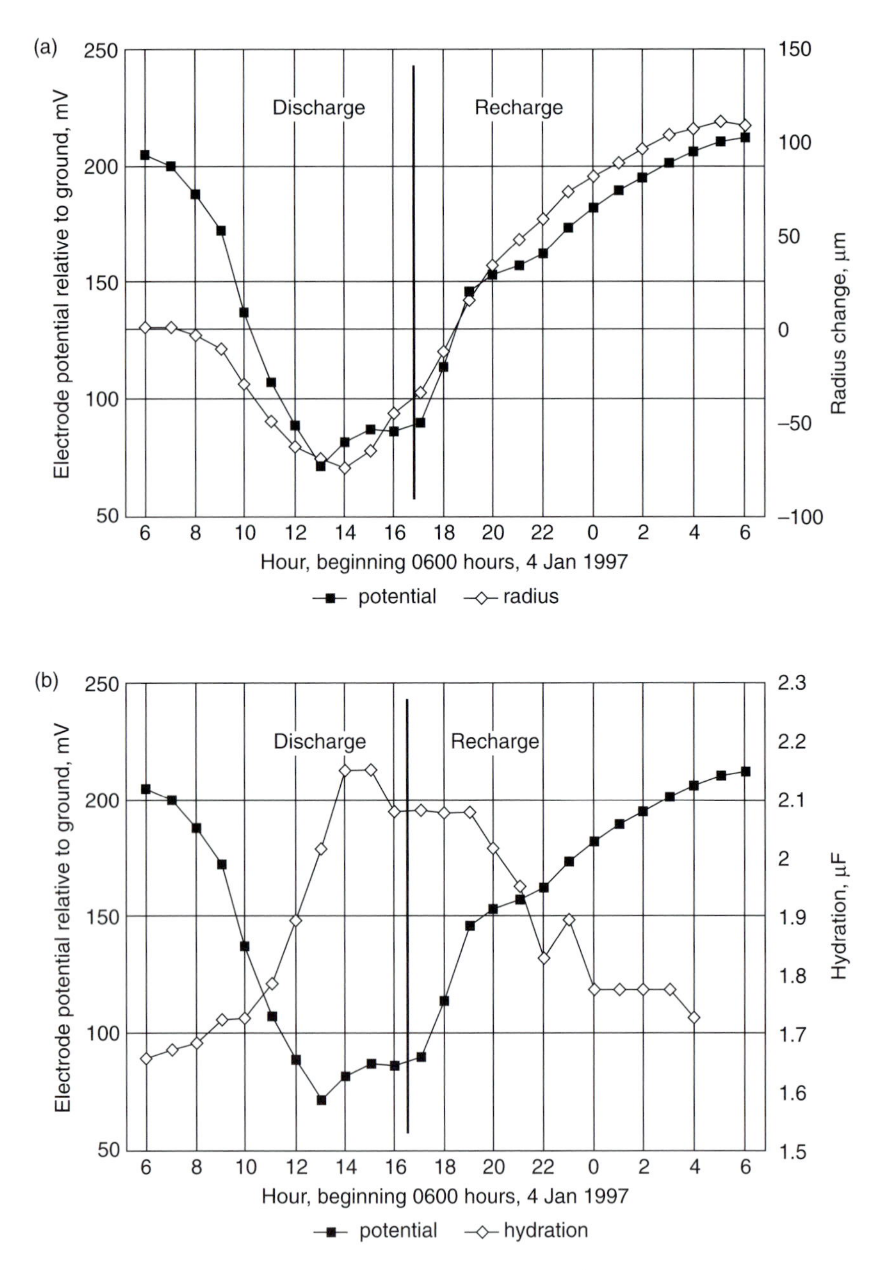

Fig. 3.8. (a) Diurnal pattern of potential change and radius change, Lewisham, 4 January 1997. Tree 5, Upper dendrometer B20: north aspect, height: 650 cm. Electrode P43, height 700 cm. Sampling rate: once per hour. The net growth after this 24-h period was 115 µm. (b) Diurnal pattern of potential change and hydration change, Lewisham, 4 January 1997. Tree 5, Electrode P43, north aspect, height 700 cm. Sampling rate: once per hour. Temperature maximum/minimum was 31.8/13.6°C.

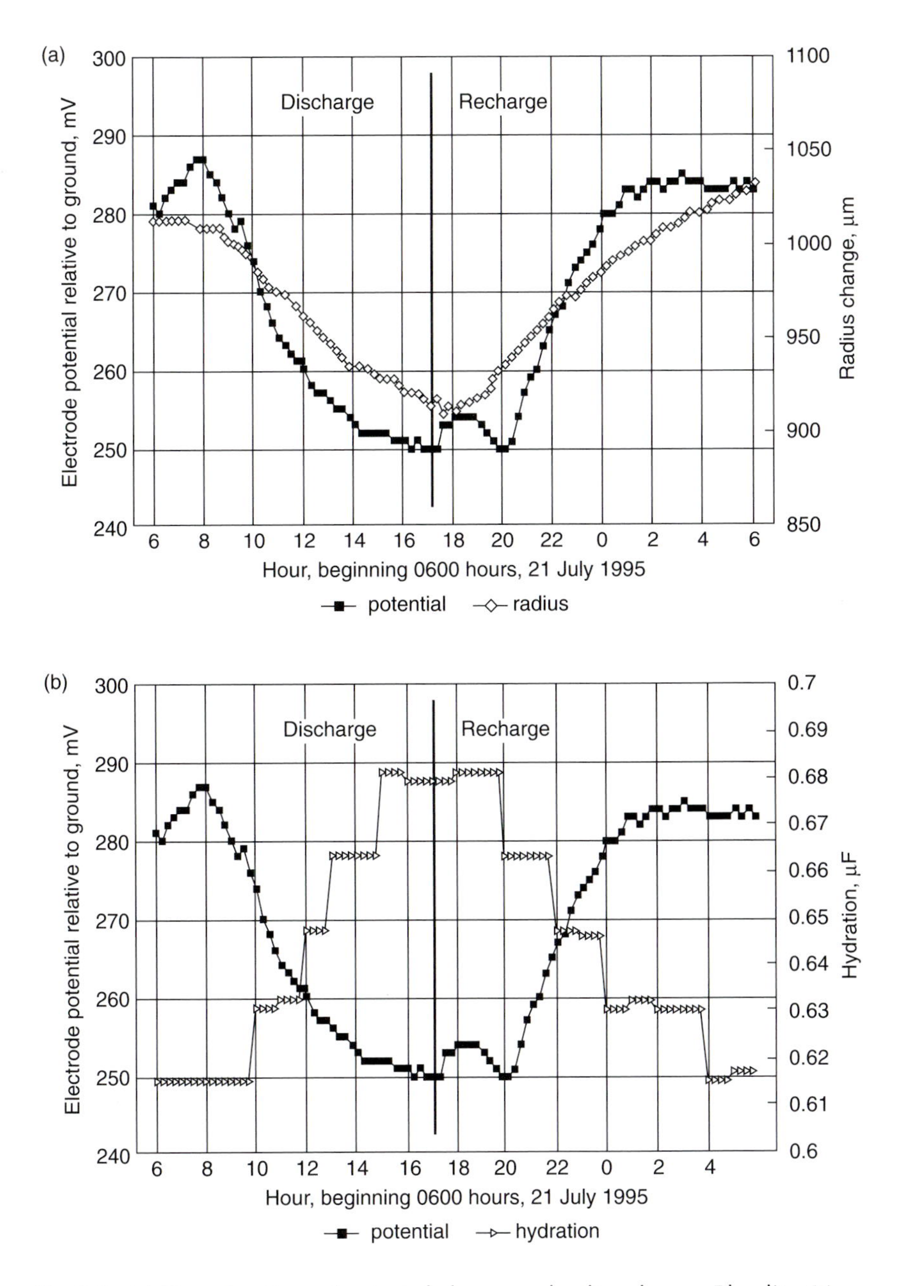

Fig. 3.9. (a) Diurnal pattern of potential change and radius change, Rhyolite, 21 July 1995. Tree 6, Lower dendrometer A1, height 272 cm. Electrode P20, south aspect, height 280 cm. Potential and radius sampling rate: once per 15 min. Temperature maximum/minimum for this day was 32.2/11.7°C. (b) Diurnal pattern of potential change and hydration change, Rhyolite, 21 July 1995. Tree 6, Electrode P20, south aspect, height 280 cm. Potential sampling rate: once per 15 min, hydration sampling rate: once per hour.

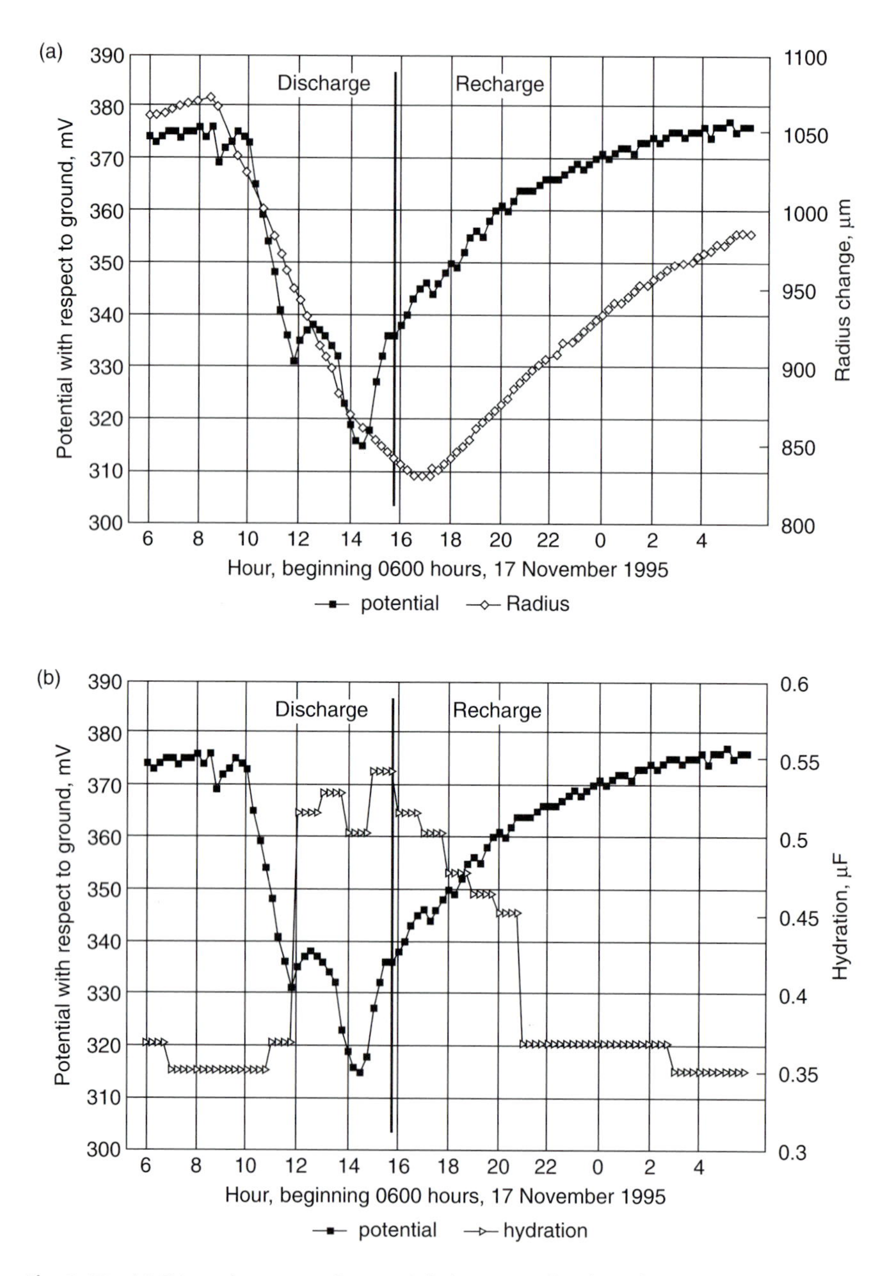

Fig. 3.10. (a) Diurnal pattern of potential change and radius change, Rhyolite, 17 November 1995. Tree 6, Lower dendrometer A1, height 293 cm. Electrode P20, south aspect, height 300 cm. Potential and radius sampling rate: once per 15 min. Temperature maximum/minimum was 17.0/1.0°C. (b) Diurnal pattern of potential change and hydration change, Rhyolite, 17 November 1995. Tree 6, Electrode P20, south aspect, height 300 cm. Potential sampling rate: once per 15 min, hydration sampling rate: once per hour.

Table 3.2. Values of T_0, T_1, T_2, T_f and M for the summer and winter potential, hydration and radius pattern at Lewisham and Rhyolite.

		T_0	T_1	T_2	T_f	M
Lewisham, 4 Jan 1997	Potential	0600	1300	1700	0600	−123
	Radius	0700	1400	1500	–	−60
	Hydration	0700	1300	1300	0300	3.3 (45)
Rhyolite, 21 Jul 1995	Potential	600	1400	2000	0200	−46
	Radius	900	1800	1800	0600	−110
	Hydration	1000	1500	2000	0400	0.065 (11)
Rhyolite, 17 Nov 1995	Potential	1000	1430	1430	0200	−60
	Radius	0830	1630	1700	–	−240
	Hydration	1100	1200	1600	0300	0.180 (51)

Values of T are in hours. Values of M are in mV for potential, μm for radius and μF for hydration. Numbers in parentheses are percentage change based on the equilibrium value as 100%. See p. 67 and Fig. 3.7 for the definition of the T values and M.

period; (ii) the radius pattern is similar with the prominent exception of a lack of an extended constant radius period during the midday; (iii) the hydration pattern is characterized by a rise in the morning, a midday plateau and then a slow decline back to the equilibrium level; and (iv) slope of the change for all three variables is more rapid in the morning than the slope in the late afternoon.

Electrochemical and Physical Interpretation of the Potential, Hydration and Radius Pattern

Energetically, the period, T_0 to T_1, in which the proton level is dropping, is interpreted as an energy discharge period concomitant with protons moving across the cell membrane and into the cell as shown in Fig. 3.3 (Flowers and Yeo, 1992). Conversely, the period, T_2 to T_f, in which the proton level is rising, is interpreted as an energy recharge period as protons are actively pumped out of the cell and into the liquid bridge. In other words, an increase in proton concentration in the liquid bridge is an energy storage.

Another interpretation is to consider the protons located in the liquid bridge as possessing a variable level of potentiality much like water has a certain potentiality in tracheids. The electrode output in terms of the interfacial potential can be looked upon as a quantitative measure of the energy storage of the protons per unit of charge. Protons function as 'metabolic money'. The electrode senses the variable storage level of this money placed in the extracellular 'bank'. Quantitatively, the chemical potential of protons in the liquid bridge is (Nobel, 1991, equation 2.4 applied to protons and setting gravitational and pressure contributions equal to zero)

$$\mu_{\mathrm{p}} = \mu_{\mathrm{p}}^{*} + R\,T \ln a_{\mathrm{p}} + F\,E \qquad (3.13)$$

where μ_{p} is the chemical potential of the protons in the liquid bridge [J mol^{-1}]; μ_{p}^{*} is the standard chemical potential [J mol^{-1}]; R is the gas constant [J mol^{-1} K]; T is Kelvin temperature; a_{p} is the proton activity [mol cm^{-3}]; F is the Faraday constant [C mol^{-1}]; and E is the electric potential [J C^{-1}].

For example, in Fig. 3.8b, the change in electric potential of the electrode from dawn to mid-afternoon was −123 mV. This corresponds to a change of 123/36 log base ten units or 7.85 log base e units. Assume that the change took place under conditions of constant temperature of 25°C and no net charge density change in the liquid bridge, then the change in the electric potential in the liquid bridge is zero. Assume further that the proton activity is equal to the concentration. The decrease in chemical proton potential of protons in the liquid bridge was then 8.31 by 298 by 7.85 which equals 19,439 in J mol^{-1}. Presumably, this energy was used to pump water out of the cells.

While it is possible to interpret increases and decreases of proton concentration in the liquid bridge in terms of energy, it is not possible to directly ascertain the use of the energy as the protons move back into the cells. Nor is it possible to determine any changes in cell membrane potential due to these proton transfers. In other words, the question of cellular energy partitioning is moot. Energy can be used for different tasks other than water movement (Raven and Smith, 1974; Smith and Raven, 1979; Giaquinta, 1983). The presence of the diurnal cycling on days of temperature stress and their absence on days of little or no temperature stress gives support to the usage of the energy for water movement. Regression analysis of the potential and hydration patterns yields correlation coefficients as high as 0.95 on some days but as low as 0.06 on other days. The higher values appear to occur on higher temperature stress days. Furthermore, there is a distinct phase relation in the patterns of potential and hydration which can be explained in terms of a cause and effect relation between potential and hydration, but this phase delay would lead to very low values of linear regression coefficients when computed as if no phase delays were present.

The rise in hydration is interpreted as a movement of water out of the cells which in turn increases the wetted area of the electrode surface. For example, in the diurnal pattern of hydration for 21 July 1995 shown in Fig. 3.9b, the equilibrium value of hydration was 0.615 μF. Using the calibration curve in Fig. 3.6, this would correspond to a wetted area of electrode surface of 13.4 mm^2. During the day, the capacitance increased 0.065 μF or an increase in wetted area of 1.51 mm^2.

Comparison of the Summer and Winter Patterns at Rhyolite

Comparison of the potential patterns from July (Fig. 3.9) and November (Fig. 3.10) indicate the general form of the pattern is the same, that is, a drop in potential in the morning and early afternoon, followed by a rise in the late afternoon which extends until dawn the next day. The time of onset of the potential decrease, T_0, is later in November than in July. The midday equilibrium period (T_2–T_1) is greater in July than in November. The time of the onset of the recovery, T_2, is earlier in November than in July. The magnitude of the potential drop, M, is greater in November. There is virtually no midday equilibrium period in November compared with a 6-hour period in July. The time of onset of the potential recovery, T_2, is much earlier in November than in July. Table 3.2 quantifies these differences. In general, one can conclude the potential pulse is shorter in duration and greater in magnitude in November than in July.

With regard to hydration, the time of onset of the hydration increase, T_0, is considerably later in November than in July. The residence time (T_2–T_1), is approximately the same in July and November, 4 hours. The onset of the recovery is earlier in November than in July. A major difference between the July and November hydration patterns is the magnitude of the hydration increase over equilibrium. In general, the hydration rise above equilibrium in November is greater in magnitude but shorter in duration than in July. The maximum hydration in November is less than in July because auto-dehydration leads to lower hydration equilibrium values.

With regard to radius, the time of onset of the radius decrease, T_0, is approximately the same. The time of onset of the radius recovery is earlier in November than in July. There is a much greater decrease in the magnitude of the radius, M, in November (240 μm) than in July (110 μm).

The radius midday equilibrium, (T_2–T_1), is zero in July and only half an hour in November. This is a reasonable result since a zero change in radius over time would require that at each and every $C_j(t)$ along the radius vector is zero. The only exception to this would be in the case of algebraic cancellation wherein some of the $C_j(t)$ are increasing in volume and some of the $C_j(t)$ are decreasing in volume at the exact same absolute rate at the same time. This is highly unlikely.

In general, the magnitude of the radius change in November is greater than in July. This appears contradictory since the temperature stress in July is much greater than in November. The reason for this probably lies in the auto-dehydration that occurs in the autumn and winter. The bark has a reduced quiescent hydration level during these months. This gives rise to the large percentage increase in extracellular hydration during the diurnal cycle in November. Concomitant with this large-magnitude hydration pulse is an equally large magnitude radius decrease. In other words, the magnitude of the diurnal pulse in radius is influenced by the hydration equilibrium level and the ensuing hydration pulse magnitude. The reduced winter hydration equilibrium

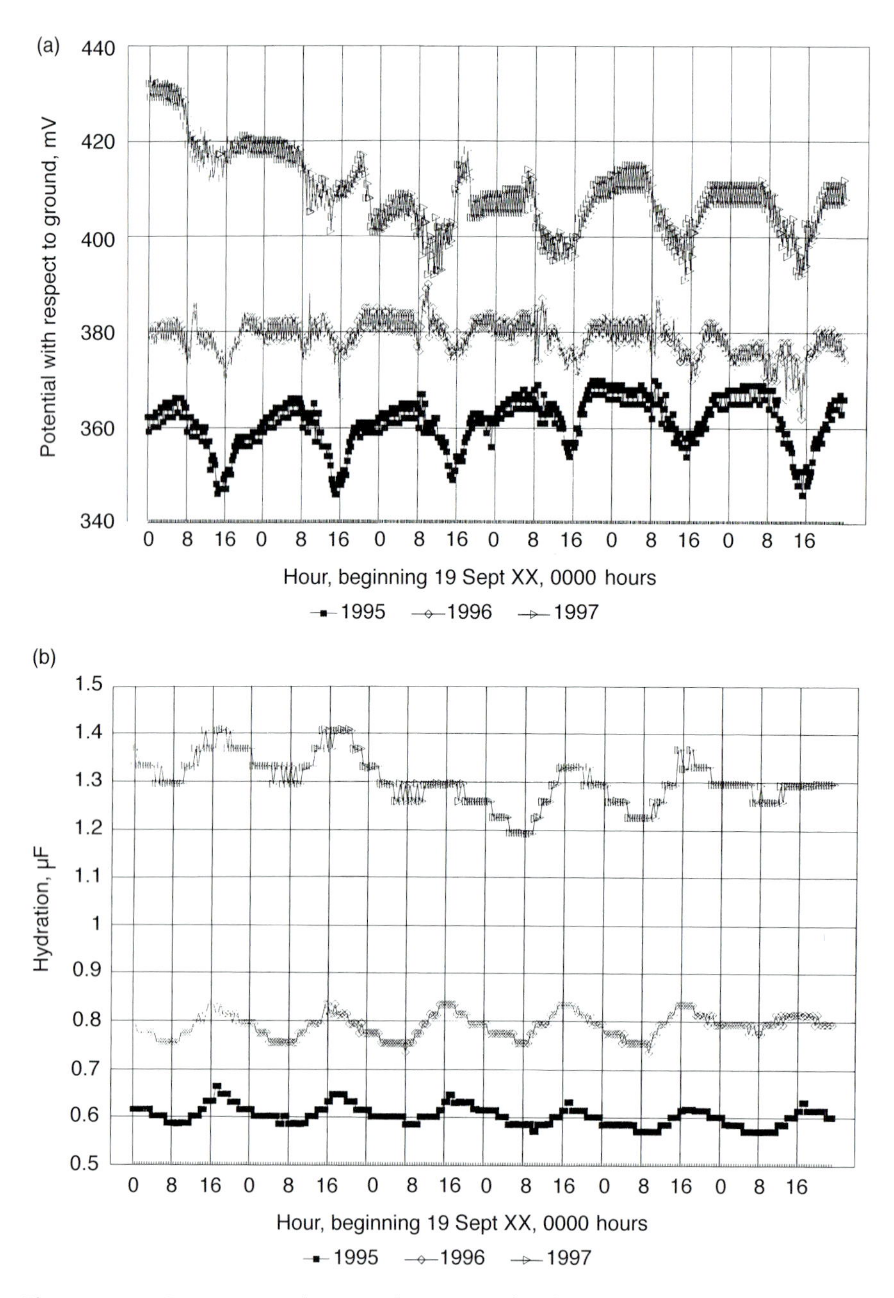

Fig. 3.11. (a) Comparison of potential patterns, Rhyolite, 19–24 September 1995, 1996, 1997. Tree 5, Electrode P34, south aspect, height 300 cm. Sampling rate: once per 15 min. (b) Comparison of hydration patterns, Rhyolite, 19–24 September 1995, 1996, 1997. Tree 5, Electrode P34, south aspect, height 300 cm. Sampling rate: once per hour 1995, 1996, once per half hour 1997.

(*Continued opposite*)

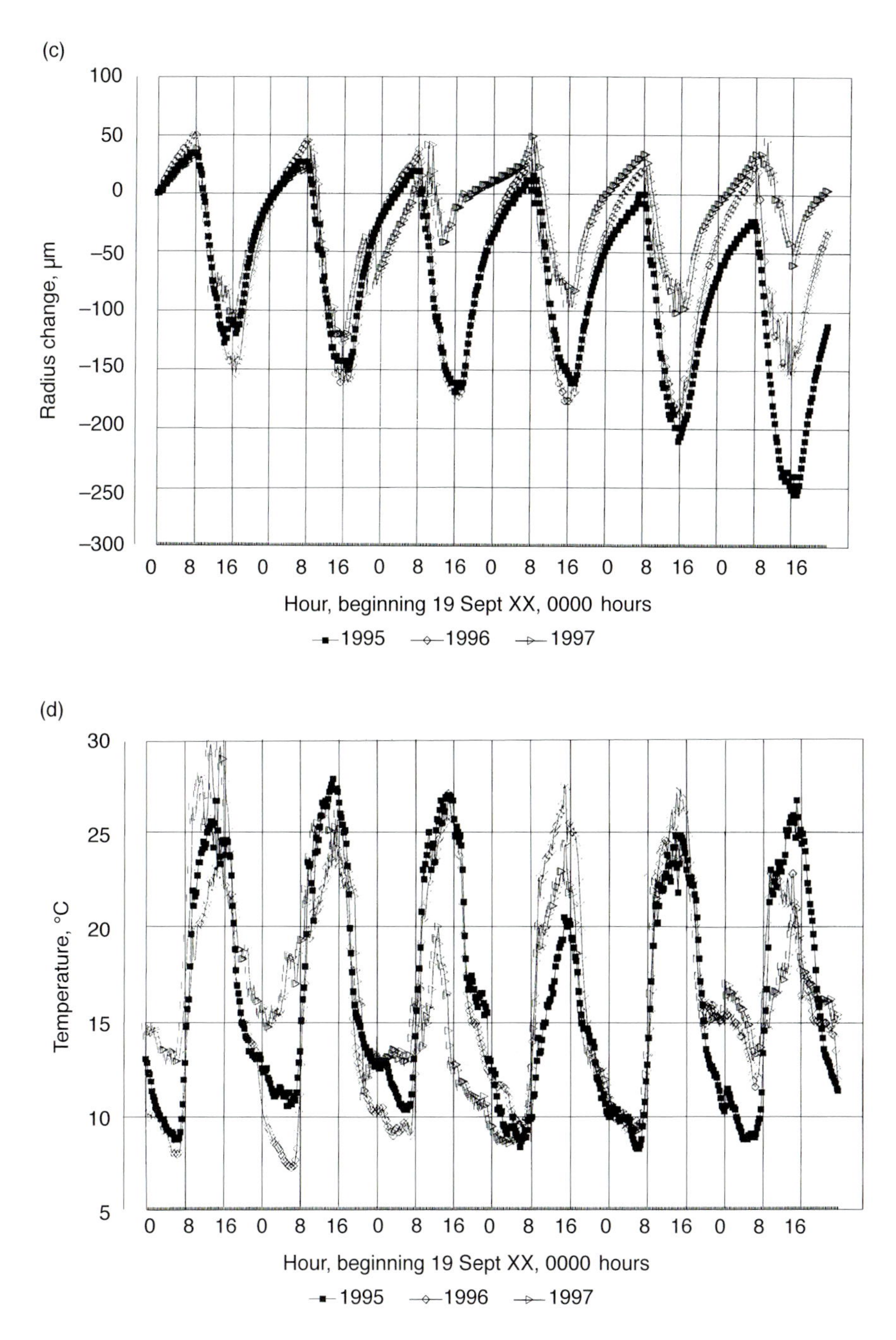

Fig. 3.11. (c) Comparison of radius patterns, Rhyolite, 19–24 September 1995, 1996, 1997. Tree 5, Lower dendrometer A1, south aspect, height 293 cm. Sampling rate: once per 15 min. (d) Comparison of temperature patterns, Rhyolite, 19–24 September 1995, 1996, 1997. Sampling rate: once per 15 min.

level strongly influences the form of the diurnal cycle of hydration *and* the magnitude of the radius cycle.

Rhyolite: Comparison of Patterns 4, 14 and 20 Months after Implant

An important question concerns the location of the electrode as the tree grows. In other words, where is the electrode located in the long term. This question can best be answered by anatomical analysis. A second analysis technique is to test the functionality of the electrode by comparing the potential and hydration patterns at the same time during the yearly cycle. Figure 3.11a, 3.11b, 3.11c and 3.11d give the potential, hydration, radius and temperature patterns for the period from 19–24 September 1995, 1996 and 1997, respectively. This corresponds to a period 4 months, 16 months and 28 months after implant of the electrode in the cambial region, respectively. This time interval was selected because it follows the end of the summer rainy season and a period of relatively stable weather from year to year. The radius curves have been normalized to zero on 19 September for each year. Absolute values of radius increase from May to September of 1995, from September 1995 to September 1996, and from September 1996 to September 1997 were 843, 1023 and 2913 μm, respectively. In other words, the radius increase was a total of 4779 μm over the 28-month interval.

The similarity of weather in September 1995 and 1996 is apparent from the temperature plots. Weather in the September 1997 period was relatively unsettled. But the low temperature on 21 September 1997 caused a distinct lack of radius change. This response was matched in both the potential and hydration curves for that day. By contrast, the similarity of the weather on 23 September for the three years resulted in a similar potential and hydration response. This strongly suggests that the functionality of the electrode is unimpaired after 28 months of residence in the tissue.

Division of Hydration Pattern into Cell Water Discharge and Cell Water Recharge Periods

The hydration pattern in Fig. 3.8b can be divided into two periods: a period in which water movement is out of the cell and into the apoplast and a period in which water movement is from the apoplast into the cell. Since the wetted area returns to an equilibrium value at the end of a 24-h cycle, the assumption is made that the amount of water leaving the cells is the same as the amount of water returning to the cells. This assumption permits a determination of the transition time, T_t, whereupon the direction of water transfer changes. Now the volume of the aereoplast is much smaller than the volume of the cells themselves. Furthermore, the net tissue volume decreases as this hydration level rises as seen by the decrease in radius. This suggests that when the water moves out of the cells it must be further transported out of the apoplast because of the limited volume of the aereoplast.

With these assumptions, discharge of water from the cells into the apoplast can be described by the time integral:

$$\text{Discharge from cells to apoplast} = \int_{T_0}^{T_t} K\,C(t)\,\mathrm{d}t \tag{3.14}$$

The lower limit of integration, T_0, is the onset of the hydration rise in the early morning. The upper limit of integration, T_t, is the time at which discharge ceases and recharge begins, or what might be termed the transition time. The integrand is the absolute value of the rate of water transfer given by a proportionality constant, K, times the measured capacitance, C.

The recharge of water from apoplast into the cells can be described by a similar time integral:

$$\text{Recharge from apoplast to cells} = \int_{T_t}^{T_f} K\,C(t)\,\mathrm{d}t \tag{3.15}$$

The lower limit of integration, T_t, is the beginning of the recharge period or the transition time which occurs in the afternoon. The upper limit of integration, T_f, is the time at which recharge ceases and the hydration level remains constant. The integrand is the absolute value of the rate of water transfer given by a proportionality constant, K, times the measured capacitance, C.

Notice that the measured capacitance is used in both integrals even though the direction of water transfer changes. This is possible because the absolute value of water transfer removes directionality from the relation between capacitance and water transfer. Notice further that the total capacitance value is used, not just the increment of capacitance above the equilibrium level. The total capacitance is used because water movement is assumed to occur over the entire wetted area, not just the daytime increase in wetted area.

Since the hydration level during stable weather before and after the hydration rise is the same, discharge equals recharge. This means the two integrals can be equated. The transition time can then be determined from the relation:

$$\int_{T_t}^{T_f} K\,C(t)\,\mathrm{d}t = 2\int_{T_0}^{T_t} K\,C(t)\,\mathrm{d}t \tag{3.16}$$

If linearity is assumed, the proportionality constant is independent of the capacitance value and can be cancelled out of the equation. The onset time T_0 and the time to return to equilibrium time T_f are known. The transition time can be determined by evaluation of the integrals.

Applications of these mathematical relations are given in the diurnal hydration patterns in Figs 3.8, 3.9 and 3.10. The transition time, T_t, occurs earlier in the afternoon in November compared to July as one would expect. The lateness of the transition time in both seasons indicates that storage water is moving out of the bark region for a large part of the daytime period.

Chain of Causality

The proton pattern is consistently phase advanced over the hydration pattern. Furthermore, the sequence of cell membrane transfers between protons and water described in the literature strongly suggests that the proton transfer precedes the water transfer (Nobel, 1991). The increase in pulse magnitude of both hydration and radius in the winter further suggests that the radius shrinkage follows from the movement of water out of the cells. These results indicate the chain of causality is: proton transfer leads to hydration transfer which leads to radius change.

The timing and magnitude of the coupling is strongest in the summer under conditions of high transpiration. In the winter, the radius change in Fig. 3.10a showed a distinct mismatch in the timing of the afternoon transition time, T_2. This is probably due to the location of the electrode near the cambium and the fact that the measured radius change is a summation of volume changes along the entire radius vector. The radius continued to shrink in the presence of a putative shift in water transfer in the cells near the electrode. This suggests that the cells more radial than the electrode were still shrinking while the cells near the cambium were expanding. Mathematically, some of the $C_j(t)$ in equation 3.11 at lower values of j were increasing while some of the $C_j(t)$ at higher values of j were decreasing.

Klepper *et al.* (1971) reported a phase delay in the diurnal cycle of radius with respect to the water content cycle in cotton and Brough *et al.* (1986) reported a similar but slight phase delay in apple.

SUMMARY

The PHYTOGRAM is explained in terms of the characteristics of an invasive electrode resident in the tissue. A liquid bridge is formed between the surface of the electrode and the cells adjacent to the electrode surface. The interfacial potential of the electrode–liquid interface is employed to sense changes in proton concentration in the liquid bridge. Electrode capacitance is employed to sense changes in hydration in the apoplast. Point dendrometers are used to provide a physical variable to cross-correlate with the electrochemical measures of proton change and hydration change.

The instrumentation was applied to *Eucalyptus globulus* in Tasmania and *Pinus engelmannii* in Arizona. Consistent diurnal patterns of potential, hydration and radius were obtained. The proton pattern is interpreted in terms of energy buildup and release. The hydration pattern is interpreted in terms of sequential periods of discharge and recharge of cell water into the apoplast. Similar patterns in potential, hydration and radius after 28 months of electrode residence indicate functionality is unimpaired. A chain of causality for the three variables based on the interpretation of the patterns is: proton transfer into the cells leads to water transfer out of the cells. Radius changes follow from the transfer of water into and out of the cells.

REFERENCES

Armstrong, W. (1979) Aeration in higher plants. In: Woolhouse, H. (ed.) *Advances in Botanical Research*, Vol. 7. Academic Press, New York, pp. 226–232.

Bard, A.J. and Faulkner, L.R. (1980) *Electrochemical Methods, Fundamentals and Applications*. Wiley, New York, pp. 488–519.

Bockris, J.O'M. and Reddy, A.K.N. (1970) *Modern Electrochemistry*, Vol. 2. Plenum Publishing Company, New York, pp. 623–841.

Brough, D.W., Jones, H.G. and Grace, J. (1986) Diurnal changes in water content of the stems of apple trees, as influenced by irrigation. *Plant, Cell and Environment* 9, 1–7.

Canny, M.J. (1995) Apoplastic water and solute movement. *Annual Review of Plant Physiology and Plant Molecular Biology* 46, 215–236.

Davis, W., Shigo, A. and Weyrick, R. (1979) Seasonal changes in electrical resistance of inner bark in red oak, red maple, and eastern pine. *Forest Science* 25, 282–286.

Fensom, D.S. (1959) The bio-electrical potentials of plants and their functional significance III. The production of continuous potentials across membranes in plant tissue by the circulation of the hydrogen ion. *Canadian Journal of Botany* 37, 1003–1026.

Fensom, D.S. (1966) On measuring electrical resistance *in situ* in higher plants. *Canadian Journal of Plant Science* 46, 169–175.

Flowers, T.J. and Yeo, A.R. (1992) *Solute Transport in Plants*. Blackie Academic and Professional, Glasgow, pp. 49–73.

Gensler, W. (1978) Tissue electropotentials in *Laenchoe blossfeldiana* during wound healing. *American Journal of Botany* 65, 152–157.

Gensler, W. (1979) Electrochemical healing similarities between animals and plants. *Biophysical Journal* 27, 461–466.

Gensler, W. (1994) Measurement of micro-oxygen supply and demand. *Encyclopedia of Agricultural Science*, Vol. 3. Academic Press, San Diego, pp. 11–15.

Giaquinta, R.T. (1983) Phloem loading of sucrose. *Annual Review of Plant Physiology* 34, 347–387.

Goldstein, A. (1982) Interface between cotton tissue and a penetrating noble metal probe. *American Journal of Botany* 64, 513–518.

Grignon, C. and Sentenac, H. (1991) pH and the ionic conditions of the apoplast. *Annual Review of Plant Physiology and Plant Molecular Biology* 42, 103–128.

Hoare, J.P. (1965) Oxygen overpotential measurements on bright platinum in acid solution, I. Bright platinum. *Journal of the Electrochemical Society* 112, 602–607.

Hook, D.D., Brown, C.L. and Wetmore, R.H. (1972) Aeration in trees. *Botanical Gazette* 133, 443–454.

Hze, H. (1985) H^+ translocating ATPases: Advances using membrane vesicles. *Annual Review of Plant Physiology* 36, 175–208.
Klepper, B., Browning, V.D. and Taylor, H.M. (1971) Stem diameter in relation to plant water status. *Plant Physiology* 48, 683–685.
Lekas, T.M., MacDougall, R.G., MacLean, D.A. and Thompson, R.G. (1990) Seasonal trends and effects of temperature and rainfall on stem electrical capacitance of spruce and fir trees. *Canadian Journal of Forest Research* 20, 907–977.
MacDougall, R.G., MacLean, D.A. and Thompson, R.G. (1988) The use of electrical capacitance to determine growth and vigor of spruce and fir trees and stands in New Brunswick. *Canadian Journal of Forest Research* 18, 587–594.
Matthews, C. and Van Holde, K. (1990) *Biochemistry*. Benjamin Cummings Publishing Co., Redwood City, CA.
Nobel, P.S. (1991) *Physiological and Environmental Plant Physiology*. Academic Press, San Diego.
Raven, J.A. (1977) The evolution of vascular land plants in relation to supracellular transport processes. In: Woolhouse, H. (ed.) *Advances in Botanical Research* 5. Academic Press, New York, pp. 153–219.
Raven, J.A. and Smith, F.A. (1974) Significance of hydrogen ion transport in plant cells. *Canadian Journal of Botany* 52, 1035–1048.
Raven, J.A. and Smith, F.A. (1976) Nitrogen assimilation and transport in vascular land plants in relation to intracellular pH regulation. *New Phytologist* 76, 415–431.
Rugenstein, S. (1982) Tissue response to palladium microprobe as observed in *Gossypium hirsutum*, L. (*Malvaeceae*). *American Journal of Botany* 64, 519–528.
Silva-Diaz, F.W., Gensler, W. and Sechaud, P. (1983) *In vivo* cyclic voltammetry in cotton under field conditions. *Journal of the Electrochemical Society* 30, 1464–1468.
Smith, F.A. and Raven, J.A. (1979) Intracellular pH and its regulation. *Annual Review of Plant Physiology* 30, 289–311.
Smith, K.T. and Ostrofsky, W.D. (1993) Cambial and internal resistance of red spruce trees in eight diverse stands in the northeastern United States. *Canadian Journal of Forest Research* 23, 322–326.
Zimmermann, M.H. (1983) *Xylem Structure and Function*. Springer-Verlag, Berlin.

Diurnal Variation and Radial Growth of Stems in Young Plantation Eucalypts

Geoffrey M. Downes, Chris Beadle, William Gensler, Daryl Mummery and Dale Worledge

INTRODUCTION

Until recently, tree improvement programmes have focused on traits such as stem volume and form. Tree volume increment remains the major factor driving the economics of plantation productivity within the pulp and paper industry (Greaves *et al.*, 1997). However, as technologies become available for cost-effective breeding for wood quality traits (Downes *et al.*, 1997), more emphasis will be placed on selection for wood quality.

Wood properties are the expression of the activities of the vascular cambium in cell division, expansion and secondary wall development. These activities are dictated by the interaction between the genetic structure of the individual tree and the silvicultural and environmental factors that determine growth. Genetic differences have a physiological expression, and understanding the physiological basis of particular wood quality traits is important in optimizing the growth environment of plantations. This is particularly true if the tree species is sensitive to environmental variation or change.

The lack of studies relating water stress in trees to wood property development was highlighted by Zahner (1968), and attributed to the difficulty in regulating water availability in older stands. This observation appears to be still valid. The data presented in this chapter were collected as part of a larger study to explore the effects of climate and water availability on the development of wood properties. Many intra-ring wood properties can now be measured accurately using automated analysis technology (Evans *et al.*, 1995, 1996). These measurements are expressed as a function of radial distance from the pith. In contrast, environmental variation is measured chronologically on an hourly, daily or weekly scale. To investigate accurately the effects of climate and water availability on wood properties, a template is required to relate temporal measurements of environment to spatial measurements of wood properties (Downes *et al.*, 1994). Point dendrometers were used in this study to generate a template of stem growth with a high spatial and temporal resolution that could be related to climatic and wood property data.

This chapter describes differences in the pattern of stem growth of two species of eucalypts grown under two differing regimes of water availability. One regime, the irrigated treatment, acted as a control indicating maximum attainable growth when available water was not limiting. Diurnal changes in stem radius are described in relation to patterns of sap flow and environmental variation, and electrochemical measurements made in the cambial region. In this investigation the cambial region is defined as those tissues between the mature wood and the bark including the living phloem, and the zones of cell division, cell expansion and secondary wall formation.

MATERIALS AND METHODS

Site Description

The investigation was undertaken in a 2 ha plantation of 6-year-old *Eucalyptus globulus* Labill. and *E. nitens* (Deane and Maiden) Maiden located in south-eastern Tasmania, which is a temperate region. Tree growth (height, stem diameter and water use) has been monitored intensively since establishment in August 1990. The plantation was established to investigate the effects of water stress on the comparative growth and physiology of the two species. The stocking rate was 1428 stems ha^{-1}. Further site characteristics have been described elsewhere (Honeysett *et al.*, 1996; White, 1996). The soil consists of a shallow red–brown loam A horizon and a light brown, medium clay B horizon. Mean soil depth to bedrock was 0.6 m. The annual rainfall was low (*c.*515 mm) and not normally suited to plantation establishment. The site was chosen as it allowed trees to be taken from an irrigated to severely drought-stressed condition in 2 to 4 weeks. The investigation did not address fertilizer effects. However, fertilizer was applied prior to this study as described in Honeysett *et al.* (1996).

Monitoring Tree Growth

Point dendrometers (Agricultural Electronics Corporation (AEC), Tucson, Arizona) were installed on 12 trees (Fig. 4.1) at *c.*25% tree height (*c.*3 m) in March 1995. The dendrometers were mounted on three stainless steel threaded rods of 4 mm diameter which were screwed 40 mm into the stem. The rods were arranged in a triangle about the dendrometer's sensing head and never immediately above or below it. The sensing head was at least 50 mm laterally

Fig. 4.1. Point dendrometers were installed on three stainless steel rods inserted 40 mm into the stem, and sensitive to movements of greater than 4 μm.

and 70 mm vertically from the rods. Subsequent tree growth resulted in the rods becoming more deeply embedded over time and had minimal effect on stem growth. The diameter of the rods was considerably less than nearby branch stubs.

The experimental design was a 2 by 2 factorial comparing the two eucalypt species in two treatments of varying water availability (irrigated and rain-fed). In each species and treatment combination a plot of 30 trees had been established for long-term monitoring of growth (Honeysett *et al.*, 1996). Three trees in each growth plot were randomly selected from the upper 25th percentile of tree height and diameter to ensure that dendrometers were installed on dominant trees.

A further six dendrometers were used to determine the component of stem movement that could be attributed to the contraction and expansion of mature wood alone. Dendrometers were mounted on two *E. globulus* trees, located outside the two treatments, at three heights (0.5, 1.3 and 3.0 m). For three of these dendrometers, a 10 × 10 mm area of bark and cambium was removed from beneath the dendrometer's sensing head such that the head was resting on mature wood. On tree 1, bark was removed from beneath the heads of the dendrometers at 0.5 and 3.0 m. Bark was removed beneath the dendrometer at 1.3 m only on tree 2. On tree 2 a sapflow sensor was installed at 1.5 m in August 1996.

Sapflow was estimated using Greenspan sapflow sensors (Greenspan Technology, Warwick, Queensland, Australia). Two probe sets were installed at 1.5 m on opposite sides of the stem (north–south orientation) and shielded from external radiative heat sources. Each probe set consists of three parallel stainless steel tubes, two containing a pair of thermistors 5 mm apart, and the remaining tube a line heater. One pair of thermistors is located 10 mm downstream from the heater element and the other pair of thermistors 10 mm upstream. Probe sets were inserted so that one thermistor pair was 10 to 15 mm under the cambium and the second pair 20 to 25 mm.

At 15 min intervals a heat pulse of 0.8 s was generated and sap flux calculated using techniques developed by Marshall (1958), Swanson and Whitfield (1981), Edwards and Warwick (1984) and Hatton *et al.* (1990).

As a part of the dendrometer system, ten PHYTOGRAM electrodes (see Gensler, Chapter 3, this volume) were installed in the cambial tissue adjacent to the dendrometers. Changes in electrode potential and capacitance, which are reported to relate to changes in extracellular proton level and hydration, respectively, were monitored hourly.

Monitoring Water Stress and Scheduling Irrigation

Soil moisture variability was monitored fortnightly throughout the site using a neutron moisture probe (CPN 503 Hydroprobe) (Honeysett *et al.*, 1996). An automatic weather station was located approximately 500 m from the site. Weather data, including precipitation, temperature, relative humidity, solar radiation and vapour pressure deficit, were monitored automatically every 5 min, and hourly and daily averages determined. A class A pan evaporimeter was used to schedule irrigation events in conjunction with the neutron moisture probe data (Worledge *et al.*, 1998).

Irrigation was applied through micro-sprinklers. Soil water deficit in the irrigated treatment was not allowed to fall below approximately −40 mm (Honeysett *et al.*, 1996). The soil water deficit is defined as the amount of water required to return the soil to field capacity. The soil at field capacity held between 120 and 130 mm of available water. Irrigation was applied frequently in small amounts to avoid large changes in water content between fortnightly monitoring events (Worledge *et al.*, 1998). Supplemental irrigation of the rain-fed treatment was necessary to prevent tree death. Monitoring of soil water content and evaporation allowed the applied irrigation in this treatment to be controlled such that soil water could be raised to prevent tree death but not alleviate water stress sufficiently to allow growth. The soil water deficit at which irrigation was applied varied over the 2-year period but was commonly greater than −110 mm. Irrigation applied in the winter months in the irrigated treatment was reduced (e.g. see June 1996, Table 4.3) to encourage root development lower down the soil profile.

Predawn leaf water potentials were measured at intervals throughout the growth period as described by Honeysett *et al.* (1996).

RESULTS

The trees used in this study grew quickly enough to be available for commercial pulp wood harvesting between 8 and 10 years of age. Table 4.1 shows the average tree height and breast height diameter over bark (DBHOB) of the three trees in each treatment at the start of the dendrometer study. Predawn leaf water potentials showed the levels of water stress experienced by the rain-fed treatment were often severe, in contrast to those in the irrigated treatment (Table 4.2). The ability of the site to sustain severe water deficits in winter is apparent, with the rain-fed *E. globulus* experiencing potentials of −2.23 MPa in June 1996.

Over the two growing seasons the irrigated treatment received approximately 1500 mm precipitation (irrigation plus rainfall) annually, while the rain-fed treatment received 900 mm (Table 4.3). The average radial increment measured by the dendrometers over the 2-year period for irrigated *E. globulus* and *E. nitens* was 9.5 mm $year^{-1}$ and 11.2 mm $year^{-1}$, respectively. In the rain-fed treatment, the increments were 7.4 mm $year^{-1}$ and 8.9 mm $year^{-1}$, respectively (Table 4.3). These differences were not significant between treatments or species due to the large variability between trees (see Fig. 4.2). However, a significant difference existed between the years, with 1995/96 producing more growth than 1996/97.

Table 4.1. Average heights and diameters of each of three trees in the treatments.

	Tree height (m)		Breast height diameter (cm)	
	E. globulus	*E. nitens*	*E. globulus*	*E. nitens*
Rain-fed	11.9	10.5	12.5	13.6
Rain-fed to irrigated	10.5	9.4	12.1	11.9
Irrigated to rain-fed	12.6	10.5	12.2	12.8
Irrigated	14.7	11.8	15.1	16.0

Table 4.2. Predawn water potential (MPa) over the period July 1995 to June 1997, with standard errors in parentheses. Significant water stress was generated in the rain-fed trees.

	Irrigated		Rain-fed	
	E. globulus	*E. nitens*	*E. globulus*	*E. nitens*
29 Nov 95	−0.48 (0.03)	−0.41 (0.04)	−0.44 (0.02)	−0.46 (0.03)
6 Mar 96	−0.47 (0.03)	−0.49 (0.05)	−2.26 (0.08)	−1.97 (0.05)
13 Jun 96	−0.60 (0.02)	−0.58 (0.05)	−2.23 (0.3)	−1.73 (0.21)
4 Sep 96	−0.47 (0.03)	−0.39 (0.02	−2.79 (0.1)	−2.04 (0.17)
23 Oct 96	−0.44 (0.06)	−0.34 (0.03)	−0.90 (0.05)	−0.92 (0.05)
4 Dec 96	−0.43 (0.02)	−0.19 (0.02)	−1.37 (0.08)	−1.55 (0.07)
24 Feb 97	−0.47 (0.02)	−0.43 (0.06)	−1.73 (0.05)	−1.54 (0.05)
1 May 97	−0.51 (0.03)	−0.49 (0.03)	−2.14 (0.02)	−1.57 (0.08)

Daily Data

E. nitens consistently grew more slowly during winter (June–July) than *E. globulus* (Fig. 4.2a). The growth rate of *E. globulus* was more uniform across the seasons, due predominantly to greater growth in winter and slower growth in spring and autumn than *E. nitens*. In the rain-fed treatment *E. globulus* maintained stem growth under conditions of moderate water stress (Fig. 4.2b). *E. nitens* was much more reactive to water stress, as can be seen by the more exaggerated stem contraction during periods of water stress (e.g. in August 1996, February 1997 in Table 4.3).

Irrigation resulted in a much smoother growth pattern compared to that of the rain-fed trees (Fig. 4.2a,b). The irregular growth pattern of rain-fed trees included many periods of long-term (days to months) stem contraction. During the dry spring in 1995 (September−October), trees ceased growth for 6–7 weeks until a large rainfall in early November resulted in rapid radial growth rates of up to 3 mm per month (Fig. 4.2b). Figure 4.2b also shows the difference in the rate of onset of water stress across different seasons. Stem contraction in winter, evident from July to August 1996, was more gradual than that in December 1995 and December 1996. The effect of rainfall events in initiating radial expansion was readily apparent.

When soil water deficits were not limiting growth, rain-fed trees often exhibited a greater rate of growth than irrigated trees (Fig. 4.2c). A positive slope in Fig. 4.2c indicates a faster growth rate in the irrigated trees (e.g. September–October 1995) and a negative slope a faster growth rate in the rain-fed trees (e.g. November 1995 following rain). The changes in slope from positive to negative correspond to large rainfall/irrigation events. After 92 weeks there was a net difference of *c.* 5 mm of radius between the irrigated and rain-fed trees.

Table 4.3. Monthly data from the 1995–97 growing seasons. Monthly precipitation was significantly correlated with monthly radial increment in both the rain-fed *E. globulus* and *E. nitens* trees ($r^2 = 0.76$ and 0.61, respectively) but not in the irrigated *E. globulus* and *E. nitens*.

Month	Irrigated increment (µm)		Rain-fed increment (µm)		Monthly rainfall (mm)	Irrigated precipitation (mm)	Rain-fed precipitation (mm)
	E. globulus	*E. nitens*	*E. globulus*	*E. nitens*			
Jul 95	529	68	588	685	47	47	47
Aug 95	1,042	870	955	1,282	67	67	67
Sep 95	983	1,235	0	128	32	109	32
Oct 95	1,951	1,986	14	78	32	149	32
Nov 95	1,467	1,626	2,511	3,045	49	149	181
Dec 95	1,061	1,061	1,202	1,695	102	214	102
Jan 96	739	1,442	878	1,201	92	170	92
Feb 96	842	1,323	169	705	68	145	68
Mar 96	745	1,165	443	885	33	132	33
Apr 96	487	793	1,120	1,787	123	129	123
May 96	525	517	165	167	11	57	11
Jun 96	439	247	282	−10	29	29	59
Jul 96	589	154	525	221	62	72	62
Aug 96	617	669	−317	−557	25	85	25
Sep 96	1,203	1,378	1,083	1,488	54	155	109
Oct 96	1,218	1,705	1,344	1,193	42	170	142
Nov 96	811	1,065	756	452	41	186	55
Dec 96	871	850	1,440	1,249	25	194	159
Jan 97	741	380	734	452	85	236	130
Feb 97	329	504	−466	−987	39	172	39
Mar 97	740	1,409	904	1,707	62	187	131

Continued over

Table 4.3. (*Continued*)

	Irrigated increment (μm)		Rain-fed increment (μm)		Monthly rainfall (mm)	Irrigated precipitation (mm)	Rain-fed precipitation (mm)
Month	*E. globulus*	*E. nitens*	*E. globulus*	*E. nitens*			
Apr 97	565	1,058	4	246	15	99	15
May 97	278	530	198	364	24	89	50
Jun 97	375	407	196	309	21	62	21
1995/96	10,810	12,333	8,327	11,648	685	1,139	846
1996/97	8,337	10,109	6,401	6,137	494.6	1,706	937
Total	19,147	22,442	14,728	17,785	1,179.6	2,848.66	1,789.48

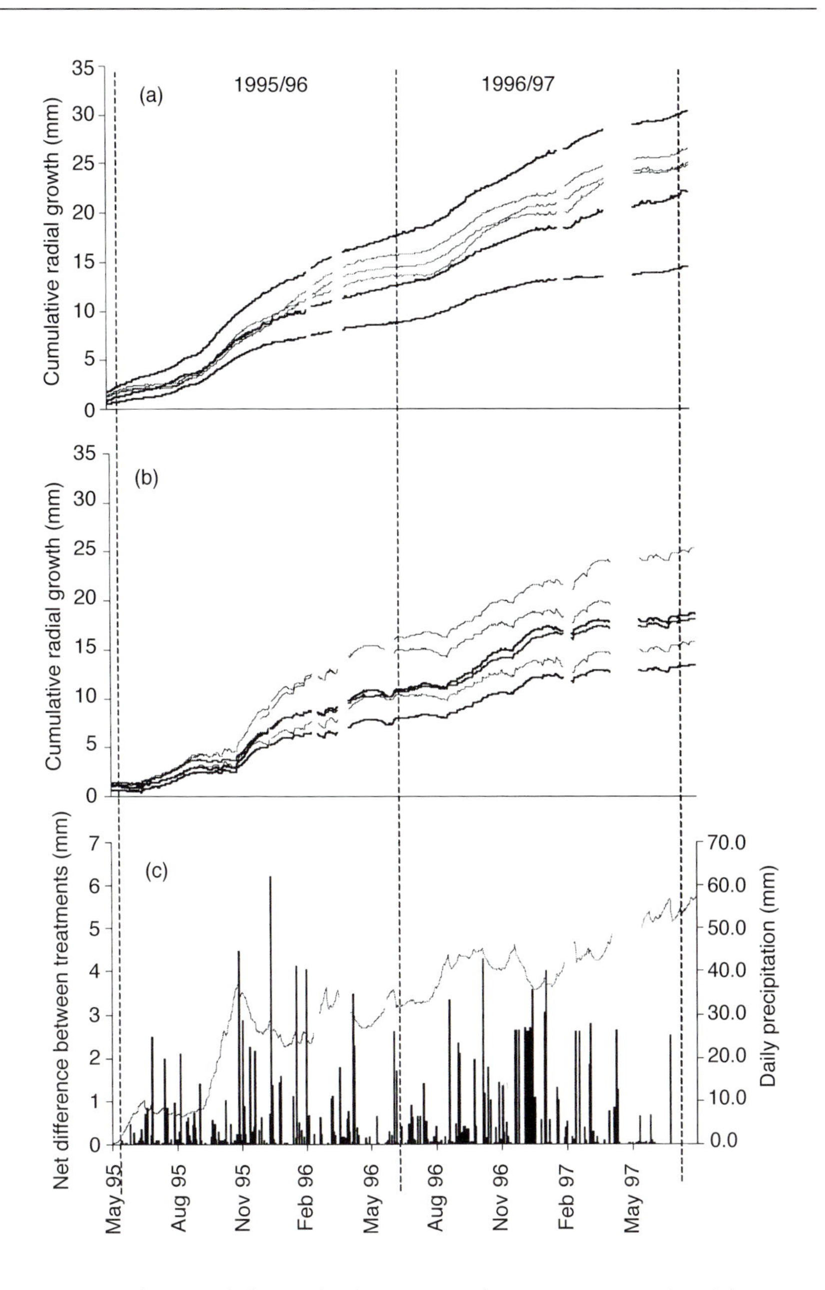

Fig. 4.2. Cumulative radial growth of young eucalypts in (a) irrigated and (b) rain-fed treatments. Three trees each of irrigated *E. globulus* (thick line) and *E. nitens* (thin line) were monitored by point dendrometers at 15 min resolution. Daily averages are shown here. The net difference between the two treatments for *E. globulus* is shown in (c). A positive slope indicates a faster growth rate in the irrigated trees.

Hourly Data

Stem contraction began at sunrise and was coincident with the opening of stomata and the commencement of transpiration. Contraction typically ceased between early and mid afternoon. If water stress was not severe, the stem then began to expand, and this continued until contraction began the following day.

Rainfall and irrigation resulted in a marked increase in stem radius (Fig. 4.3), the extent of which was dependent upon the existing level of water stress. Responses by trees to these events were observed within 30 min. The magnitude of the diurnal variation in stem radius was reduced following the event and commonly took several days to re-establish. When water stress was severe (> -2 MPa Ψ_{pd}), the dendrometers detected some stem expansion in response to rainfall as low as 1 mm. In such cases rain does not reach the ground and only affects the canopy.

The rain received on 11 October 1996 was sufficient to rehydrate the stem but not sufficient to allow further radial expansion, as indicated by the negative growth trend between 12 and 23 October (Fig. 4.3). This was a typical response when stress was alleviated but soil water deficit remained below -50 to -60 mm.

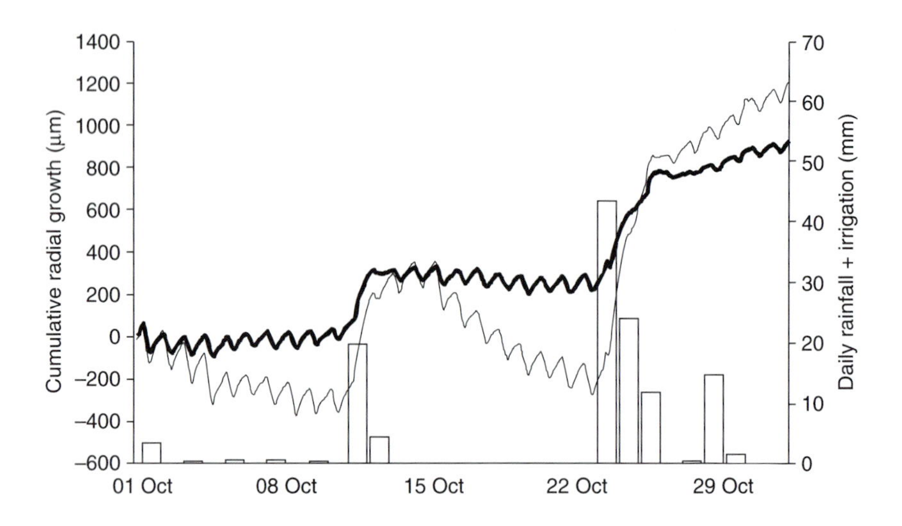

Fig. 4.3. The hourly growth pattern during October 1996 of an *E. globulus* (thick line) and an *E. nitens* (thin line) is shown. The response of the trees to both rainfall and irrigation events is evident by the changes in slope. The greater sensitivity of *E. nitens* is also evident in the greater shrinkage of the tree over the first 3 weeks. The response to rainfall is almost immediate.

Species Differences

The magnitude of diurnal variation differed significantly between the two species during summer (Fig. 4.4). *E. nitens* exhibited significantly ($P<0.001$) greater diurnal contraction (up to 200 μm) than *E globulus* (up to 140 μm), when the normal diurnal pattern was not affected by rainfall (e.g. as on 22 January). The stem expansion due to rainfall on 22 January resulted in an interaction between species and treatment with the rain-fed *E. globulus* exhibiting significantly less expansion than both the rain-fed *E. nitens* and the irrigated *E. globulus*; however, it was not different from the irrigated *E. nitens*. The patterns shown were representative of species differences in these treatments across the two summers monitored. During winter, diurnal variation was similar in both species (*c.*50 μm). In general, *E. nitens* had a more pronounced diurnal cycle.

The response of trees of both species to climatic events was similar (Fig. 4.4). Stem expansion followed by contraction was a feature often observed following rainfall (e.g. 22 and 28 January), particularly in the irrigated treatment. It was as if the trees initially took up more water than they needed. The greater sensitivity of *E. nitens* to water stress is again evident, as indicated by the greater long-term stem contraction recorded by the dendrometers in the rain-fed treatment between 12 and 23 October (Fig. 4.3) and 8 and 21 January (Fig. 4.4).

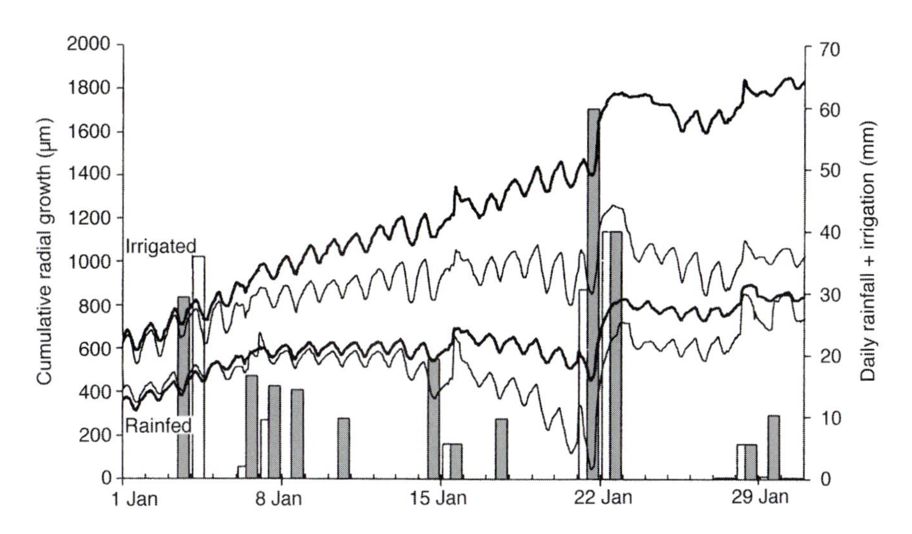

Fig. 4.4. During summer both species exhibit more diurnal variation than in winter. However, *E. globulus* (thick line) has a lower magnitude (*c.*140 μm) than *E. nitens* (*c.*200 μm) (thin line). The magnitude was more a function of species than of treatment. The precipitation received in the irrigated (shaded) and rain-fed (open) treatments are shown as columns.

Partitioning Growth between Wood and Cambium

A large component of diurnal stem movement was attributable to the dendrometer sensing movement in mature wood (Fig. 4.5). Comparing the magnitude of the diurnal variation between dendrometers over a 6-week period, it was found that 40–50% of the total daily contraction/expansion could be attributed to the dendrometer data from wood. This did not include days on which large expansion occurred in response to rainfall/irrigation. Longer-term contraction and expansion, following periods of rainfall and irrigation, were attributable to movement in the cambial region. Data from other researchers using similar dendrometer systems (Zweifel, personal communication[1]) suggested that a large proportion of this wood variation, although not all, is an artefact of the effect of temperature on the dendrometer system itself. Thus, subtracting the wood component from the dendrometer sensing bark movement is in effect removing the temperature effect on the dendrometer system. After further testing these temperature effects appear to be minimal in the data presented here. Figures 4.5b, 4.6a and 4.7a show the movement attributable to the cambial region alone when wood movement is subtracted from the bark movement.

The diurnal pattern of stem contraction corresponded closely, though inversely, with that of sapflow (Fig. 4.6). Over a 6-day period the cambial region accounted for *c.*70% of diurnal movement (Fig. 4.6a). Stem contraction generally lagged the commencement of sapflow by one to several hours, depending on environmental conditions. For example, at high vapour pressure deficit (*D*, on 25 November, see Fig. 4.6b) stem contraction commenced 1 h earlier than on other days when *D* was lower. Sapflow matched solar radiation closely when *D* was low to moderate (23 and 26–28 November), but when *D* was higher the effect of stomatal closure on reducing sapflow was evident (25 November). The large, but brief, reduction in solar radiation on 28 November was related to a small amount of precipitation, and resulted in a brief reduction in sapflow and a small amount of stem expansion. Reduced sapflow in response to high *D* was associated with reduced contraction in the cambial region (on 25 November, Fig. 4.6). However, the contraction in wood continued for several more hours than contraction in the cambial region.

Electrochemical Monitoring of Cambial Region

Stem contraction (Fig. 4.7a) was coincident with a decrease in electrode potential (Fig. 4.7b). Electrode capacitance increased as the stem contracted and, conversely, decreased during stem expansion. This pattern was consistent across electrodes and matched the patterns of sapflow (Fig. 4.7b). However, the

[1] Roman Zweifel, Swiss Federal Institute for Forest, Snow, and Landscape Research (WSL), CH-8903 Birmensdorf, Switzerland.

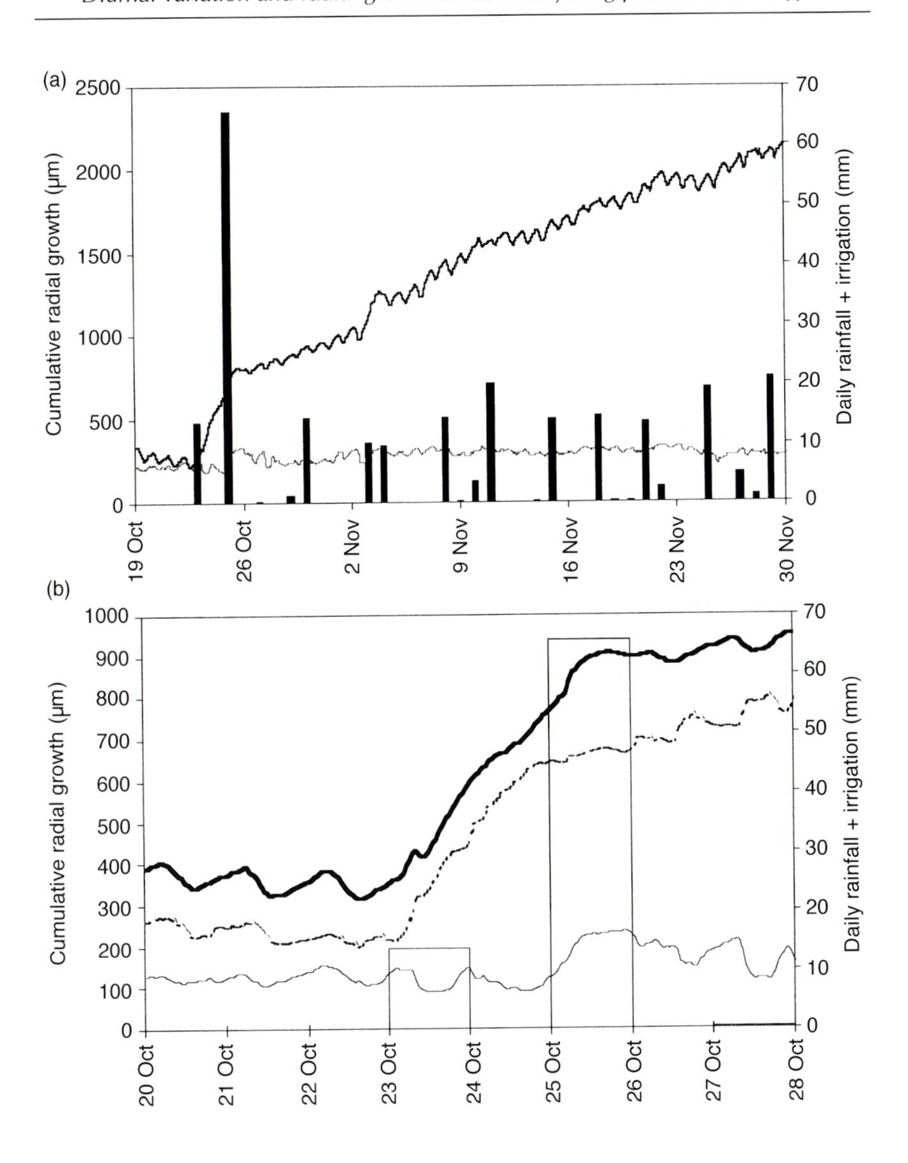

Fig. 4.5. During October/November 1996 stem movement was partitioned between the wood (thin line) and cambial region (thick line). In (a) the pattern over 6 weeks is shown, at the start of which the tree was under considerable water stress that was relieved on 23 October. In (b) a closer view of days 2 to 10 is given. The wood movement has been subtracted from the bark movement to give an estimate of the movement attributable to the cambial region alone (dashed line). Daily precipitation plus irrigation is shown as columns in both plots.

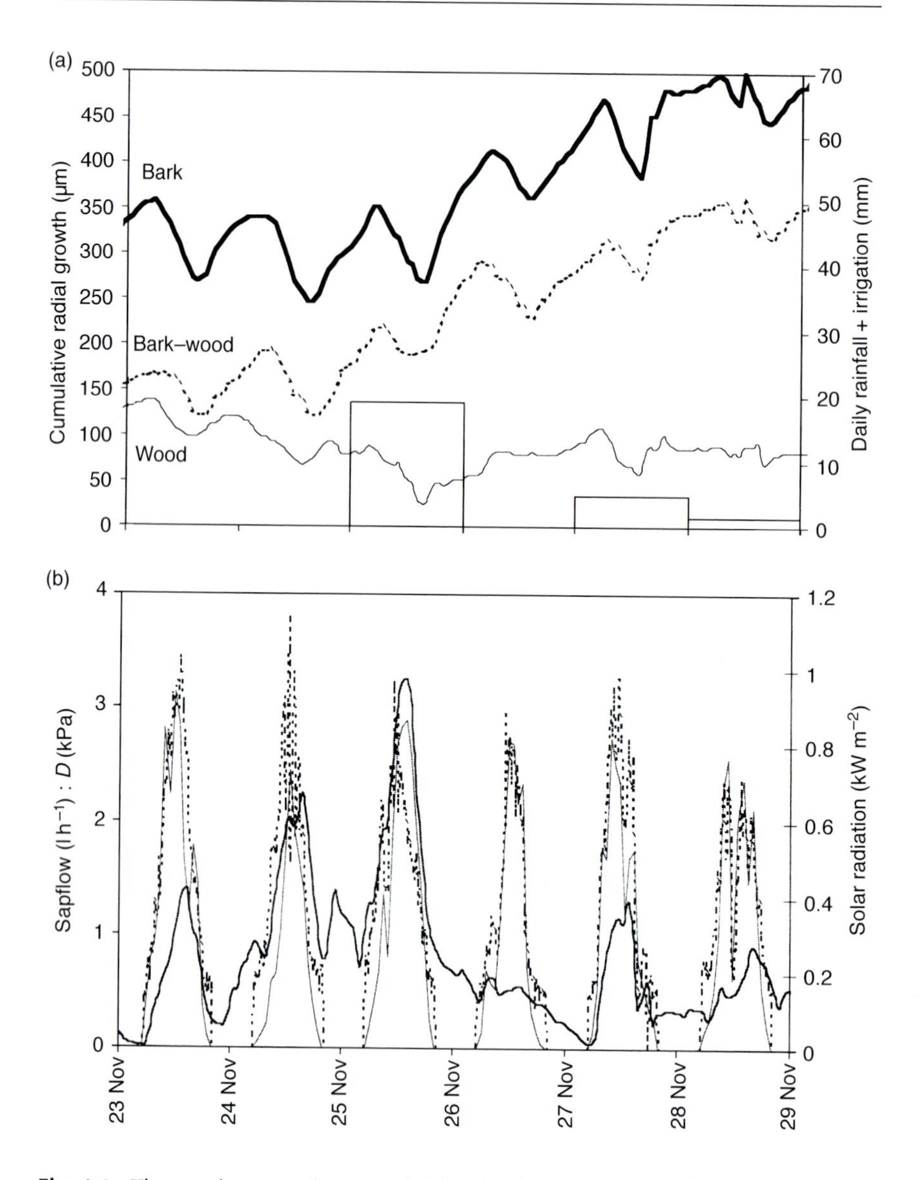

Fig. 4.6. The synchronicity between (a) the dendrometer data and (b) sapflow, vapour pressure deficit (VPD) and solar radiation is apparent. In (a) diurnal movement in stem radius (thick line), wood radius (thin line) and the cambial region (dashed line) is shown, along with precipitation + irrigation (column). In (b) the variation in sapflow (dashed line), VPD (thick line) and solar radiation (thin line) is from the same 6-day period.

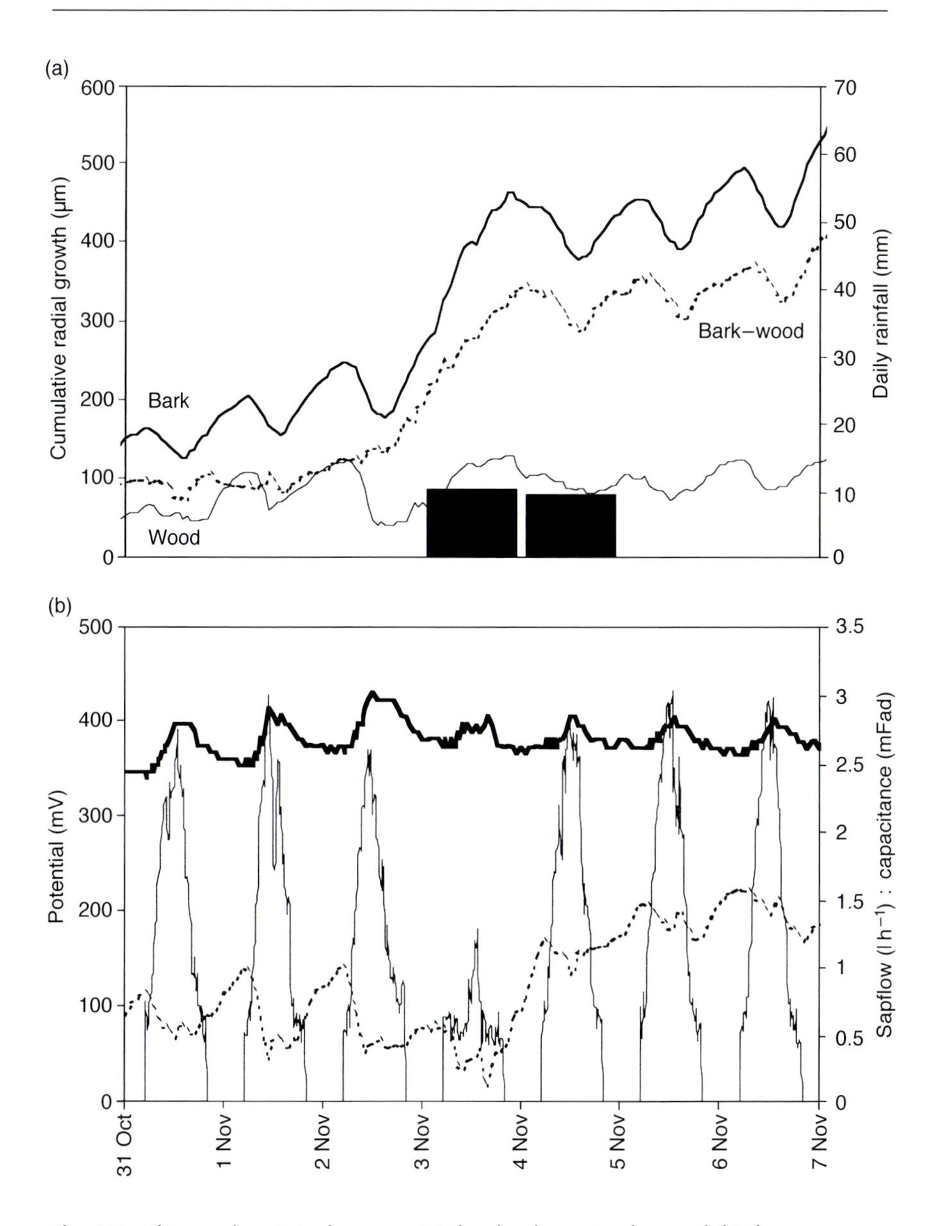

Fig. 4.7. The synchronicity between (a) the dendrometer data and (b) the PHYTOGRAM electrode potential (dashed line) and capacitance (thick line) together with the sapflow data (dashed line) is shown. The shaded columns in (a) represent average daily precipitation.

magnitude of the variation differed between electrodes and appeared to be influenced, in part, by the placement of the electrode. It was not possible to determine the exact location of the electrode in the cambial region. Aperiodic positive and negative spikes were another unusual feature of the data and were often coincident with rainfall recorded in the climatic data. The synchrony between the sapflow and the electrode data confirms that the patterns were real, as these measurements were completely independent. The increase in capacitance commenced as sapflow began. Similarly, the periods of precipitation on 3 November were coincident with an increase in the baseline of the electrode potential. These relationships are typical of those obtained over the monitoring period.

DISCUSSION

This investigation has shown that point dendrometers can be used to provide a chronological template of radius change with a high temporal resolution. There were differences in the pattern of change between the two eucalypt species, with *E. nitens* showing a slower rate of stem radius increase in winter, and *E. globulus* a more uniform increase throughout the year. *E. nitens* was more reactive to water stress, but *E. globulus* maintained a faster growth rate at a given level of water stress. The magnitude of the diurnal variation varied between winter and summer, with *E. nitens* exhibiting greater diurnal variation in summer. Diurnal contraction and expansion were coincident with variation in sapflow rate as well as with changes in the potential and capacitance of electrodes inserted into the cambial region.

Dendrometers have often been used to investigate patterns of radial growth (Fritts, 1976), although point dendrometers are used less than band dendrometers. A comparison of the Fritts dendrograph (Fritts, 1976), the most extensively used mechanical form of point dendrometer, with that used in this study was conducted and showed that the former was less sensitive to diurnal variation than the latter (Fritts and Gensler, personal communication). This is believed to be a consequence of the force applied to the cambium by the dendrometer. The AEC point dendrometer uses a cantilever action to hold the sensing head against the bark and applies only 0.2 N of constant force. The use of springs, as in the mechanical form, which leads to an increasing pressure as the tree expands and the spring compresses, is thus avoided.

The radial growth rates recorded by the dendrometers were similar to those observed in the growth plots from which the trees were selected. Diameter growth decreases with stand age during this period of growth and mean values, based on diameter tape measurements, were <20 mm $year^{-1}$ across species and treatments (D. Worledge, personal communication). The longer-term effect of these treatments on stem volume and height growth, and pulp and wood properties of trees harvested from the treatments, have been described previously (Honeysett *et al.*, 1996; Beadle *et al.*, 1997).

The more uniform growth pattern of *E. globulus*, compared with that of *E. nitens*, over the year has not, to our knowledge, been reported previously. However, it has been observed that in Tasmania *E. nitens* generally has a more clearly defined (i.e. sharper) annual ring boundary, whereas the boundary of *E. globulus* is more blurred (Dadswell, 1972; J. Ilic, personal communication[1]). This is consistent with the observed patterns of growth. *E. nitens* is a species more tolerant of frost and is generally planted on higher elevation sites. Neither species, unlike most northern hemisphere species, displays a period of winter dormancy. If conditions are favourable, the trees grow.

The dendrometer data showed that radial shrinkage in *E. nitens*, in response to water stress, was larger than in *E. globulus*. The growth of *E. nitens* has been shown to be more sensitive to moderate levels of water stress than *E. globulus* (Honeysett *et al.*, 1996). Both species respond to drought by stomatal closure but, at any given level of water stress, there was a greater reduction of stomatal conductance in *E. nitens* (White, 1996). *E. globulus* has also been shown to maintain leaf turgor over a wider range of relative water contents than *E. nitens* (White *et al.*, 1996). Maintenance of positive turgor is an essential prerequisite for any expansion growth.

The daily stem radius changes were characterized by periods of expansion and contraction. Within these periods, real cambial growth occurred but was confounded by them. Previous studies had determined that tree growth stopped at soil water deficits <-50 to -60 mm (Honeysett *et al.*, 1996). Under irrigated conditions, the smooth radial changes were probably indicative of steady rates of cell division and irreversible cell expansion. The irregular growth pattern under rain-fed conditions, and the capacity of rain-fed trees to have greater rates of radial expansion after rewatering, may reflect an imbalance between these two processes. It may also reflect, in part, a greater root surface area in the rain-fed trees and, hence, a greater potential rate of water extraction from the soil. The diurnal pattern of stem shrinkage reflects the changing pattern of internal stem water stress. Zahner (1968) indicated that cell enlargement in the cambium may only occur for several hours per day, when internal conditions are favourable. Ultimately, the objective of studies like the one described here is to define, if possible, the conditions when real growth (cell division and cell wall production) occurs.

Zahner (1968) discussed the difficulty of studying cambial activity by both instruments monitoring tree growth and via interpretation of intra-annual ring patterns. The former do not provide direct observations of tissue morphogenesis. The latter cannot accurately relate the resultant within-ring anatomy to production conditions. The use of dendrometers to provide a means of relating the anatomy to production conditions was examined prior to this study (Downes and Evans, 1994) and will be pursued in later papers. However, anatomical

[1] Jugo Ilic, CSIRO Forestry and Forest Products, Private Bag 10, Clayton South MDC, Victoria, 3168, Australia.

observation of the cambium indicated that when trees were water stressed the width of the zones of cell division and enlargement, but not the numbers of cells in them, were reduced (Herbert and Downes, 1997, unpublished data). Thus, when the stress is alleviated these cells probably expand and continue growth, facilitated by a greater root water uptake.

Others have noted differences in the magnitude of diurnal shrinkage between species and between seasons (see Kozlowski, 1972). Differences across seasons probably relate both to changes in the amount of water transpired and in the width of the cambial region. The difference between these closely related species was surprising. These differences can arise from several causes. They may reflect a difference in the structure of the cambial zone in terms of the width of the dividing and enlarging zones. Alternatively they may reflect differences in the physiology of the trees and the role of the stem as a capacitor buffering water loss via transpiration against water uptake from the roots. Conceptually, differences in the pattern of loss and uptake may result in changing magnitudes of diurnal variation. Lassoie (1973) observed that soil moisture levels had an impact on the magnitude of diurnal shrinkage. It is also evident that, as water stress became severe, the magnitude of diurnal shrinkage reduced, particularly in *E. globulus*.

The cambial region was defined here as all living tissue between the mature wood and the bark. The question arose as to how much of the diurnal movement in the dendrometer data could be attributed to this tissue, and how much to expansion/contraction of mature wood. Long-term drought has often been reported to result in stem shrinkage that can negate previous radial increases in that season (Kozlowski, 1972). This must involve shrinkage of the mature wood. The point dendrometers allowed this question to be addressed by removing a small portion of the bark and cambium immediately beneath the sensing head such that it was resting directly on mature wood. A second dendrometer mounted nearby, with the sensing head resting normally on bark, recorded the combined movement. A large proportion of the diurnal variation was attributed to movement in the stem wood as distinct from the cambial region. This has also been reported by others (MacDougal, 1924).

The sensitivity of the diurnal pattern of sapflow to factors affecting transpiration has been examined in other species (Herzog *et al.*, 1995; Moreno *et al.*, 1996). The data presented here showed similar relationships. The patterns of diurnal shrinkage observed were consistent with the five phases described by Herzog *et al.* (1995), although as water became limiting the second phase became much reduced.

The PHYTOGRAM system was examined for its potential to obtain additional information about the sensitivity of the cambial region to environmental variation. Although there is still much to be done in regard to validating the physiological basis of the technique, the data indicated that diurnal variation in the electrode potential and capacitance occurred in concert with the diurnal changes in stem radius and sapflow. An interpretation of these changes is based on movement of protons and water into and out of the cell (see Gensler, Chapter 3, this volume). During the night protons are pumped out of the cell into the extracellular space, thus lowering the pH of the cell wall, and effectively acting as an energy store. At the same time water moves into the cell, lowering the water content (capacitance) of the extracellular space. As the sun rises, and coincident with the commencement of radial contraction, water moves from the cell as protons move into it. The transpiration stream then removes the water released into the extracellular space.

SUMMARY

Eighteen point dendrometers were installed in a 6-year-old plantation of *Eucalyptus globulus* Labill. and *E. nitens* (Deane and Maiden) Maiden on a low rainfall site (*c.* 500 mm year^{-1}) in southeastern Tasmania. The site was divided into two blocks comparing rain-fed and irrigated treatments.

Radial stem growth was recorded in 12 trees across treatments at 15 min intervals. In addition three dendrometers were mounted at different heights on each of two *E. globulus* trees, and bark and cambium removed from beneath half of them to partition stem movement between the cambial region and wood. PHYTOGRAM™ electrodes collected data on electrochemical changes in the cambial region.

Radial growth rates up to 3 mm month^{-1} (15 mm year^{-1}) over 2 years were observed in irrigated trees. Radius changes were smoother in irrigated than rain-fed trees. Rain-fed trees experienced predawn water potentials (Ψ_{pd}) down to −2.4 MPa. The magnitude of diurnal variation in tree diameter varied between trees, treatments and season. Within trees it ranged from *c.* 50 to 200 μm over a year, was less in irrigated than rain-fed trees, and was negligible following large rainfall events or during severe water stress. Stem contraction began at sunrise and ceased between 1 and 4 h after midday.

Electrode potential and capacitance (extracellular pH and hydration) showed diurnal changes synchronous with dendrometer data. Potential decreased and capacitance increased with stem contraction and vice versa. Preliminary interpretation is that water movement out of the cambial region into the transpiration stream drives contraction. As transpiration decreases, water moves into the cambial region and stem expansion occurs. Marked changes in electrode potential were evident during high rainfall.

ACKNOWLEDGEMENTS

Financial support for the establishment of the plantation used here was provided by the Australian Centre for International Agricultural Research. Funds for the equipment and maintenance of the study were provided by Australian Newsprint Mills, Forest Management, and the Cooperative Research

Centre for Hardwood Fibre and Paper Science. We thank Drs Don White and Richard Benyon for valuable comments on the manuscript.

REFERENCES

Beadle, C.L., Banham, P.W., Worledge, D., Russell, S.L., Hetherington, S.L., Honeysett, J.L. and White, D.A. (1997) Irrigation increases growth and improves fibre quality of *Eucalyptus globulus* and *E. nitens*. *Proceedings of the IUFRO Conference on Silviculture and Improvement of Eucalypts*, Salvador, Brazil, Vol. 4. Embrapa, Colombo.

Dadswell, H.E. (1972) Anatomy of eucalypt woods. *CSIRO Division Applied Chemistry Technical Paper* No. 66.

Downes, G.M. and Evans, R. (1994) Effects of environment on tracheid dimensions. *Proceedings of the Workshop 'Modelling of Tree-ring Development – Cell Structure and Environment'*, Freiburg, 5–9 September 1994. Institut für Waldwachstum, University of Freiburg, Germany, pp. 58–68.

Downes, G.M., Evans, R., Benson, M. and Myers, B. (1994) Application of a new wood microstructure analyser to the assessment of environmental effects on radiata pine tracheid dimensions. *Proceedings of the Appita 48th Annual General Conference*, 2–6 May 1994, Melbourne, Australia, pp. 461–466.

Downes, G.M., Hudson, I.L., Raymond, C.A., Dean, G.H., Michell, A.J., Schimleck, L.S., Evans, R. and Muneri A. (1997) *Sampling Plantation Eucalypts for Wood and Fibre Properties*. CSIRO Publishing, Melbourne, 132pp.

Edwards, W.R.N. and Warwick, N.W.M. (1984) Transpiration from a kiwifruit vine as estimated by the heat pulse technique and the Penman–Monteith equation. *New Zealand Journal of Agricultural Research* 23, 3–27.

Evans, R., Downes, G.M., Menz, D. and Stringer, S. (1995) Rapid measurement of variation in tracheid transverse dimensions in a radiata pine tree. *Appita Journal* 48, 134–138.

Evans, R., Downes, G.M. and Murphy, J.O. (1996) Application of new wood characterization technology to dendrochronology. In: Dean, J.S., Meko, D.M. and Swetnam, T.W. (eds) *Tree Rings, Environment and Humanity. Radiocarbon* special issue. Department of Geosciences, University of Arizona, pp. 743–749.

Fritts, H.C. (1976) *Tree Rings and Climate*. Academic Press, New York.

Greaves, B.L., Borralho, N.M.G. and Raymond, C.A. (1997) Breeding objective for plantation eucalypts grown for production of kraft pulp. *Forest Science* 43, 465–472.

Hatton, T.J., Catchpole, E.A. and Vertessy, R.A. (1990) Integration of sapflow velocity to estimate plant water use. *Tree Physiology* 6, 201–209.

Herzog, K.M., Hasler, R. and Thum, R. (1995) Diurnal changes in the radius of a subalpine Norway spruce stem: their relation to the sap flow and their use to estimate transpiration. *Trees* 10, 94–101.

Honeysett, J.L, White, D.A., Worledge, D. and Beadle, C.L. (1996) Growth and water use of *Eucalyptus globulus* and *E. nitens* in irrigated and rain-fed plantations. *Australian Forestry* 59, 64–73.

Kozlowski, T.T. (1972) Shrinking and swelling of plant tissues. In: Kozlowski, T.T. (ed.) *Water Deficits and Plant Growth*, Vol. 1. Academic Press, New York, pp. 1–64.

Lassoie, J.P. (1973) Diurnal dimensional fluctuations in a Douglas fir stem in response to tree water status. *Forestry Science* 19, 251–255.

MacDougal, D.T. (1924) *Dendrographic Measurements*. Carnegie Institute, Washington, Publication 350, pp. 1–88.

Marshall, D.C. (1958) Measurement of sap flow in conifers by heat transport. *Plant Physiology* 33, 385–396.

Moreno, F., Fernandez, J.E., Clothier, B.E. and Green, S.R. (1996) Transpiration and root water uptake by olive trees. *Plant and Soil* 184, 85–96.

Swanson, R.H. and Whitfield, D.W.A. (1981) A numerical analysis of the heat-pulse velocity theory and practice. *Journal of Experimental Botany* 32, 221–239.

White, D.A. (1996) Physiological responses to drought of *Eucalyptus globulus* and *Eucalyptus nitens* in plantations. PhD thesis, University of Tasmania, 166pp.

White, D.A., Beadle, C.L. and Worledge, D. (1996) Leaf water relations of *Eucalyptus globulus* ssp. *globulus* and *E. nitens*: seasonal, drought and species effects. *Tree Physiology* 16, 469–476.

Worledge, D., Honeysett, J.L., White, D.A., Beadle, C.L. and Hetherington, S.J. (1998) Scheduling irrigation of *Eucalyptus globulus* and *E. nitens*: a practical guide. *TasForests* 10, 91–101.

Zahner, R. (1968) Water deficits and growth of trees. In: Kozlowski, T.T. (ed.) *Water Deficits and Plant Growth*, Vol. 2. Academic Press, New York, pp. 191–254.

Zimmermann, M.H. (1983) *Xylem Structure and the Ascent of Sap*. Springer-Verlag, Berlin.

B

Analysing Anatomical and Structural Features

5

Significance of Vertical Resin Ducts in the Tree Rings of Spruce

Rupert Wimmer, Michael Grabner, Giorgio Strumia and Paul R. Sheppard

INTRODUCTION

Resin ducts are normal features of the wood and bark of many conifers, particularly of *Pinus*, *Picea*, *Pseudotsuga* and *Larix* (Fig. 5.1, LaPasha and Wheeler, 1990). These species develop resin ducts axially as well as horizontally, forming an interconnecting system, the extent of this system varying between genera. In addition, wounding (pressure wounds, frost, wind) may result in the formation of traumatic resin ducts, even in *Abies* or *Tsuga* where ducts are not present in healthy tissues, giving the potential for further resin production in the injured area. Resin ducts never occur in the wood of some conifers, e.g. *Juniperus* and *Cupressus* (Fahn and Zamski, 1970). The duct diameters are largest in *Pinus* species (60–300 μm) and smaller in *Larix* (40–80 μm), *Picea* (40–70 μm) and *Pseudotsuga* (40–45 μm) (Larson, 1994). Both vertical and radial resin ducts are found in these genera, and both types occur in traumatic as well as normal duct systems.

The frequency of resin ducts in tree rings, their position within tree rings and their horizontal and vertical variability within a tree are poorly investigated. Reid and Watson (1966) report that vertical ducts form during the latter half of the seasonal growth period. According to their results, resin ducts are located mainly in the latewood portion of tree rings. For lodgepole pine it was found that 88% of the resin ducts are in the last 40% of the tree ring. But number and position of vertical resin ducts might also be influenced by external conditions (Wodzicki, 1961; Larson, 1994). In a laboratory experiment using 2-year-old *Pinus halepensis* Mill., Zamski (1972) showed that temperature and photoperiod changes influence resin duct formation with a time lag of several months. The study also proved temperature to be of higher importance than the photoperiod.

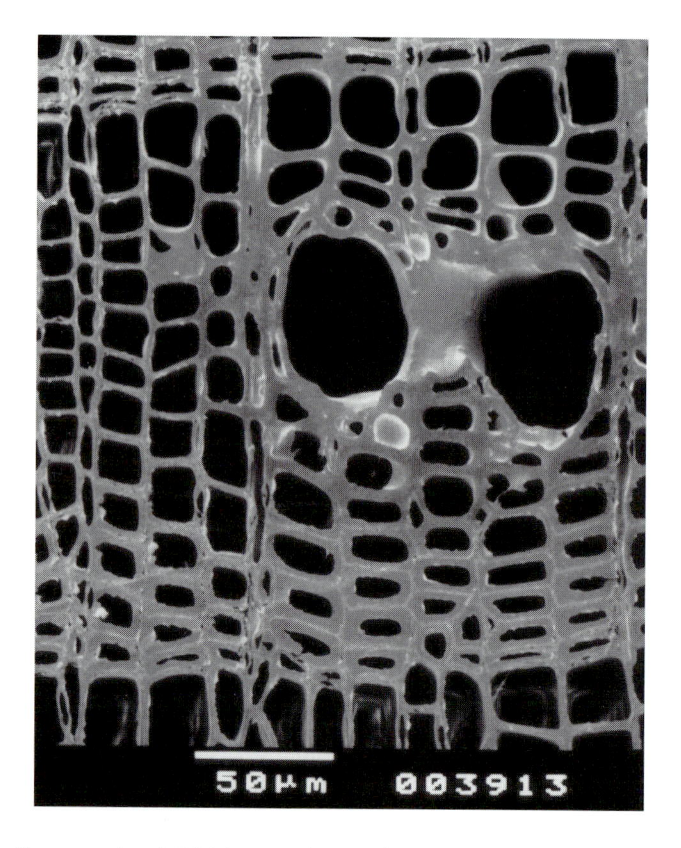

Fig. 5.1. Cross-sectional SEM-image of vertical resin ducts in a spruce tree ring.

Ruden (1987) measured ring width and resin duct frequencies in Scots pine and was able to show that both variables are related to climatic data. Wimmer and Grabner (1997) demonstrated with mature spruce trees that the number of resin ducts per unit area in tree rings is positively linked to summer temperatures. The current study presents two new aspects of resin ducts in spruce tree rings: first, a complete resin duct stem-analysis and, second, the periodicity of resin duct series.

MATERIAL AND METHODS

The first sampling site is located in the 'Lachforst', Ranshofen, Upper Austria, approximately 60 km north from the city of Salzburg, Austria. The site is located on a fluvial-terrace of the Inn River at about 380 m a.s.l. The natural forest community is submontane mixed oak–beech and oak–hornbeam. The actual forests are dominated by even-aged conifers with spruce as the dominant species. At this site, one mature spruce tree (*Picea abies* (L.) Karst.) with a straight, unbroken trunk and a regular-shaped crown was selected. The sample tree was felled in January 1995 and the stem was cut into pieces and sawn into halves by cutting lengthwise through the pith. This procedure was necessary to accurately identify all terminal shoots. From each terminal shoot, disc sections were cut with a chain saw. In total, 81 sections were obtained which corresponded with the number of rings on the disc taken from the base of the tree. The discs were sanded using conventional sandpaper ranging from 100 to 1000 grit. On the sanded discs the resin ducts were counted along radii in each tree ring using a regular stereo-microscope equipped with a video-system. With this procedure it was possible to resolve clearly single tracheids.

The numbers were related to an area that was obtained by multiplying the given tangential window by the ring width, resulting in the parameter 'resin duct density'. In addition, the relative positions of the resin ducts within the tree rings were determined by using the five-step ordinate scale: initial, earlywood–latewood transition, latewood, terminal and dispersed.

The second site is located in the forest district Seyde, Eastern Erzgebirge, about 50 km south of Dresden in Germany, close to the Czech border. On this site, even-aged and approximately 80-year-old Norway spruce trees were sampled. The natural forest community is a mixed beech–fir–spruce forest at a maximum altitude of 800 m a.s.l. Spruce trees cover 80% of this area and on the quartz-porphyric bedrock the predominant brown soils are partially podzolized. In April 1993, 20 dominant and codominant trees were felled and 5-cm-thick stem discs were removed at the 4-m tree height. Discs were transported to the laboratory for the preparation of continuous series of blocks from pith to bark. Transverse sections 20 μm thick were cut from these blocks using a sledge microtome and sections were dehydrated, stained with methylene blue and mounted permanently in Malinol on slides (Gerlach, 1984). The tree rings formed between 1941 and 1987 were analysed. Three radii facing to north, south and east were prepared from each disc and sections were observed through a transmitting light microscope with a CCD-camera connected to a standard video system. Resin ducts were counted in each individual ring from pith to bark. Numbers were also related to an area that was obtained by multiplying the given tangential window by the ring width. We tried to differentiate between regular and traumatic resin duct formations and excluded the latter from the investigation. The final resin duct density is substantial and contains 58 series.

The Eastern Erzgebirge is located in a transition from atlantic to continental climate types. This area experiences especially high fluctuations in temperature with cold winters and little precipitation. Total annual precipitation for these sites is 965 mm, 38% of which is snow. Annual mean temperature is 5.5°C, with −23°C as the lowest temperature measured. Homogenized climatic data representative for the Seyde area were used (Deutscher Wetterdienst, Wetteramt Dresden).

Samples from both sites were crossdated according to the procedure outlined in Swetnam *et al.* (1985) and ring widths were measured to the nearest 0.01 mm using a regular incremental measuring machine. Data were further checked for dating and measurement errors using COFECHA (Holmes, 1983). Figure 5.2 demonstrates that the parameters resin duct density and ring widths are independent and Fig. 5.3 illustrates that ring widths generally follow a negative exponential curve while resin duct density trend is basically linear. Therefore, the resin duct data set from the Erzgebirge site was not detrended, only a horizontal line was fitted through the mean. To investigate the frequency properties of the resin duct density series over time, a spectral analysis was performed using the SPSS (1997) computer software. With a power spectral analysis the variance of each wavelength is estimated through a Fourier transformation of the autocorrelation function (Fritts, 1976).

RESULTS AND DISCUSSION

Resin Duct Stem Analysis

The resin duct densities are shown in Fig. 5.4a. In total, 3280 tree rings are included in each figure. Two trends can be extracted from the tree. First, the resin duct densities for each year are averaged and plotted in Fig. 5.5. Second, the series are sorted by their cambial age and re-plotted in Fig. 5.6. Particular years show extremely high resin duct densities, such as the dry year 1976 (Figs

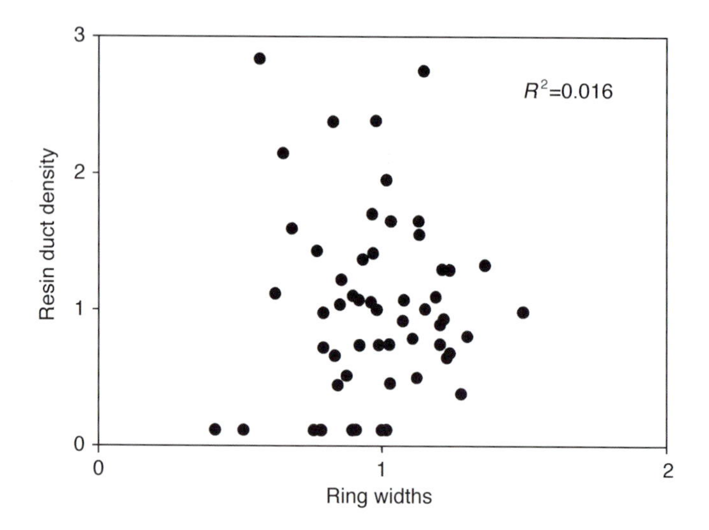

Fig. 5.2. Scatterplot of ring widths and resin duct density. There is no significant relationship between these two parameters.

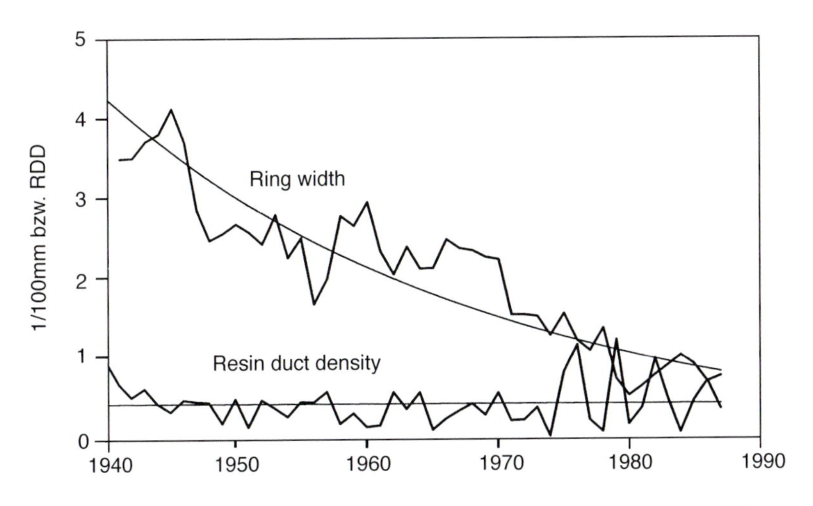

Fig. 5.3. Radial trends of ring widths and resin duct density. Ring widths follow basically a negative exponential curve while resin duct density follows a nearly horizontal line through the mean.

5.4a and 5.5). These years show continuous bands along the stem. Other years have low resin duct densities, which might be due to regionally favourable climatic conditions. Figure 5.6 shows the trend caused by cambial ageing. The 10 years around the pith have numerous resin ducts (Fig 5.4a). Tree rings with a cambial age above 10 show a slight but constant increase for resin duct density. Trendelenburg and Mayer-Wegelin (1955) have already reported such an increase of the proportion of resin ducts from pith to bark.

Results for the relative positions of resin ducts are shown in Fig. 5.4b. According to this figure the majority of the resin ducts are located in the transition between earlywood and latewood. In Fig. 5.7 the 'position' trends averaged over the tree height through time are shown. In the early part, 1930–50, resin ducts are more likely to occur in the latewood, but since 1950 they fluctuate around the earlywood–latewood transition. In Fig. 5.8 the resin duct positions are plotted against their cambial age and this figure is showing basically the same tendency: the resin ducts in tree rings of young cambial age are more likely in the latewood, then they stay constant in the earlywood–latewood transition up to an age of 60 years, and then shift later slightly towards the earlywood. From Fig. 5.4b it can be seen that higher up in the stem resin ducts are located more frequently in earlywood. Comparisons with literature are difficult because only a few reports were focusing on the position of resin ducts in coniferous trees. A general statement was made by Reid and Watson (1966) mentioning that vertical resin ducts are located preferentially in the latewood of tree rings. Their results were based on young experimental trees. In mature spruce tree rings the position of resin ducts is most likely the

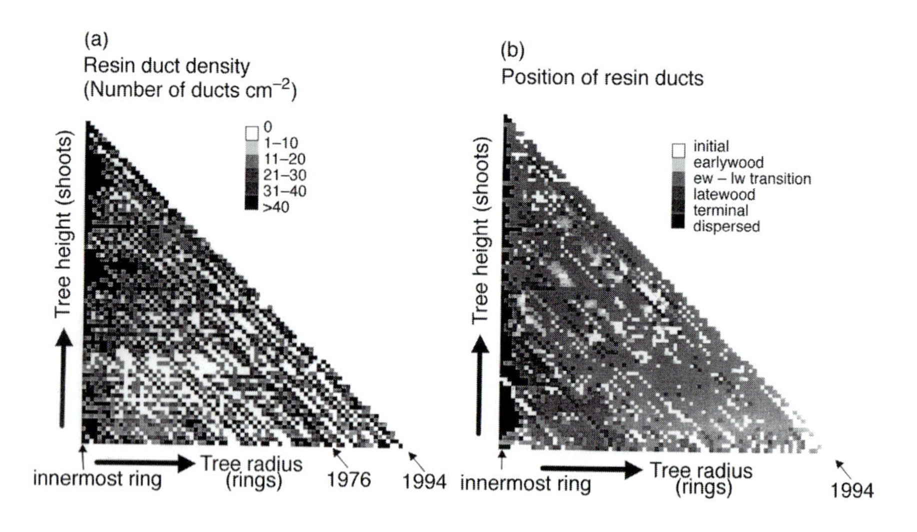

Fig. 5.4. Resin duct patterns showing the variation within the stem. (a): Resin duct density (unit: resin ducts cm^{-2}), (b): Resin duct position within tree rings (ordinate scale).

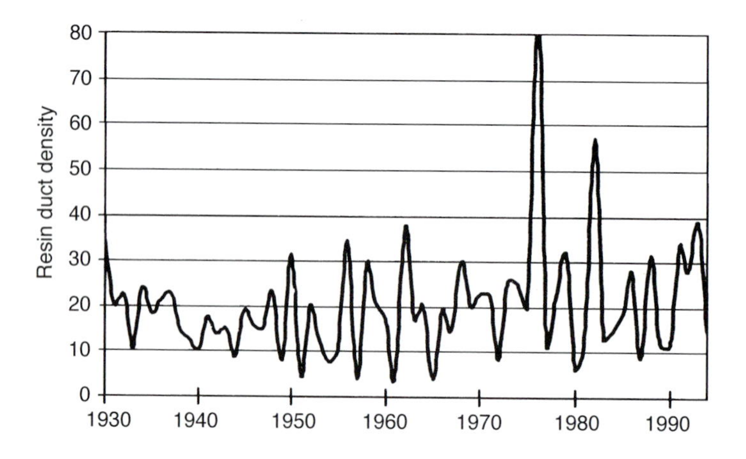

Fig. 5.5. Radial profile of resin duct density (unit: resin ducts cm^{-2}). The curve is an average over all profiles measured in each annual shoot.

transition from earlywood to latewood. Due to environmental stresses that occur earlier in the season, resin ducts could be formed even in the quickly differentiating earlywood. Werker and Fahn (1969) and Fahn (1979) found resin ducts concentrated mainly in the earlywood–latewood transition whereas

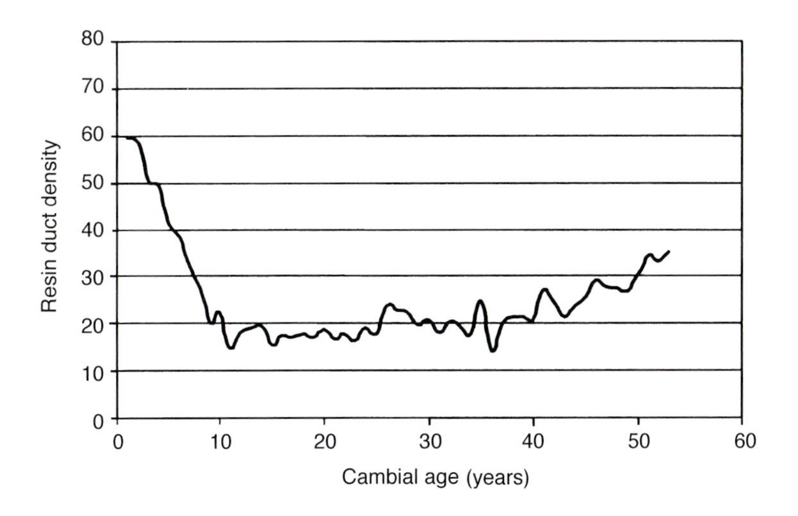

Fig. 5.6. Resin duct density profile and cambial age. The curve is an average over all profiles measured in each annual shoot sorted for their cambial age.

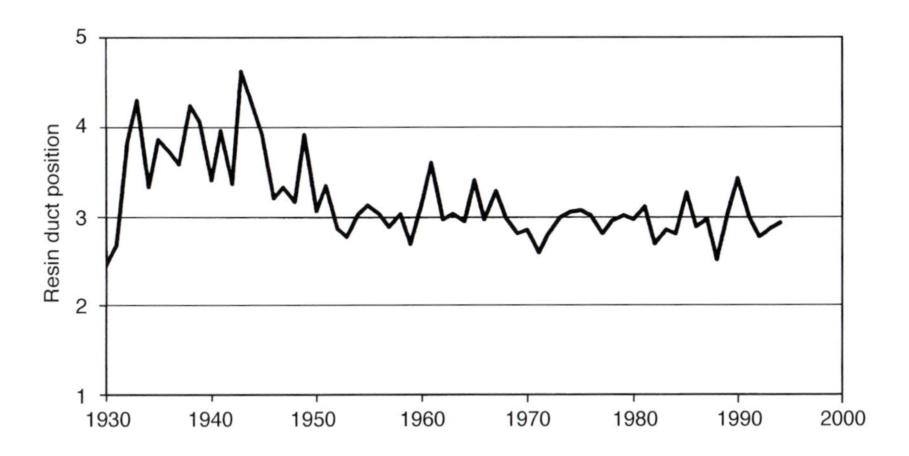

Fig. 5.7. Resin duct position profile. The curve is an average over all position profiles determined in each annual shoot. Relative duct position scale: 1 = initial, 2 = earlywood, 3 = transition earlywood–latewood, 4 = latewood, 5 = terminal.

Stephan (1967), Alfieri and Evert (1968) and Zamski (1972) concluded that resin ducts are concentrated in the latewood. A review of the literature gives the impression that vertical resin ducts are more likely to be confined to the latewood in *Pinus* than in *Picea* and *Larix*.

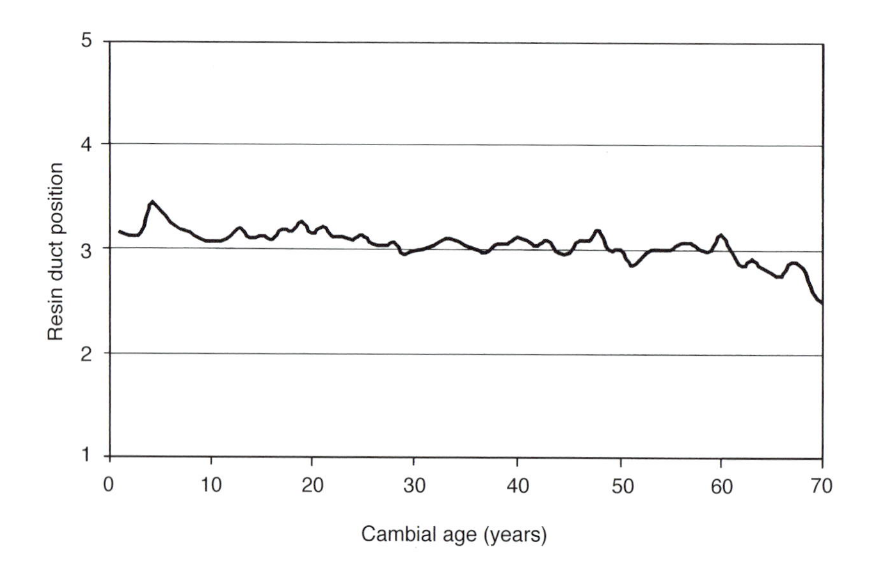

Fig. 5.8. Resin duct position profile and cambial age. The curve is an average over all position profiles determined in each annual shoot which were sorted by their cambial age. Relative duct position scale: 1 = initial, 2 = earlywood, 3 = transition earlywood–latewood, 4 = latewood, 5 = terminal.

Periodicity and Climate Signal in Resin Duct Series

The Erzgebirge data set with 58 resin duct series coming from 20 spruce trees was analysed earlier by Wimmer and Grabner (1997). Resin duct density data were symmetrically distributed with relatively high mean sensitivity and standard deviation but lower signal-to-noise ratio than the corresponding growth rate. The power spectrum analysis estimates the variance of each wavelength (power) and expresses it as a continuous distribution of wavelength throughout the entire spectrum. Figure 5.9 shows a peak at a period of approximately 3 years, which means that the resin ducts behave like a sine wave at this period length. This basic wavelength seems to be the only significant one as no other peak at wavelengths corresponding to one or more higher harmonics of the basic wavelength are seen.

In the previous study (Wimmer and Grabner, 1997), a correlation analysis between monthly climatic data and resin duct density in tree rings showed significant positive relationships with June and July temperature of the current year. In contrast, ring width showed a significant negative response to above-normal precipitation from June to August but no response with temperature. This was also found by others such as Reid and Watson (1966) who suggested circumstantial evidence for a direct relationship between high summer temperatures and large numbers of vertical ducts. Ruden (1987) used resin duct

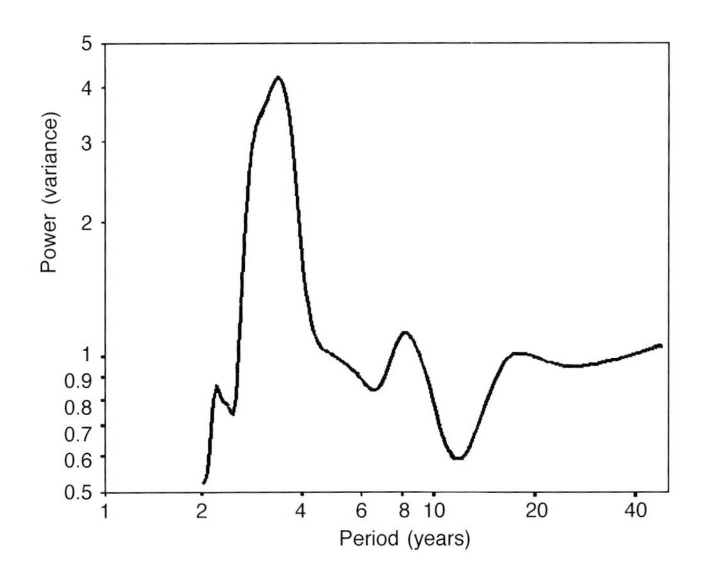

Fig. 5.9. Result of power spectra analysis of resin duct density series from the Erzgebirge (58 series from 20 trees). The maximal peak is at an approximately 3-year period.

frequency and ring width of seven old Scots pines to reconstruct climate for a certain period. He also found resin duct frequency correlated highly with summer drought.

In a cross-power spectrum analysis, the covariance expressed in the frequency domain between the resin duct series and the June–July temperature was computed. Coherency curves are obtained as a measure of similarity between the two series at each frequency. In Fig. 5.10 the highest coherency of 0.8 is at a period of about 3 years, which indicates that there is a causal relationship between resin ducts and summer temperature. A smaller peak can be seen at around 12 years (coherence = 0.6). Most spectral analyses in past dendrochronological research have dealt with radial increment as, for example, Stockton (1975), Fritts (1976), Stahle and Cleaveland (1988), Cook and Kairiukstis (1990) and Cook *et al.* (1998). Time series of various tree-ring parameters (density, anatomical features, ...) should be investigated for their frequency domain as an insight into biological and statistical relationships with regard to possible periodicities of factors controlling tree growth.

CONCLUSIONS

This study deals with two aspects of resin ducts in tree rings. First, a resin duct stem analysis demonstrates the within-tree variability and the effects of

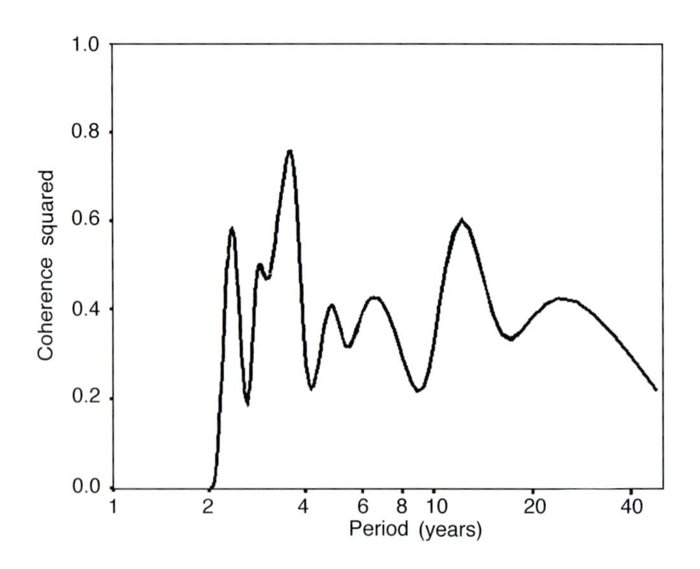

Fig. 5.10. Result of cross-power spectra analysis (coherence (= analogous to R^2) spectra) for resin duct density from the East Erzgebirge, Germany (58 series from 20 trees) and June/July temperature means.

cambium age on resin duct density. Second, the periodicity of resin duct series are explored statistically using a set of samples extracted from 20 spruce trees. The parameter 'resin duct density' is independent of ring width. Particular years show extreme high resin duct densities, such as the dry year 1976, and these years show continuous bands throughout the stem. Tree rings with a cambial age above 10 years show a slight but constant increase in resin duct densities. The majority of the resin ducts are located at the transition between earlywood and latewood. The resin ducts of tree rings at young cambial age are more likely to be in latewood. Summer temperature affects the formation of vertical resin ducts most. At a high frequency of about 3 years in the resin duct series a causal relationship can be seen with summer temperature. The current results should encourage researchers to investigate vertical resin ducts in tree rings of other coniferous species across different site and climate conditions. Although measurement procedures of resin ducts are much more time consuming than measuring ring widths, the results will provide new environmental information and an additional parameter useful in dendroclimatic reconstructions. New insights into the biological relationships can be achieved by investigating factors that control tree growth.

SUMMARY

A resin duct stem-analysis demonstrates the within-tree variability and the effects of cambium age on resin duct density. Particular years show extreme high resin duct densities, such as the dry year 1976, and these years show continuous bands throughout the stem. Summer temperature affects the formation of vertical resin ducts most. The periodicity of resin duct series are explored statistically using a set of samples extracted from 20 spruce trees. At a high frequency of about 3 years in the resin duct series a causal relationship can be seen with summer temperature. The current results should encourage researchers to investigate vertical resin ducts in tree rings.

ACKNOWLEDGEMENTS

Research was conducted with funding from the Austrian Science Foundation (P9200–BIO).

REFERENCES

Alfieri, F.J. and Evert, R.F. (1968) Seasonal development of the secondary phloem in *Pinus*. *American Journal of Botany* 55, 518–528.

Cook, E.R. and Kairiukstis, L.A. (eds) (1990) *Methods of Dendrochronology. Applications in the Environmental Sciences*. Kluwer Academic Publishers, Dordrecht.

Cook, E.R., D'Arrigo, R.D. and Briffa, K. (1998) A reconstruction of the North Atlantic Oscillation using tree-ring chronologies from North America and Europe. *The Holocene* 8, 9–17.

Fahn, A. (1979) *Secretory Tissues in Plants*. Academic Press, New York, 302pp.

Fahn, A. and Zamski, E. (1970) The influence of pressure, wind, wounding and growth substances on the rate of resin duct formation in *Pinus halepensis* wood. *Israel Journal of Botany* 19, 429–446.

Fritts, H.C. (1976) *Tree Rings and Climate*. Academic Press, New York.

Gerlach, D. (1984) *Botanische Mikrotechnik*. Georg Thieme Vlg., Stuttgart.

Holmes, R.L. (1983) Computer-assisted quality control in tree-ring dating and measurement. *Tree-Ring Bulletin* 43, 69–78.

LaPasha, C.A. and Wheeler, E.A. (1990) Resin canals in *Pinus taeda*. Longitudinal canal lengths and interconnections between longitudinal and radial canals. *IAWA Bulletin n.s.* 11, 227–238.

Larson, P.R. (1994) *The Vascular Cambium. Development and Structure*. Springer-Verlag, Berlin.

Reid, R.W. and Watson, J.A. (1966) Sizes, distributions, and numbers of vertical resin ducts in lodgepole pine. *Canadian Journal of Botany* 44, 519–525.

Ruden, T. (1987) Hva furuårringer fra Forfjorddalen kan fortelle om klimaet i Vesterålen 1700–1850. Rapport, *Norsk Institutt for Skogforskning* 4, 1–12.

SPSS (1997) *SPSS for Windows*. Version 7.5, Statistical Products and Science Solutions, Inc. Chicago.

Stahle, D.W. and Cleaveland, M.K. (1988) Texas drought history reconstructed and analyzed from 1698 and 1980. *Journal of Climate* 1, 59–74.
Stephan, G. (1967) Untersuchungen über die Anzahl der Harzkanäle in Kiefern (*Pinus sylvestris*). *Archiv für Forstwesen* 16, 461–470.
Stockton, Ch.W. (1975) *Long-Term Streamflow Records Reconstructed from Tree Rings.* Papers of the Laboratory of Tree-Ring Research Number 5, The University of Arizona Press, Tucson, Arizona, 111pp.
Swetnam, T.W., Thompson, M.A. and Sutherland, E.K. (1985) *Using Dendrochronology to Measure Radial Growth of Defoliated Trees.* USDA Forest Service Handbook 639, Washington, DC.
Trendelenburg, R. and Mayer-Wegelin, H. (1955) *Das Holz als Rohstoff.* Verlag Carl Hanser, München, 541pp.
Werker, E. and Fahn, A. (1969) Resin ducts of *Pinus halepensis* Mill. – their structure, development and pattern of arrangement. *Botanical Journal of the Linnean Society* 62, 379–411.
Wimmer, R. and Grabner, M. (1997) Effects of vertical resin duct density and radial growth of Norway spruce [*Picea abies* (L.) Karst.]. *Trees* 11, 271–276.
Wodzicki, T. (1961) Investigation on the kind of *Larix polonica* Rac. wood formed under various photoperiodic conditions. II. Effect of different light conditions on wood formed by seedlings grown in greenhouse. *Acta Societatis Botanicorum Poloniae* 30, 111–131.
Zamski, E. (1972) Temperature and photoperiodic effects on xylem and vertical resin duct formation in *Pinus halepensis* Mill. *Israel Journal of Botany* 21, 99–107.

6

Practical Application of Annual Rings in the Bark of *Magnolia*

Zhong-Zhen Zhao, Mei Hu, Yutaka Sashida, Xiao-Jun Tang, Hiroko Shimomura and Hiroko Tokumoto

INTRODUCTION

Over the decades, very limited studies have been carried out on annual rings in the bark. In a temperate climate, the age of a tree is usually estimated by counting the annual rings in the xylem. However, this is not practical for standing trees. So far, there has not been any non-destructive and convenient method to determine the age of a standing tree. On the other hand, many crude drugs originate from species of the genus *Magnolia*, and the quality and content of drugs in the crude bark is traditionally evaluated based on the bark thickness. However, the effective compounds in the bark are related to the age of the tree and no standards have been set for a comparison of the descriptions of microscopic characteristics.

Esau (1969, 1977) presented a detailed discussion of the annual growth in the phloem. Her work is based on references such as DeBary (1877), Strasburger (1891), Huber (1939, 1949), Holdheide (1951), Holdheide and Huber (1952) and Evert (1960). Some of these articles comment directly on the number of phloem fibre bands per year and its variation (Courtot and Baillaud, 1956). For example, Holdheide (1951) found annual rings in tree bark of 55 species. Kosicenko (1969) also reported annual rings in the tree bark. Courtot and Baillaud (1958) observed and described the growth rhythms in softwood bark. Roth (1981) investigated the bark of 168 tropical species and discussed their bark growth rings. One conclusion of this study was the need to check whether periodic phloem growth zones correspond to xylem rings. Shimomura *et al.* (1988) conducted a preliminary study on *Magnolia* and reported annual ring-like structures in the bark. Trockenbrodt (1990) has worked on the structural changes of bark tissue during its development in *Quercus robur, Ulmus*

(eds R. Wimmer and R.E. Vetter)

glabra, *Populus tremula* and *Betula pendula*. Growth rings of the bark were also examined. Although growth rings in bark have been studied, the practical applications of annual rings in the bark are not yet fully explored. The major objectives of this study are: (i) to determine whether annual rings in the bark of *Magnolia* are present or not; (ii) to explore the potential of the age determination and (iii) to apply the method in the field of pharmacognosy for the identification of crude bark and for evaluating its quality.

MATERIAL AND METHODS

A total of 141 samples of various ages from 17 species belonging to the genus *Magnolia* were collected in China and Japan (Table 6.1). In order to further investigate the continuous formation of annual rings in the bark, the species observed in 1987 were re-sampled in 1992. Samples were identified based on the taxonomic classification of *Magnoliaceae* by Johnstone (1955). Stem discs were collected as well as branches. The number of annual rings in the bark, branch and stem discs was compared with the number of annual rings in the xylem. As a control, comparisons were also made with trees from botanical gardens and with other living trees where historical records proved the exact tree age (Gourlay, 1995).

Cross sections of the barks were observed microscopically. To observe the fibres, potassium iodide solution was used for the staining and a polariscopic lens was used for a bright yellow contrast (Zhao *et al.*, 1996, 1997).

RESULTS AND DISCUSSION

Annual Rings in the Bark of *Magnolia*

The family *Magnoliaceae* includes trees and shrubs of more than 200 species belonging to 12 genera. *Magnolia*, with about 70 species, is the largest genus in the family. This genus is classified into two subgenera, *Magnolia* and *Pleurochasma*. According to the investigation, many species of *Magnolia* have been used as crude drugs in China and Japan (Zhao and Tang, 1991). In the present study, annual rings were found in all of the observed barks of the 17 *Magnolia* species, which included both of the two subgenera and six sections. The cambium forms one layer of fibre bundles in each subsequent year and the pattern is a series of concentric rings. We drew this conclusion by comparing the number of layers of fibre bundles in tree bark with the number of annual rings in the xylem. *Magnolia kobus* DC. provides a good example (Fig. 6.1).

The periderm consists of phellem (cork), phellogen (cork cambium) and phelloderm. The rhytidome is basically a parenchyma, commonly associated with sclerenchyma and secretory cells. The phloem consists basically of sieve elements, namely sieve tubes, phloem fibres, phloem parenchyma cells and rays.

Table 6.1. Experimental material of *Magnolia*.

No.	Species	Part*	Sampling date	Location	Age (years)**
Subgen. Pleurochasma					
Sec. Yulania					
1	*M. denudata* Desr.	B	Jun. 1987	Tokyo, Japan	1,2,3,4,5,6,7
		T	Jun. 1987	Tokyo, Japan	12
		T	Apr. 1984	Anhui, China	34
		T	Apr. 1984	Anhui, China	>100
		T	Apr. 1990	Beijing, China	3
		T	Apr. 1990	Beijing, China	11
2	*M. denudata* Desr. var.	B	Apr. 1984	Anhui, China	2,3,4,5,6,7,8
	Dilutipurpurascens	T	Apr. 1984	Anhui, China	31
	Xie et Zhao	T	Apr. 1984	Anhui, China	>100
3	*M. elliptimba* Law. et Gao	T	Oct. 1989	Henan, China	43
4	*M. sprengeri* Pamp.	B	Feb. 1983	Shannxi, China	5
		T	Feb. 1983	Shannxi, China	>100
Sec. Buergeria					
5	*M. biondii* Pamp.	B	Feb. 1983	Henan, China	1,2,3
		T	Feb. 1987	Henan, China	20
		T	Oct. 1989	Henan, China	43
		T	Oct. 1989	Henan, China	46
		T	Nov. 1991	Henan, China	88–92***
6	*M. cylindrica* Wils.	T	Oct. 1989	Anhui, China	17
7	*M. kobus* DC.	B	Oct. 1989	Anhui, China	1,2,3,4,5,6,7,8
		T	Oct. 1989	Anhui, China	15,16,17
		T	Apr. 1992	Anhui, China	20
		T	May 1997	Chiba, Japan	50–55
8	*M. pilocarpa* Zhao et Xie	B	Mar. 1983	Hubei, China	3,4,5
		T	Mar. 1983	Hubei, China	20
9	*M. solangeana* Soul.-Bod.	B	Mar. 1990	Beijing, China	2,3,4
		T	Mar. 1990	Beijing, China	39,40,41
10	*M. salicifolia* Maxim.	B	Jul. 1987	Tokyo, Japan	3,4,5
		T	Jul. 1987	Tokyo, Japan	15
Sec. Tulipsdtrum					
11	*M. liliflora* Desr.	T	Feb. 1983	Shannxi, China	11
		T	Jul. 1987	Tokyo, Japan	13
		T	Apr. 1992	Tokyo, Japan	18
Subgen. Magnolia					
Sec. Rytidospermum					
12	*M. obovata* Thunb.	B	Jun. 1987	Tokyo, Japan	1,2,3,4,5,6,7,8
		T	Jun. 1987	Tokyo, Japan	15,20
		T	Jun. 1987	Tokyo, Japan	25
13	*M. officinalis*	B	Jun. 1988	Sichuan, China	2,3,4,5,6,7
	Rehd. et Wils.	T	Jun. 1988	Sichuan, China	15,16,30
		T	Jun. 1988	Sichuan, China	31,40
14	*M. officinalis* Rehd. et	B	Jun. 1988	Sichuan, China	1,2,3,4,5,6,7,8
	Wils. var. *biloba*	T	Jun. 1987	Jiangxi, China	11,17,33
	Rehd. et Wils.	B	Jul. 1989	Jiangxi, China	3,4,18
		T	May 1990	Jiangxi, China	21,22,23,24,25
		T	May 1990	Jiangxi, China	26,27,28,29,30
15	*M. rostrata* W.W.Sm.	B	Jun. 1988	Yunnan, China	1,2,3,4,5,6,7,8
		B	Jun. 1988	Yunnan, China	9,10,11,12,13
		T	Jun. 1988	Yunnan, China	12,15,19
		T	Jun. 1988	Yunnan, China	25,31,36
		T	Jun. 1992	Yunnan, China	51
Sec. Gwillimia					
16	*M. championii* Benth.	B	May 1990	Guangxi, China	1,2,3
		T	May 1990	Guangxi, China	5
Sec. Theorbodon					
17	*M. grandiflora* L.	B	Jun. 1987	Tokyo, Japan	1,2,3,4,5,6,7,8
		T	Jun. 1987	Tokyo, Japan	15

*B = branch, T = trunk.
**Bark samples with 1,2,3, etc. counted rings taken from the same tree. Annual rings in the bark matched with the rings in the corresponding xylem.
***Estimated aged between 88 and 92 years.

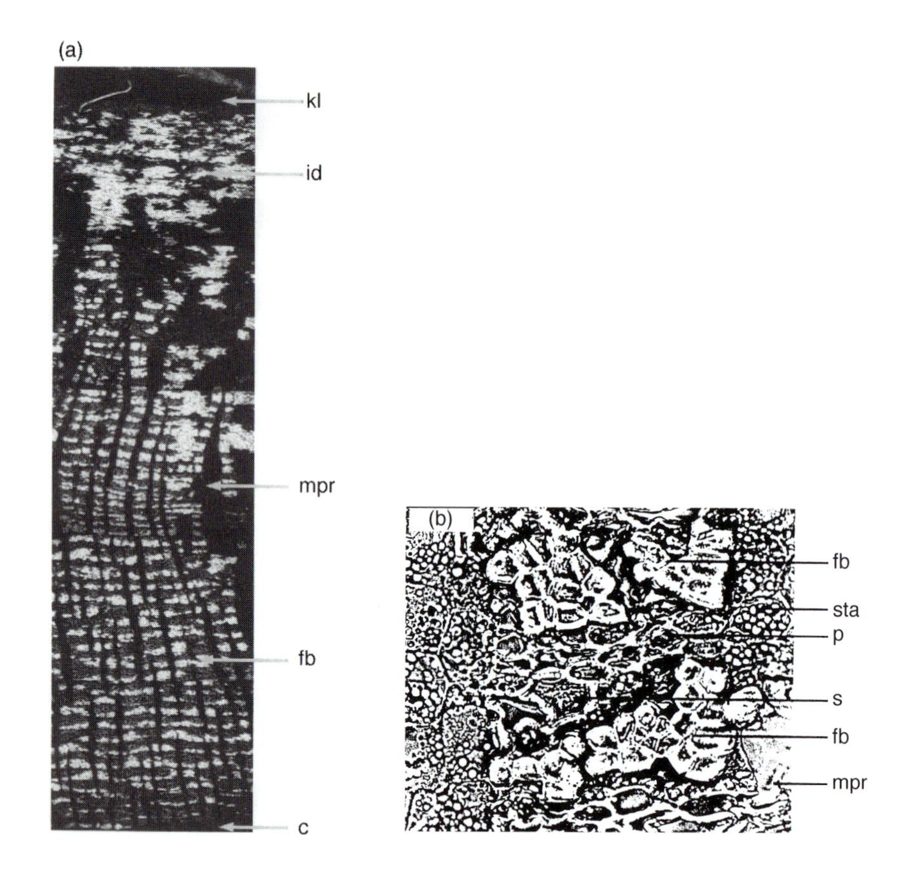

Fig. 6.1. *Magnolia kobus* DC. aged 50–55 years. (a) Cross-section of the bark. Annual rings in the bark are indicated by rings of phloem fibre bundles, photographed under polarized light. (b) Magnification of (a) under light microscope. c = cambium; fb = fibre bundle; id = idioblast; kl = cork layer; mpr = primary medullary ray; p = parenchyma (cell); s = sieve tube; sta = starch grain.

Medullary rays are usually composed of two to four cells which are wide and V-shaped. The sieve tubes are usually separated from the thick-walled phloem fibres by the tangentially aligned phloem parenchyma. The fibre bands in contiguous radial panels of the axial system occur at the same distance from the cambium and give the secondary phloem a stratified aspect. A comparison of the bark with a xylem section of the tree indicated that the number of these fibre bundle layers corresponds to the number of the annual rings in the xylem and thus can be considered as an indicator of the number of annual rings (growth rings) in the xylem. Strictly speaking, annual rings also include a certain amount of sieve elements and parenchymatous tissue, but these are easily damaged during the growth of a tree and thus can be neglected when counting the layers.

Practical Application of Annual Rings in Age Determination of *Magnolia*

Fibre bundles are mature tissue with thick walls that can easily be observed after chemical staining or polarizing microscopy. Characteristic distribution patterns of this tissue in different species provide valuable information on the changes in the growing rhythm. Since only a piece of bark is needed for microscopic observation, this is an easy and almost non-destructive technique to determine the tree's age (Zhao *et al.*, 1991a).

Identification of annual rings in the bark can also be used for age determination of old trees. The age of an old tree of *Magnolia biondii* Pamp. growing in Nanzhao County, Henan Province, China, was determined by this method. The tree is more than 20 m high and showed 1.38 m in diameter. The microscopic cross sections were stained with a potassium iodide solution to make the phloem fibre bundles more visible. The annual rings in the bark, as indicated through the phloem fibre bundles, gave an age of 88–92 years. This result coincided with the historical records provided by local people (Fig. 6.2, Zhao *et al.*, 1990).

Climatic and ecological factors (e.g. temperature, rainfall and habitat) affect the cambium activity and thus the distribution of phloem fibre bundles in the bark. However, the number of layers of fibre bundles still corresponds with the actual age. In the old trees of various species of *Magnolia*, the error in the age determination is less than 5%. Even with longitudinal fissures that appear on the bark surface with higher age, the annual rings in the bark can still be used.

Microscopic Identification of Crude Drug Cortex Magnoliae

In China, 'Hou po', Cortex Magnoliae, is derived from *Magnolia officinalis* Rehd. et Wils. and *Magnolia officinalis* Rehd. et Wils. var. *biloba* Rehd. et Wils., while in Japan, 'Hou po' is derived from *M. obovata* Thunb. It is difficult to distinguish them with macroscopic characteristics.

With the present study it was possible to identify these commercial drugs by comparing the bark thickness of samples with the same age. Here, the age determination is very important in the field of pharmacognosy (e.g. quality evaluation of crude drugs originating from tree bark). The thickness of *Magnolia* bark differs considerably from species to species, assuming that the number of phloem fibre layers in the sample are the same.

In 20-year-old bark of *Magnolia officinalis* Rehd. et Wils., the cork layers consist of more than 20 cell layers, and there are many oil cells but few sclerenchyma cells. The secondary cork cambium is already formed and is

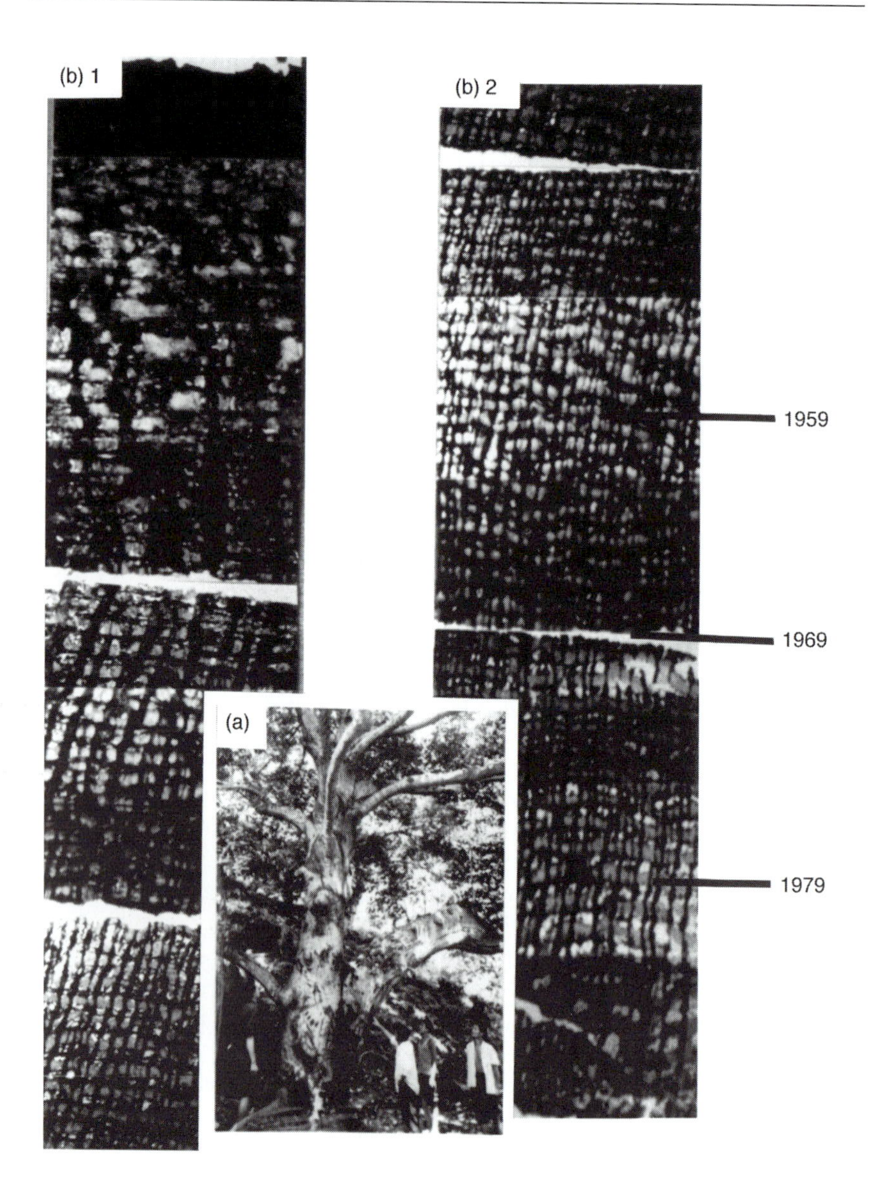

Fig. 6.2. (a) *Magnolia biondii* Pamp. in Henan Province, China, 20 m high, 1.38 m in trunk diameter, together with local villagers who verified its age. (b) Micrograph of cross section of the trunk bark. Annual rings in the bark indicated by rings of phloem fibre bundles gave an age of 88–92 years. (1: outer part, 2: inner part; 1959, 1969 and 1979 indicate the years when fibre bundles were formed.)

producing a new cork layer in the outer cortex. In *Magnolia officinalis* Rehd. et Wils. var. *biloba* Rehd. et Wils., the structure is similar to the bark of *M. officinalis* Rehd. et Wils., but the formation of secondary cork cambium is delayed compared with the former. In *Magnolia obovata* Thunb. the cork layers consist of about 20 cell layers. There are few oil cells but many sclerenchyma cells (Fig. 6.3.).

Quality Estimation of Bark Material Used for Medicine in Different Ages of Tree

Magnolol (I) and honokiol (II) are considered to be the main effective components of Hou po, Cortex Magnoliae. Normally, the content of these two components are determined in the bark to evaluate the commercial value of various Hou po samples coming from different geographical sources and different productions. The relationship between the I and II contents and the age of the bark was not well known. Traditionally, the quality of crude drug bark is evaluated just by comparing thicknesses. In fact, the effective compounds are

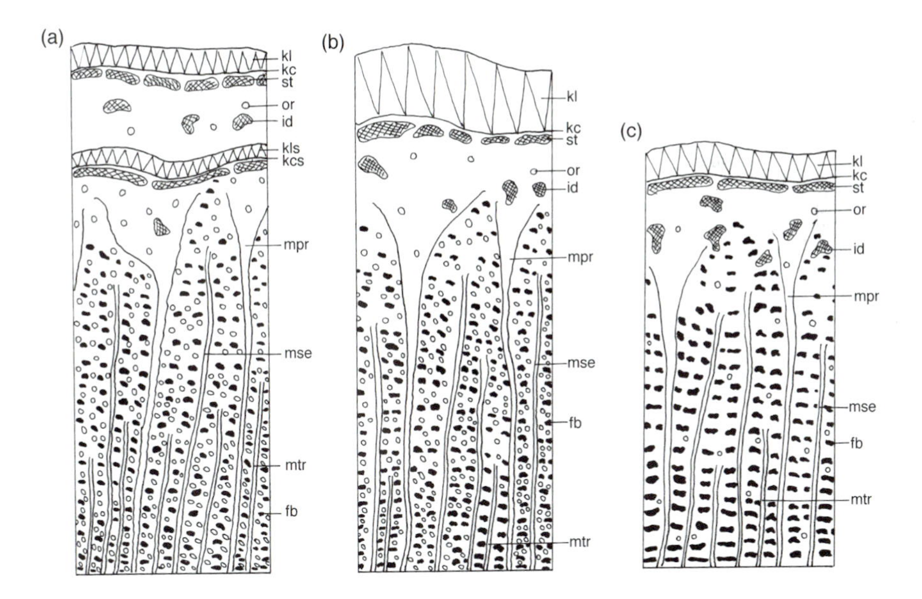

Fig. 6.3. Drawing of transverse sections of Cortex *Magnoliae*. (a) *M. officinalis* Rehd. et Wils.; (b) *M. officinalis* Rehd. et Wils. var. *biloba* Rehd. et Wils.; (c) *M. obovata* Thunb. fb = fibre bundle; id = idioblast; kc = cork cambium or phellogen; kcs = secondary cork cambium or phellogen; kl = cork layer; kls = secondary cork layer; mpr = primary medullary ray; mse = secondary medullary ray; mtr = tertiary medullary ray; or = oil cavity or oil reservoir; st = stone cell.

related to the tree age. Better methods to determine the best drug collection period have long been sought. The method of tree age determination presented here, combined with chemical analysis delivers a better basis for quality evaluation. We first estimated the age of the bark by examining the annual rings in the bark and then determined the I and II contents in the bark of different age parts from a single tree of *Magnolia officinalis* Rehd. et Wils. The reliability of this method was confirmed by running a parallel thin-layer chromatography scan for quantitative evaluation of magnolol and honokiol in some *Magnolia* species. The results demonstrate that the amount of these two bark components is related to the age of the bark and varies as the tree gets older. For *Magnolia officinalis* Rehd. et Wils. the peak content occurs in a bark of about 25 to 30 years of age (Fig. 6.4). Therefore, this would be the optimum age to collect crude drug from this tree (Zhao *et al.*, 1991b).

SUMMARY

A total of 141 samples were taken from 17 tree species belonging to the genus *Magnolia* collected from different regions of China and Japan to examine annual growth rings in the bark. The number of fibre bundle layers in the phloem was successfully compared with the number of annual rings in the corresponding

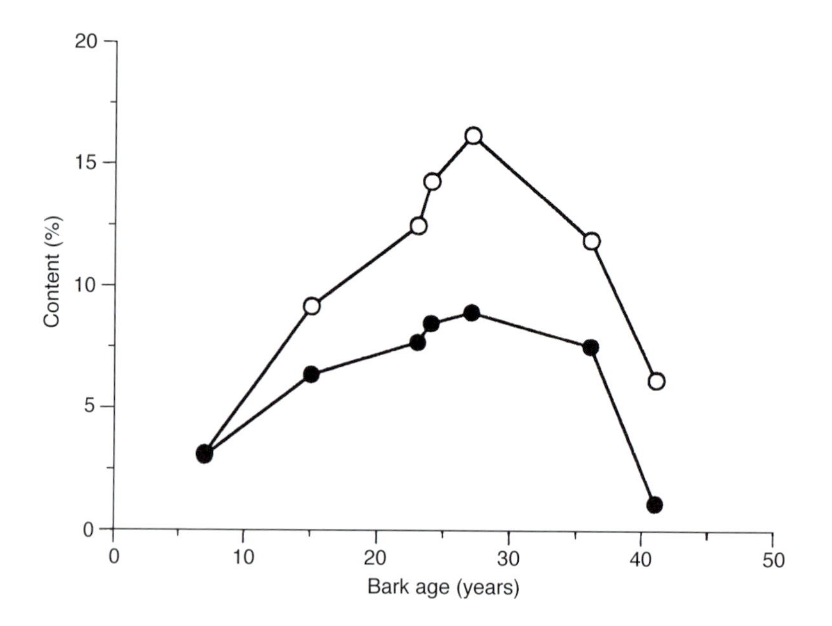

Fig. 6.4. Magnolol (○) and honokiol (●) content in dried bark of different ages from *Magnolia officinalis* Rehd. et Wils.

xylem. Thus, the annual rings in the bark can serve as a relatively easy method to determine the age of standing *Magnolia* trees. In addition, this chapter discusses applications of age determination of bark material for medicinal purposes. Age determination proved to be useful for increasing the quality of crude drugs originating from tree barks.

ACKNOWLEDGEMENTS

The investigations were carried out in cooperation with Mr Izumi, H., Medicinal Plant Garden of Tokyo University of Pharmacy and Life Science, Mr Hu, Z.-X., Beijing Bureau of Forestry, China, Mr Wu, J.-L., Botanical Bureau of Emei Mountain, China, and Mr Huang, H.-X., Henan Bureau of Forestry, China. The authors wish to thank Dr S.Y. Zhang, Forintek Canada Corp., Canada, for his assistance.

REFERENCES

Courtot, Y. and Baillaud, L. (1956) Sur la periodicite des differenciations liberiennes chez le Tilleul. *Annual Science University Besancon Series* 2. Bot. 8, 73–87.

Courtot, Y. and Baillaud, L. (1958) Quelques modalites du rythme de la differenciation dans le liber des Cupressacees. *Annual Science University Besancon Series* 2. Bot. 12, 53–57.

DeBary, A. (1877) *Vergleichende Anatomie der Vegetationsorgane der Phanerogamen und Farne*. W. Engelmann, Leipzig.

Esau, K. (1969) The phloem. *Handbuch der Pflanzenanatomie*. Vol. V/2. Gebrüder Borntraeger, Berlin-Stuttgart.

Esau, K. (1977) *Anatomy of Seed Plants*, 2nd edn. Wiley, New York.

Evert, R.F. (1960) Phloem structure in *Pyrus communis* L. and its seasonal changes. *University of California Publications in Bot*any 32, 127–194.

Gourlay, I.D. (1995) Growth ring characteristics of some African *Acacia* species. *Journal of Tropical Ecology* 11, 121–140.

Holdheide, W. (1951) Anatomie mitteleuropäischer Gehölzrinden. In: Freund, H. (ed.) *Handbuch der Mikroskopie in der Technik*, Vol. V/1. Frankfurt a. M., Umschau, pp. 193–367.

Holdheide, W. and Huber, B. (1952) Ähnlichkeiten und Unterschiede im Feinbau von Holz und Rinde. *Holz als Roh- und Werkstoff* 10(7), 263–268.

Huber, B. (1939) Das Siebröhrensystem unserer Bäume und seine jahreszeitlichen Veränderungen. *Jahrbuchet für Wissenschaftlicher Botanik* 88, 176–242.

Huber, B. (1949) Zur Phylogonie des Jahrringbaues der Rinde. *Svensk. Bot. Tidskr.* 43, 376–382.

Johnstone, G.H. (1955) *Asiatic* Magnolia *in Cultivation*. The Royal Horticultural Society Press, London.

Kosicenko, N.E. (1969) The annual rings of the bark of woody species. *Lesnoi Zhurnal Arhangelsk* 12(1), 33–37 (in Russian).

Roth, I. (1981) Structural patterns of tropical barks. *Handbuch der Pflanzenanatomie*, Vol. IX/3. Gebrüder Borntraeger, Berlin.

Shimomura, H., Sashida, Y., Zhao, Z.-Z., Tokumoto, H. and Kobayashi, H. (1988) Pharmacognostical studies on the *Magnolia* bark (1) A preliminary study on the annual ring-like structure in the bark of *Magnolia*. *The Japanese Journal of Pharmacognosy* 42, 220–227.

Shimomura, H., Sashida, Y., Zhao, Z.-Z., Tokumoto, H. and Tang, X.-J. (1989) Pharmacognostical studies on the *Magnolia* bark (2) Morphological and histological studies of crude drug cortex Magnoliae. *The Japanese Journal of Pharmacognosy* 43, 148–158.

Strasburger, E. (1891) *Ueber den Bau und die Verrichtung der Leitungsbahnen. Histologische Beiträge* 3. G. Fischer, Leipzig.

Trockenbrodt, M. (1990) Qualitative structural changes during bark development in *Quercus robur, Ulmus glabra, Populus tremula* and *Betula pendula*. *IAWA Bulletin*, n.s. 11, 141–166.

Zhao, Z.-Z. and Tang, X.-J. (1991) Investigation of medicinal plant resources in *Magnolia*. *Journal of Chinese Materia Medica* 5, 44–48.

Zhao, Z.-Z., Tang, X.-J. and Hu, M. (1990) A study of annual rings in tree bark and their significance. *Acta Beijing Forest University* 12, 127–131.

Zhao, Z.-Z., Tang, X.-J. and Hu, M. (1991a) A new method for identifying the age of the tree. Chinese Bureau (ed.) *Bulletin of Chinese Patents*. No. 90,1,03136.4, Beijing (in Chinese).

Zhao, Z.-Z., Hu, M., Sashida, Y. and Tang, X.-J. (1991b) Pharmacognostical studies on the *Magnolia* bark (3) Determination of Magnolol and Honokiol in cortex Magnoliae prepared from the bark of different age. *The Japanese Journal of Pharmacognosy* 45, 145–147.

Zhao, Z.-Z., Sashida, Y., Tang, X.-J., Tokumoto, H. and Shimomura, H. (1991c) Pharmacognostical studies on the *Magnolia* bark (4) Morphological and histological studies on the bark of *Magnolia rostrata* W.W. Smith and *Magnolia grandiflora* L. *The Japanese Journal of Pharmacognosy* 45, 190–198.

Zhao, Z.-Z., Shimomura, H., Sashida, Y., Thujino, T., Okamoto, T. and Kazami, T. (1996) Identification of traditional Chinese patent medicines by a polariscope (1) Polariscopic characteristics of starch grains and calcium oxalate crystals. *Natural Medicines* 50, 389–398.

Zhao, Z.-Z., Shimomura, H., Sashida, Y., Thujino, T., Okamoto, T. and Kazami, T. (1997) Identification of traditional Chinese patent medicines by a polariscope (2) Polariscopic characteristics of stone cells, vessels and fibres. *Natural Medicines* 51, 504–511.

Tropical Tree-Ring Analysis

C

7

Remarks on the Current Situation of Tree-Ring Research in the Tropics

Roland E. Vetter and Rupert Wimmer

THE DIFFICULTIES OF TREE-RING RESEARCH IN THE TROPICS

For the tropical region, it is still widely assumed that trees grow continuously throughout the year and, as a consequence, tree-ring studies would be unlikely to yield results. This apparent absence of growth rings in the tropics was mentioned in a letter by Antonie van Leeuwenhoek to the Royal Society of London dated 12 January 1680. In this letter he described the wood anatomy of Mauritius ebony and comments on the 'continuous' growth in its tropical habitat (Baas and Vetter, 1989).

One still gets the impression that nobody feels an urgent need to investigate the growth dynamics and absolute tree age in natural untouched or even in already disturbed tropical forests. It happens from time to time that trees are discovered which are older than anticipated, and this brings into question the well-propagated claims for sustainability of forest management plans. Those plans frequently calculate with a arbitrary rotation cycle of about 30 years (Boxman *et al.*, 1987). These 30 years might be a long period for financial investment plans in tropical developing countries which usually have unstable landownership, but are too short for maintaining a sustainable natural forest resource. A reliable and objective method for age determination of harvested trees seems not to be getting the attention it deserves, even among foresters. However, the increasing interest of climatologists in using old trees of the tropics to study past climatic conditions with reference to the ongoing climate change discussion may turn out to be a streak of light on the horizon of tree-ring analysis in the tropics.

Classical references to climate–environment relationships and tree rings include those of Douglass (1919, 1928, 1936), with his compendium *Climatic*

Cycles and Tree Growth, and also those of Fritts (1976) with his book *Tree Rings and Climate*. Schweingruber (1993) has added many new aspects to tree-ring analysis, and in his book *Trees and Wood in Dendrochronology* he listed the tree species that are most frequently used in tree-ring studies.

Observations and studies on growth periodicity in tropical trees have been reported for several tropical regions over the decades (Coster, 1927; Alvim, 1964; Amobi, 1973). In recent years, the topic 'age and growth periodicity in tropical trees' has become the subject of several international meetings. The first workshop on this subject, held in 1980 at Harvard Forest in Petersham, Massachusetts, USA, identified new research directions (Borman and Berlyn, 1981). During a second meeting in São Paulo, Brazil (Baas and Vetter, 1989), and a third workshop in Kuala Lumpur, Malaysia (Eckstein *et al.*, 1995), results of progressing research activities were presented and discussed.

TREE-RING RESEARCH STILL AT ITS BEGINNING

However, tree-ring analysis in the tropics is still at its beginning and 'tropical' tree-ring chronologies exist so far only for *Tectona grandis* in South East Asia (Jacoby and D'Arrigo, 1990; Pumijumnong, chapter 10, this volume). The knowledge about tropical tree biology is rather limited, and questions such as how growth periodicity is related to environmental factors (in order to find out whether growth rings reflect annual or climatic/seasonal cycles in tree growth or not) are not yet solved (Prevost and Puig, 1981; Vetter and Botosso, 1989a).

Tropical climate does not mean constant environmental conditions without any seasonal changes. There are specific environmental factors prevalent in each tropical forest type. To name examples, tropical mountain forests cover different sites with locally very special conditions such as dryness in the rain shadow of mountains or low temperatures on the shadow side of a valley. Monsoon climates in deciduous forests are characterized by distinct dry seasons. Tropical rain forests at the Brazilian Atlantic coast or in Western Amazonia usually grow under wet conditions, but locally sites also experience extended dry periods.

In contrast to temperate regions, the current status of tree-ring analysis in the tropics does not go further than the identification of growth ring boundaries in species from some limited regions in South East Asia (Coster, 1927), Latin America (Schulz, 1960; Alvim, 1964; Détienne and Barbier, 1988; Vetter and Botosso, 1988, 1989b; Worbes, 1989; Vetter, 1995) and Africa (Mariaux, 1969; Détienne 1989; Gourlay, 1995). Exceptionally, research activities go further, for example, with analysis of growth patterns–climate relationships (Devall *et al.*, 1995) or dendrochronological investigations in Thailand (Buckley *et al.*, 1995). These results are not yet recognized by silviculturists.

NUMEROUS METHODS AND APPROACHES ARE NEEDED

Besides traditional methods of tree-ring analysis as utilized successfully in temperate regions, we should not hesitate to apply new and unorthodox methods to understand ring formation in tropical environments. But the analysis has to start with basic wood anatomical approaches before moving on to other techniques. As an example, in *Scleronema micranthum* the parenchyma bands which are important for the detection of growth periodicity cannot be detected with X-ray densitometry (Vetter, 1995). The relatively new methods of sandblasting (Lesnino, 1989) and drill resistance (Rinn *et al.*, 1989) in connection with the wood anatomical description may furnish a new understanding of growth behaviour of tropical trees. For instance, structural differences appearing in the drill resistance measurements and after sandblasting cannot be explained by density variations alone. These methods are still in their initial phase but seem to be promising for the future.

Cambium marking and increment measurements are indispensable dynamic methods for documenting growth periodicity previously detected in new species by traditional methods (wood anatomy). The cambium should be marked at short – at least monthly – intervals. With cambium marking the undesired influence of callus tissue formation should be avoided to allow the almost undisturbed development of a tree ring. Punctual cambium marking introduced by Wolter (1968) using nails or needles, and the 'pinning method' (Shiokura, 1989; Jalil *et al.*, 1994; Möller, 1997), as well as the marking with knife cuts (Jalil *et al.*, 1994), are less damaging to the cambium and could already be considered as standard methods.

Special attention should be paid to periodic changes in the availability of water and nutrients to the root system of the trees (Luizão and Schubart, 1987; Wong, 1994). It is suggested that element concentrations in increment rings correlate with growth periodicity. Trace-element determination within tree rings by means of proton induced X-ray emission (PIXE) may identify periodicity in element concentrations (Harju *et al.*, 1996). Iqbal (1994) mentions the influence of phenological events, particularly the shedding and re-growth of leaves and cambium activity. For less known species, studies on phenology and tree physiology could increase the knowledge about the influence of flowering, fruit development and internal metabolism in the course of a year on growth periodicity and tree-ring formation. Future research efforts should focus more on dynamic methods of periodic measurements and physiological factors such as starch distribution and activity of acid phosphatase in correlation to cambium activity (Fink, 1982).

For large tropical trees with unknown age, radiocarbon dating could be applied for age estimation. However, radiocarbon measurements are rather expensive and therefore not very practicable. But the few results on age determination by radiocarbon dating available underline the importance of this method. For example, radiocarbon dating of harvested trees from Central Amazonia brought up tree ages between 200 and 500 years (Carmargo *et al.*, 1994; Vetter, 1995; Chambers *et al.*, 1998), and one exemplar dated by

Chambers *et al.* (1998) was as old as 1400 years. These interesting data show the high potential and need for further tree-ring analysis efforts in the tropics.

SITE SELECTION

Most areas of the vast tropical forests are unknown. Appropriate site selection in these forests is critical in order to apply tree-ring analysis successfully. Sites should be close to meteorological stations with long-term climatic records available. An indication of the climatic and geomorphological characteristics is essential before a periodicity study of growth ring structures starts in at least some species. Continuous data collection and regular monthly cambium marking require that the sites are accessible throughout the year and that periodic fieldtrips are covered by the research budget. Measurements and data have to be standardized at least for the most important factors to facilitate comparison between stands. Self-registering data acquisition of increment rates, soil and xylem/cambium moisture are important new approaches (Downes *et al.*, Chapter 4; Gensler, Chapter 3, this volume) and should be employed in the tropical forests.

STRENGTHENING MULTIDISCIPLINARY RESEARCH AND TEACHING

Coordinated multidisciplinary research programmes are necessary and should be promoted to investigate the complex relationships between environmental conditions and tree biology. Soil scientists looking at soil texture and moisture regime could explain biological behaviour of the living system 'tropical tree'. Climatologists may add their expertise as well. Research on the influence of El Niño events on tree growth in Amazonia, for example, is a result of such a multidisciplinary collaboration (Vetter and Botosso, 1989a).

University teaching should emphasize education in tree-ring analysis, its importance, techniques and applications. Special attention should be paid to the subject in the curricula in forest science and forest management at all levels. Tree-ring researchers at a senior level could be more involved in supervising MSc and PhD students, and could work with them on site on these topics.

Tropical tree-ring research has to demonstrate its potential for climatology, forest conservation and ecology, for silviculture, and for forest management and wood utilization. This potential would hopefully gain the interest of people coming from forest management and wood utilization, and thus also ensure future research funding in the tropical region.

SUMMARY

In this brief chapter the current problems of tree-ring research in tropical regions are addressed. Suggestions are made on how to promote tree-ring research to provide basic and reliable results to contribute to the sustainability of tropical forests in the future. Coordinated multidisciplinary research programmes are necessary and should be promoted.

REFERENCES

Alvim, P. (1964) Tree growth periodicity in tropical climates. In: Zimmermann, M.H. (ed.) *The Formation of Wood in Forest Trees*. Academic Press, New York, pp. 479–495.

Amobi, C.C. (1973) Periodicity of wood formation in some trees of lowland rainforest in Nigeria. *Annals Botany n.s.* 37, 211–218.

Baas, P. and Vetter, R.E. (eds) (1989) Growth rings in tropical trees. *IAWA Bulletin (Special Issue)* 10(2), 95–174.

Borman, F.H. and Berlyn, G. (eds) (1981) *Age and Growth Rate of Tropical Trees: New Directions for Research*. Bulletin No. 94, Yale University, School of Forestry and Environmental Studies, New Haven, 137pp.

Boxman, O., Graaf, N.R. de, Hendrison, J., Jonkers, W.B.J., Poels, R.L.H., Schmidt, P. and Tjon Lim Sang, T. (1987) Forest land use in Surinam. In: Beusekom, C. van, Goor, P. van and Schmidt, P. (eds) *Wise Utilization of Tropical Rain Forest Lands (Tropenbos Scientific Series 1)*, Tropenbos, Utrecht, pp. 119–129.

Buckley, B.M., Barbetti, M., Watanasak, M., D'Arrigo, R., Boonchirdchoo, S. and Sarutanon, S. (1995) Dendrochronological investigations in Thailand. *IAWA Journal* 16(4), 393–409.

Carmargo, P.B., Salomão, R.P., Trimbore, S. and Martinelli, L.A. (1994) How old are large Brazil-nut trees (*Bertholletia excelsa*) in the Amazon? *Scientia Agricola* 51, 389–391.

Chambers, J., Higuchi, N. and Schimel, J.P. (1998) Ancient trees in Amazonia. *Nature* 391(6663), 135.

Coster, C. (1927) Zur Anatomie und Physiologie der Zuwachszonen- und Jahrringbildung in den Tropen I. *Annales du Jardin Botanique de Buitenzorg* 37, 49–161.

Détienne, P. (1989) Appearance and periodicity of growth rings in some tropical woods. *IAWA Bulletin n.s.* 10(2), 123–132.

Détienne, P. and Barbier, C. (1988) Rhythmes de croissance de quelques essences de Guyane Francais. *Bois et Forêts de Tropiques* 217, 63–76.

Devall, M.S., Parresol, B.R. and Wright, S.J. (1995) Dendroecological analysis of *Cordia alliodora*, *Pseudobombax septenatum* and *Annona spraguei* in Central Panama. *IAWA Journal* 16(4), 411–424.

Douglass, A.E. (1919) *Climatic Cycles and Tree Growth*. Carnegie Institute of Washington, Publication No. 289, Volume I, 129pp.

Douglass, A.E. (1928) *Climatic Cycles and Tree Growth*. Carnegie Institute of Washington, Publication No. 289, Volume II, 166pp.

Douglass, A.E. (1936) *Climatic Cycles and Tree Growth*. Carnegie Institute of Washington, Publication No. 289, Volume III, 171pp.

Eckstein, D., Sass, U. and Baas, P. (eds) (1995) Growth periodicity in tropical trees. *IAWA Journal* 16(4), 323–442.

Fink, S. (1982) Histochemische Untersuchungen über Stärkeverteilung und Phosphataseaktivität im Holz einiger tropischer Baumarten. *Holzforschung* 36(6), 295–302.

Fritts, H. (1976) *Tree Rings and Climate*. Academic Press, New York, 567pp.

Gourlay, I.D. (1995) The definition of seasonal growth zones in some African *Acacia* species – A review. *IAWA Journal* 16(4), 353–359.

Harju, L., Lill, J.-O., Saarela, K.-E., Heselius, S.-J., Herberg, F.J. and Lindroos, A. (1996) Study of seasonal variations of trace-element concentrations within tree rings by thick-target PIXE analysis. *Nuclear Instruments and Methods in Physics Research Bulletin* 109/110, 536–541.

Iqbal, M. (1994) Periodicity of radial growth in tropical trees. *Meeting on Periodicity of Growth in Tropical Trees*, 16–18 November 1994 in Kuala Lumpur (Abstracts of papers), 2–3.

Jacoby, G.C. and D'Arrigo, R.D. (1990) Teak (*Tectona grandis* L.), a tropical species of large-scale dendroclimatic potential. *Dendrochronologia* 8, 83–98.

Jalil, N., Sari, M.H., Itoh, T. and Jusoh, M.Z. (1994) Periodicity of xylem growth of rubberwood (*Hevea brasiliensis*) in Malaysian plantations. *Meeting on Periodicity of Growth in Tropical Trees*, 16–18 November 1994 in Kuala Lumpur (Abstracts of papers), Annex.

Lesnino, G. (1989) Untersuchungen von durch Sandstrahlen erzeugten Härtestrukturen in Laubhölzern und ihre mögliche Auswirkungen auf Holzeigenschaften. Dr thesis, Forstwissenschaftliche Fakultät, Universität München, Germany, 177pp.

Luizão, F.J. and Schubart, H.O.R. (1987) Litter production and decomposition in a terra firme forest of Central Amazon. *Experientia* 43(3), 259–265.

Mariaux, A. (1969) La périodicité des cernes dans le bois de limba. *Bois et Forêts des Tropiques* 128, 39–53.

Möller, R. (1997) Untersuchungen zur Holzbildungsdynamik von Buche und Robinie in Norddeutschland mit Hilfe der 'pinning'-Methode. MSc thesis, Universität Hamburg, Germany.

Prevost, M.F. and Puig, H. (1981) Accroisement diamétral des arbes en Guyane: observations sur quelques arbres de forêt primaire et de forêt secundaire. *Adansonia, série 4*, 3(2), 147–171.

Rinn, F., Becker, B. and Kromer, B. (1989) Ein neues Verfahren zur direkten Messung der Holzdichte bei Laub- und Nadelhölzern. *Dendrochronologia* 7, 159–168.

Schulz, J.P. (1960) *The Vegetation of Surinam. Vol.II: Ecological Studies on Rain Forest in Northern Surinam*. Van Eedenfonds, Amsterdam, 276pp.

Schweingruber, F.H. (1993) *Trees and Wood in Dendrochronology*. Springer-Verlag, Berlin, 402pp.

Shiokura, T. (1989) A method to measure radial increment in tropical trees. *IAWA Bulletin n.s.* 10(2), 147–154.

Vetter, R.E. (1995) Untersuchungen über Zuwachsrhythmen an tropischen Bäumen in Amazonien. Dr thesis, Albert-Ludwigs-Universität, Freiburg, Germany.

Vetter, R.E. and Botosso, P.C. (1988) Observações preliminares sobre a periodicidade e taxa de crescimento em árvores tropicais. *Acta Amazônica* 18(1–2), 189–196.

Vetter, R.E. and Botosso, P.C. (1989a) El Niño may affect growth behaviour of Amazonian trees. *GeoJournal* 19(4), 419–421.

Vetter, R.E. and Botosso, P.C. (1989b) Remarks on age and growth rate determination of Amazonian trees. *IAWA Bulletin n.s.* 10(2), 133–145.

Wolter, K.E. (1968) A new method for marking xylem growth. *Forest Science* 14, 102–104.

Wong, A. (1994) Variations in wood elemental compositions associated with periodicity in tree growth – Previous studies and possible analysis in moist tropical forest timbers. *Meeting on Periodicity of Growth in Tropical Trees*, 16–18 November 1994 in Kuala Lumpur (Abstracts of papers), 19.

Worbes, M. (1989) Growth rings, increment and age of trees in inundation forests, savannas and mountain forest in the neotropics. *IAWA Bulletin n.s.* 10(2), 109–122.

8

Identification of Annual Growth Rings Based on Periodical Shoot Growth

Patrícia Póvoa de Mattos,
Rudi Arno Seitz and
Graciela Ines Bolzon de Muniz

INTRODUCTION

The periodicity of growth ring formation in tropical forest trees is frequently mentioned in the literature (Mariaux, 1969, 1970; Détienne and Mariaux, 1977; Détienne and Barbier, 1988; Vetter and Botosso, 1989; Pumijumnong *et al.*, 1995). In tropical regions with a well-defined wet and dry season, the flushing of shoots is seasonal and formation of the growth rings is mainly related to the water deficit periods (Ash, 1983; Détienne, 1989; Villalba and Boninsegna, 1989; Worbes, 1989).

Amobi (1973), studying tree species from the lowland rain forest in Nigeria, observed that the cambial activity re-started after the end of the dry season, and that the re-start was positively correlated with the bud break as well as the unfolding of new leaves. Worbes (1995) also mentioned that in tropical regions with seasonally alternating favourable and unfavourable growth conditions the growth periodicity of the trees is strongly linked to the leaf fall behaviour.

In trees with rhythmic radial growth, it is reasonable to expect a similar behaviour in length of shoot growth. This relationship is well known for the temperate zone, but has only been demonstrated in a few cases for trees in the tropics (Hallé and Martin, 1968; Amobi, 1974; Gil and Garcia, 1977; Venugopal and Krishnamurthy, 1987). More recently, Worbes (1995) has demonstrated with *Pseudobombax* that rhythmic radial growth and shoot extension could be related.

The objective of the current work is to verify the annual growth ring formation by the annual shoot section in tropical trees as a fast, easy and inexpensive method, aiming to help future works of age estimation and growth increment of native tropical trees.

MATERIAL AND METHODS

The species *Cedrela fissilis* Vell. was selected for this study because annual growth rings are very well defined by paratracheal parenchyma bands in combination with half-ring porosity (Boninsegna *et al.*, 1989). This species frequently presents more than one growth flush in the same vegetative period and is shedding leaves in the wintertime. The material was collected in the Curitiba metropolitan region, Parana, Brazil, where the climate is always humid. The region has a mean annual temperature of 16.5°C, with monthly means between 13 and 21°C, and annual precipitation of 1452 mm, with monthly means between 75 mm and 170 mm (Maack, 1981).

A more extended study was carried out in a semi-deciduous forest of the Pantanal Mato-grossense region, at the Nhumirim farm, in Pantanal of Nhecolandia, Corumba. The average annual precipitation varies between 1000 and 1400 mm, with about 80% of the rainfall concentrated in summer, with maximum peak in December and January. The annual average temperature is above 25°C (Cadavid Garcia, 1984). In this region the trees shed their leaves in the dry winter.

The trees were selected among individuals presenting good canopy development belonging to the species from terra firme forest listed in Table 8.1. From each tree a branch was cut from the upper part of the crown. The last three annual shoots were determined by taking the following parameters into consideration: presence of leaves, presence of scars, branch architecture and bark texture. The use of these characteristics to determine the annual shoots in tropical tree branches was mentioned earlier by Tomlinson and Longman (1981) and Worbes (1995). From the lowest part of each shoot, microscopic cross sections were prepared. Sections were stained with safranin/astra blue and mounted on slides. Annual growth ring boundaries were determined, described and counted (see Fig. 8.1 as an example).

RESULTS AND DISCUSSION

In the branch of *Cedrela fissilis* studied (Fig. 8.2), it was possible to distinguish the annual shoots for the last four years. It was verified anatomically in *Cedrela fissilis* that each shoot, although showing more than one vegetative growth flush per year, had only one growth ring in the corresponding year. The tree ring was identified by its semi-ring porosity as well as by its flat-shaped fibres. The formation of a single growth ring per growing season was also observed by Gil and Garcia (1977) with *Vallea stipularis* and by Amobi (1974) with some tree species in Nigeria.

Cedrela fissilis presents only one growth ring per year, but it may have two growth flushes during the growing season. In contrast to observations by Venugopal and Krishnamurthy (1987), the species *Albizzia lebbeck*, *Dalbergia sissoo* and *Terminalia crenulata* present two defoliations per year which

Table 8.1. Boundaries of the growth rings from branches of Pantanal Mato-grossense native trees.

Families	Species	Boundaries
Anacardiaceae	*Astronium fraxinifolium* Schott	Flattened fibres
Bignoniaceae	*Jacaranda cuspidifolia* Mart.	Flattened fibres and marginal parenchyma
	Tabebuia aurea (Manso) B. et H.	Parenchyma band with marginal boundary and flattened fibres
	Tabebuia heptaphylla (Vell.) Tol.	Semi-porous concentration and parenchyma band
	Tabebuia impetiginosa (Mart.) Standl.	Marginal parenchyma and fibre band
	Tabebuia ochracea (Cham.) Stdl.	Parenchyma band with marginal boundary and fibres
	Tabebuia roseo-alba (Rid.) Sandw.	Parenchyma band with marginal boundary and fibres
Bombacaceae	*Pseudobombax marginatum* (St. Hil.) Rob.	Flattened fibres
Boraginaceae	*Cordia glabrata* (Mart.) A. DC.	Parenchyma band with marginal boundary and fibres
Burseraceae	*Protium heptaphyllum* (Aubl.) March.	Parenchyma band with marginal boundary and fibres
Combretaceae	*Terminalia argentea* Mart. et Zucc.	Marginal parenchyma and fibre band
Flacourtiaceae	*Casearia decandra* Jacq.	Semi-ring-porous concentration and flattened fibres
	Casearia sylvestris Sw. var. *lingua*	Flattened fibres
Lauraceae	*Ocotea suaveolens* Hassl.	Flattened fibres and fibre band
Leguminosae–Caesalpinioideae	*Hymenaea stigonocarpa* (Mart.) Hayne	Marginal parenchyma
	Pterogyne nitens Tul.	Semi-ring-porous concentration, marginal parenchyma and fibre band
	Sclerolobium aureum (Tul.) Bth.	Thick wall fibres and fibre band associated with parenchyma band
Leguminosae–Faboideae	*Andira cuyabensis* Bth.	Marginal parenchyma with semi-ring-porous tendency

Continued over

Table 8.1. *(Continued)*

Families	Species	Boundaries
Leguminosae–Faboideae	*Bowdichia virgilioides* H.B.K.	Marginal parenchyma with semi-ring-porous tendency
	Dipterix alata Vog.	Marginal parenchyma with semi-ring-porous concentration and fibre band
	Vatairea macrocarpa (Bth.) Ducke	Fibre band
Leguminosae–Mimosoideae	*Albizia niopioides* (Spruce) Burk.	Semi-ring-porous tendency and flattened fibres
	Anadenanthera colubrina (var. *cebil*) Bren	Small pore concentration and parenchyma band
	Inga urugüensis H. et A.	Flattened fibres
	Stryphnodendron obovatum Bth.	Flattened fibres
Melastomataceae	*Mouriri elliptica* Mart.	Pore concentration and parenchyma band with marginal boundary
Meliaceae	*Trichilia elegans* A. Juss. ssp. *elegans*	Marginal parenchyma and tendency to pore concentration
Rhamnaceae	*Rhamnidium elaeocarpum* Reiss.	Flattened fibres and marginal parenchyma
Rutaceae	*Fagara hassleriana* Chod.	Parenchyma band with marginal boundary, flattened fibre and pore concentrations in some rings
Sapindaceae	*Dilodendron bipinnatum* Radlk.	Flattened fibres
Sapotaceae	*Pouteria ramiflora* (Mart.) Radlk.	Fibre band
Sterculiaceae	*Sterculia apetala* (Jacq.) Karst.	Pore concentration, ray enlargement and flattened fibres
Tiliaceae	*Luehea paniculata* Mart.	Rays enlargement and flattened fibres
Verbenaceae	*Vitex cymosa* Bert.	Ring-porous concentration
Vochysiaceae	*Qualea grandiflora* Mart.	Parenchyma band and fibre band

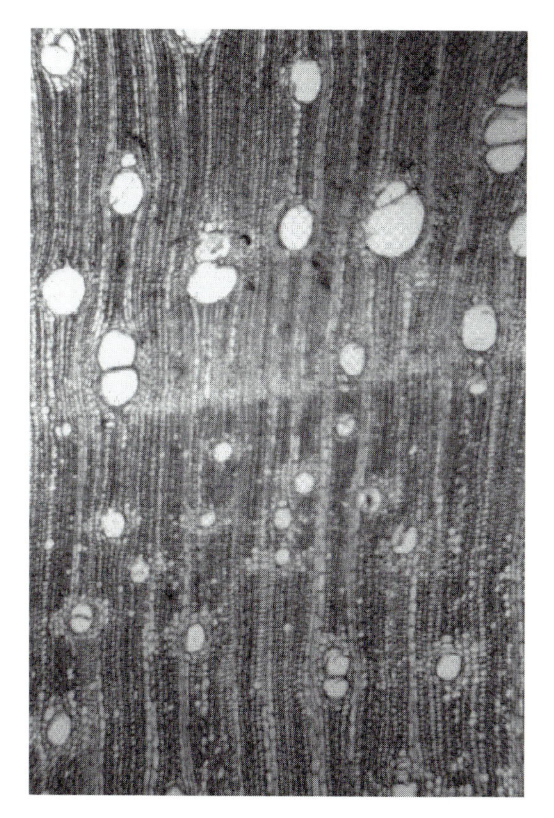

Fig. 8.1. Cross-section of a branch of *Cedrela* showing a growth ring boundary.

corresponded with the formation of two growth rings associated with new bud breaks and leaf formation. *Tectona grandis* presents only one defoliation per year which corresponds to a single period of xylem production.

In the branches of the investigated species from Pantanal Mato-grossense the shoot length growth corresponded with radial growth. Similar results were seen with *Cedrela fissilis*: the first shoot did not present a growth ring boundary, representing the growth of the year. In the following sections, there was an increment of a ring in each annual shoot (Fig. 8.2). The growth ring boundaries, delimited in most of the species by flat fibres with thick cell walls, as well as by parenchyma bands (Table 8.1), are not always distinct, probably due to the slow growth of the corresponding annual shoot.

CONCLUSIONS

It is possible to determine the annual growth ring formation by the annual growth modules of the branches. Several species from Pantanal have annual

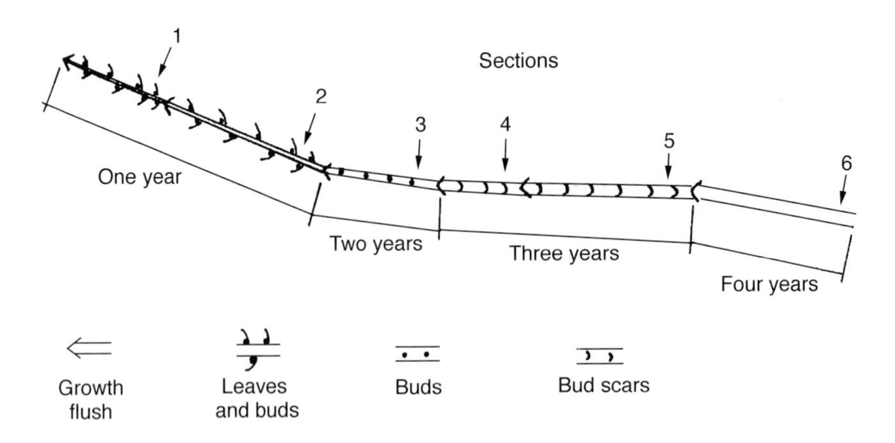

Fig. 8.2. Annual shoots of a *Cedrela fissilis* branch.

growth rings, and these may be used to estimate the growth dynamics in height and diameters of these species.

SUMMARY

Age and radial growth of trees, estimated by annual growth rings, are basic information for studies of forest dynamics. The aim of this work was to verify the correlation of growth rings with the annual periodical shoot growth. Branches were taken from the upper part of the crown. The trees were selected in regions with well-defined growth seasons in South Brazil and Pantanal Matogrossense characterized by one leaf-shedding period per year. The growth of at least the three most recent annual shoots were identified on the branches, by the presence of leaves, scars and the branch architecture. Anatomical analyses were done on cross sections and the growth ring boundaries were counted and described. In *Cedrela fissilis* and *Astronium fraxinifolium* Schott rings can be identified by the semi-ring porosity as well as the flat fibres. In *Tabebuia* spp. and *Cordia glabrata* (Mart.) A. DC., ring-porosity and parenchyma bands helped to identify growth rings. In *Pseudobombax marginatum* (St. Hil.) Rob. rings could be seen mainly through flat fibres. In these species the shoot segment from the current year does not present a ring boundary. In the following older segments, a new ring for each new segment was formed. The rhythmic growth due to the dry seasons in the Pantanal region resulted in annual growth ring formations. *Cedrela fissilis* presented only one new growth ring per year, even with two vegetative growth flushes.

REFERENCES

Amobi, C.C. (1973) Periodicity of wood formation in some trees of lowland rainforest in Nigeria. *Annals Botany n.s.* 37, 211–218.

Amobi, C.C. (1974) Periodicity of wood formation in twigs of some tropical trees in Nigeria. *Annals Botany n.s.* 38, 931–936.

Ash, J. (1983) Tree rings in tropical *Callitris macleayana* F. Muell. *Australian Journal of Botany* 31, 277–281.

Boninsegna, J.A., Villalba, R., Amarilla, L. and Ocampo, J. (1989) Studies on tree rings, growth rates and age–size relationships of tropical tree species in Misiones, Argentina. *IAWA Bulletin n.s.* 10, 161–169.

Cadavid Garcia, E.A. (1984) *O clima no Pantanal Mato-grossense*. Circular técnica 14, Corumbá, MS: EMBRAPA-UEPAE, Corumbá, MS, 39pp.

Détienne, P. (1989) Appearance and periodicity of growth rings in some tropical woods. *IAWA Bulletin n.s.* 10, 123–132.

Détienne, P. and Barbier, C. (1988) Rythmes de croissance de quelques essences de Guyane Française. *Bois et Forêts des Tropiques* 217, 63–76.

Detienne, P. and Mariaux, A. (1977) Nature et périodicité des cernes dans les bois rouges de méliacées africaines. *Bois et Forêts des Tropiques* 131, 52–61.

Gil, R.H. and Garcia, F.F. (1977) Ritmicidad en el crecimiento de la *Vallea stipularis* L. *Revista Florestal Venezolana* 27, 143–157.

Hallé, F. and Martin, R. (1968) Étude de la croissance rythmique chez l'hevea (*Hevea brasiliensis* Müll.-Arg. Euphorbiacées-Crotonoidées). *Adansonia, série 2*, 8(4), 475–503.

Maack, R. (1981) *Geografia Física do Estado do Paraná*. Livraria José Olympio, co-edição Secretaria da Cultura e do Esporte do Estado do Paraná, Rio de Janeiro, 450pp.

Mariaux, A. (1969) La périodicité des cernes dans le bois de limba. *Bois et Forêts des Tropiques* 128, 39–53.

Mariaux, A. (1970) La périodicité de formation des cernes dans le bois de l'okoumé. *Bois et Forêts des Tropiques* 131, 37–50.

Pumijumnong, N., Eckstein, D. and Sass, U. (1995) Tree-ring research on *Tectona grandis* in Northern Thailand. *IAWA Journal* 16, 385–392.

Tomlinson, P.B. and Longman, K. (1981) Growth phenology of tropical trees in relation to cambial activity. In: Bormann, F.H. and Berlyn, G. (eds) *Age and Growth Rate of Tropical Trees: New Directions for Research*. Yale University, School of Forestry and Environmental Studies, Bulletin No. 94, New Haven, pp. 7–19.

Venugopal, N. and Krishnamurthy, K.V. (1987) Seasonal production of secondary xylem in the twigs of certain tropical trees. *IAWA Bulletin n.s.* 8(1), 31–40.

Vetter, R.E. and Botosso, P.C. (1989) Remarks on age and growth rate determination of Amazonian trees. *IAWA Bulletin n.s.* 10(2), 133–145.

Villalba, R. and Boninsegna, J.A. (1989) Dendrochronological studies on *Prosopis flexuosa* DC. *IAWA Bulletin n.s.* 10, 155–160.

Worbes, M. (1989) Growth rings, increment and age of trees in inundation forests, savannas and a mountain forest in the neotropics. *IAWA Bulletin n.s.* 10(2), 109–122.

Worbes, M. (1995) How to measure growth dynamics in tropical trees. A review. *IAWA Journal* 16, 337–351.

9

Seasonal Variations of the Vascular Cambium of Teak (*Tectona grandis* L.) in Brazil

Mario Tomazello and Narciso da Silva Cardoso

INTRODUCTION

During the past decades, numerous investigations have been published on the structure of the vascular cambium, mainly on species of the temperate zones (Iqbal, 1990; Larson, 1994). However, studies on the effect of climate variations and phenological phases on the structure and activity of the cambium are scarce, especially concerning tropical species (Iqbal and Ghouse, 1987). The understanding of these influences on cambial activity is basic knowledge for any dendrochronological research, particularly the determination of age and growth increments of tree species. With this context in mind, it seems to be essential to identify potential tree species that could be used as a model for studying the complex interaction of physiological mechanisms controlling the cambial activity and the formation of the xylem. Teak (*Tectona grandis* L.), *Verbenaceae*, has the potential for such studies due to its phenological behaviour, wood anatomical structure and annual growth rings (Jacoby, 1989; Jacoby and D'Arrigo, 1990; Cardoso, 1991; Pumijumnong *et al.*, 1995).

MATERIAL AND METHODS

The present study was carried out with teak trees from a 31-year-old plantation located in Piracicaba, Brazil, originated from seeds which were collected from a population introduced in the vicinity of Araraquara, Brazil (22°42′30″ S, 47°38′00″ W). The coordinates of the site and plantation, soil and climatic conditions are presented by Mello (1963), Ranzani (1966) and Ometto (1989), respectively. Wood and bark samples for the analysis of cambial activity were

 Tree-Ring Analysis (eds R. Wimmer and R.E. Vetter)

collected in intervals of 2 weeks over a period of 16 months. The samples of 2 × 2 × 2 cm dimensions were taken from the trunks of two trees at 1.3 m height above ground. Samples were fixed in formaldehyde–acetic acid–ethanol (FAA), sectioned with a sliding microtome and the sections mounted on slides. The cell analyses were performed under a light microscope with an ocular micrometer.

RESULTS AND DISCUSSION

Cambial Zone Structure

The anatomical structure of the fusiform and ray initial cells in the cambial zone was analysed on transversal and tangential sections. The fusiform initials were longitudinally elongated, with pointed extremities and transversal septa on both ends. Mean values: length 0.31 mm, radial width 25 μm, lumen diameter 18 μm and wall thickness 6 μm. The ray initials were short, isodiametric, prismatic shape and diameter 13–30 μm (Figs 9.1 and 9.2). The xylem and phloem tissues, which resulted from periclinal divisions of the cambial cells, were also clearly visible. The structure of the cambial region of teak trees was similar to that described by Rao and Dave (1981) in India, although differences in cells dimensions exist, probably due to the different cambium age.

Seasonal Variation of the Cambial Activity

During the observation period clear seasonal differences were found in the cambial activity of teak trees characterized by the number of cell layers and the morphology as well as the thickness of the cell walls. At the beginning of the physiological activity of teak (late November–early December) the cambial zone was composed of 3–11 cell layers, reaching 12–20 layers at its highest activity (December–February). The cambial cells increased their tangential diameter with the cell walls thinned down. In the period of lower physiological activity (September–November) the number of cell layers declined again back to 3–11 and the cells showed a reduction in their tangential dimensions with a simultaneous expansion of the radial walls and a distinct wall thickening, mainly in the radial direction (Fig. 9.2, top). These seasonal variations of the cell layers and morphology were previously described for teak by Rao and Dave (1981) and for *Gmelia arborea* by Dave and Rao (1982), and for numerous hardwoods and softwoods by Aljaro *et al.* (1972) as well as by Ajmal and Iqbal (1987).

Following the division phase, similar effects were also observed during the cell differentiation phase in the xylem and the phloem. The latewood – characterized by thick-walled fibres, transversally flattened and small lumen, and vessels with smaller diameter – is formed in the period of lower cambial

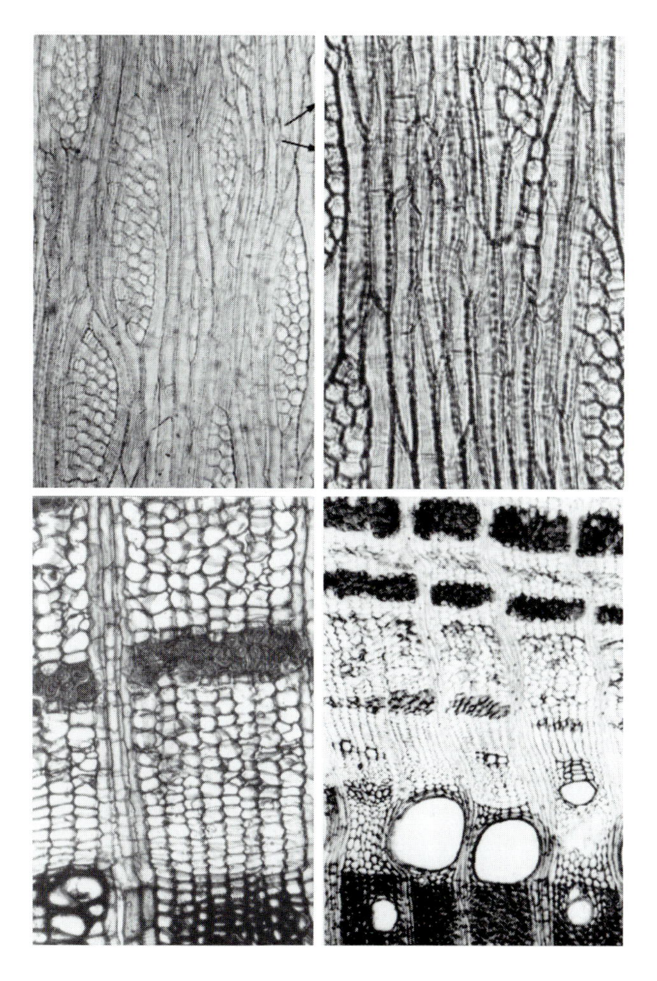

Fig. 9.1. Longitudinal and tangential sections through teak showing fusiform and radial initial cells of cambial region (top); cross-sections showing cells of cambial region, xylem and phloem (bottom).

activity, followed by a marginal parenchyma band which characterizes the next growth period (late November–early December). In this stage of intensive cell division, the differentiation and formation of the xylem cells took place. The radial parenchyma was dislocated laterally by the increasing diameter of the vessels. The largest vessels, grouped and close to the marginal parenchyma, formed the ring-porosity. The earlywood, formed in this period, presented thin-walled fibres, with larger diameter and large vessels (Fig. 9.2, bottom). The decline of cambial activity (late June–early August) corresponded, again, to the latewood formation. The cell division was more intensive at the centripetal of the cambium, producing more xylem than phloem, which increases the radius

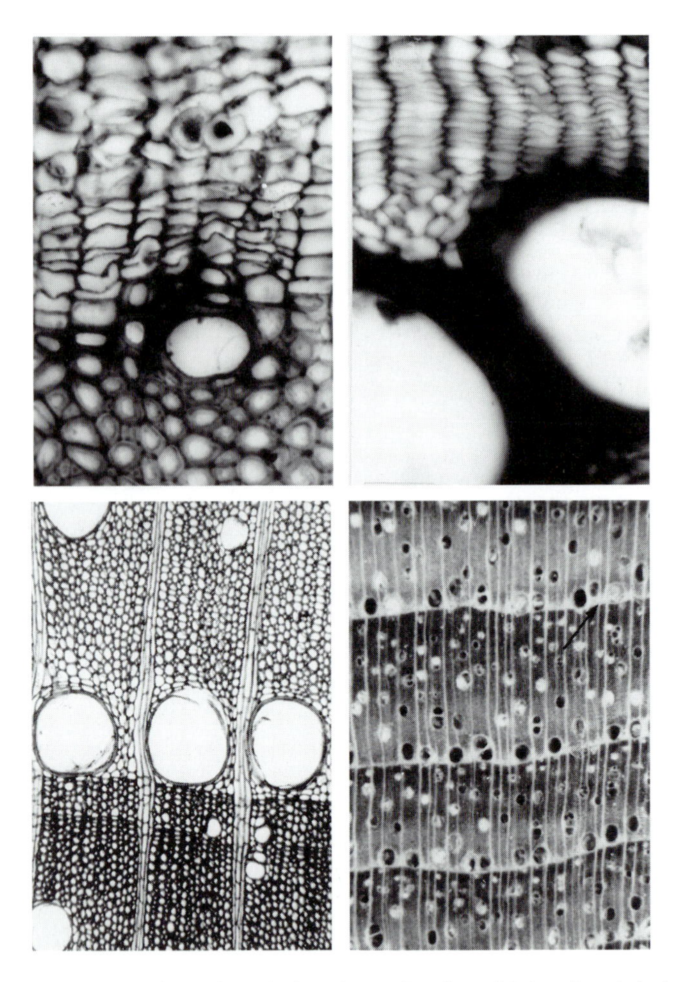

Fig. 9.2. Cross-sections through teak showing cells of cambial region, July (top, left) and December (top, right); xylem cells showing vessel elements, longitudinal and marginal parenchyma, fibres and early and latewood (bottom), indicating the formation of a growth ring. Ring-porous vessel arrangement (bottom).

of the stem. The seasonality of cambial activity and changes of cell dimensions are controlled by carbohydrates and hormones produced in axillary buds and leaves (Kozlowski, 1971).

Seasonal Variation of Cambial Activity in Relation to Phenology

Teak is a deciduous tree and showed distinct phenological phases in response to climatic variations reflecting the seasonality of cambial activity (Cardoso,

1991). The tree's growth is initiated with the bud development (November), which is followed by expansion of the leaf lamina and renewal of the shoot (December). During this time the cambium is stimulated by hormones produced in axillary buds and transported through the phloem. When the trees present the completely formed shoot, the trees start to flower as well as fruit (starts December–January, most of February). During this stage the cambium shows higher division rates, mainly forming the earlywood. When the trees enter the final phase of flowering, developing fruits and maturation (March–April), the cambium is still active, although its division rate is reduced.

The trees continue with coloration of the leaves, and the fruits became mature, dry and start to drop off the tree (May–July). During this phase the cambial activity declines. At the final phase of the vegetation period trees exhibit yellowish and senescent leaves. These start to be shed until the tree is completely defoliated (July–September), with the cambium arriving at its minimum activity, near latency. At the beginning of the next growing season the teak trees again show bud development and the cambium is activated (December).

The relationships between the phenological stages and the cambium activity level (Table 9.1) give an explanation for the reactivation of the cambial cell tissues at the trunk base. This activation takes place within 30 days after bud break. The maximum cambial activity occurs when the leaves are fully developed and high quantities of carbohydrates are available. In some species, such as *Streblus asper*, cambial activity starts 60 days after the beginning of the bud break (Ajmal and Iqbal, 1987), and in *Gmelina arborea* and *Proustia cuneifolia* (Aljaro *et al.*, 1972; Dave and Rao, 1982) activation was initiated after the first leave flush. The behaviour of the teak cambium was similar to that observed by Rao and Dave (1981) in India.

Seasonality of the Cambial Activity in Relation to Climatic Factors

The teak cambial activity began when temperature, rainfall and photoperiod started to increase (December), reaching higher cell division rates at the maximum levels of climatic factors (February). With gradual reduction of climatic parameters, the cambium decreased its activity (May–July) and entered the phase of minimum activity when growth factors were limited (August) (Table 9.1). The effect of environmental factors on cambial seasonality was already observed earlier by, for example, Waisel and Fahn (1965), who showed that temperature and photoperiod are important factors controlling cambial activation and early and latewood formation. The soil water deficit is also considered as a limiting factor, for example in *Proustia cuneifolia* (Aljaro *et al.*, 1972) and in teak trees (Rao and Dave, 1981).

Table 9.1. Relationship between phenology, cambial activity and xylem tissue in teak trees.

Month	Phenology	Cambial activity	Tissue growth
January	Leaves, young; flowering	Active (6–15 cells)	Marginal parenchyma, porous ring, formed
February	Leaves, developed; flowers and fruits	Maximum (12–20 cells)	Earlywood
March	Leaves, developed; flowering, finishing; fruits, mature	Declining (4–12 cells)	Earlywood
April	Leaves, flowers developing	Declining (4–11 cells)	Earlywood
May	Leaves, yellowing; fruits, mature	Declining (4–11 cells)	Early–latewood
June	Leaves, yellowing; fruits, dry/shedding	Declining (4–12 cells)	Vessel, small diameter fibre, tangential wall flattened
July	Leaves, yellowing; fruits, dry/shedding	Declining (4–8 cells)	Vessel, small diameter fibre, tangential wall flattened
August	Leaves, shedding; fruits, dry/shedding	Declining or dormant (4–9 cells)	Vessel, small diameter fibre, tangential wall flattened
September	Leaves, senescent and shedding	Dormant (4–10 cells)	Latewood, final phase fibre, tangential wall flattened
October	Leaves, senescent and shedding; fruits, dry/shedding	Dormant (3–6 cells)	Latewood
November	Bud, development	Dormant (4–11 cells)	Latewood
December	Leaves expanding, renewing shoot	Beginning (5–13 cells)	Earlywood marginal parenchyma and ring-porosity beginning

CONCLUSIONS

The findings of this investigation allow the following conclusions: (i) the cambium of teak trees, constituted by two cell types, present periods of minimum and maximum activities related to phenological stages and climatic variations; (ii) the cambial activity variation is expressed by alterations in number and morphology of cells; (iii) the variations of anatomical structure of the xylem, characterized by earlywood and latewood, are due to the seasonality of cambial activity; (iv) the marginal parenchyma and ring-porosity enable age dating and annual increment determination of teak trees; and (v) this species can be used in Brazil, as in the area of natural occurrence (D'Arrigo, 1997; Pumijumnong, 1997), as a model for dendrochronological studies and also as a valuable tool for silvicultural purposes.

SUMMARY

In order to study the cambial activity of teak trees in relation to seasonality, trees were selected from a 31-year-old plantation located in Piracicaba-SP, Brazil. Samples including bark, cambium and sapwood were collected during 16 months, fixed in FAA and sectioned in a sledge microtome. The slides showed the structure of fusiform and ray initials of the cambium and the variations between higher and lower cambial activity. During the cambial dormancy period (September–early November) the fusiform initial cells presented thicker radial than tangential walls. At the beginning of cambial activity (end of November–December) the fusiform cells showed differences in comparison to the dormancy period, reaching the highest activity (January–February) with significant alterations of cell wall and cell number of the cambial zone. A clear relationship was observed between cambial activity, growth ring characteristics and phenological behaviour of teak trees with local climatic conditions.

ACKNOWLEDGEMENT

The authors express their sincere thanks to Dr Martin Worbes for the revision of the English text.

REFERENCES

Ajmal, S. and Iqbal, M. (1987) Annual rhythm of cambial activity in *Strebus asper*. *IAWA Bulletin n.s.* 8, 275–283.

Aljaro, M., Avila, G., Hoffman, A. and Kummerow, J. (1972) The annual rhythm of cambial activity in two species of Chilean motorral. *American Journal of Botany* 59, 879–885.

Cardoso, N.S. (1991) Caracterização da estrutura anatômica da madeira, fenologia e relações com a atividade cambial de árvores de teca (*Tectona grandis*). MSc thesis, University of São Paulo, Piracicaba, Brazil.

D'Arrigo, R. (1997) Progress in dendroclimatic investigations in Indonesia. *Proceedings of IUFRO Conference Forest Products for Sustainable Forestry*. Washington State University, Pullman, p. 230.

Dave, Y.S. and Rao, K.S. (1982) Seasonal activity of the vascular cambium in *Gmelina arborea. IAWA Bulletin n.s.* 3, 59–65.

Iqbal, M. (ed.) (1990) *The Vascular Cambium*. Research Studies Press, Taunton, Somerset, 345pp.

Iqbal, M. and Ghouse, A.K.A. (1987) Anatomy of the vascular cambium of *Acacia nilotica* var. *telia* in relation to age and season. *Botanical Journal Linnean Society* 9, 385–397.

Jacoby, G.C. (1989) Overview of tree-ring analysis in tropical regions. In: Baas, P. and Vetter, R.E. (eds) *Growth Rings in Tropical Woods. IAWA Bulletin n.s.* 10, 99–108.

Jacoby, G.C. and D'Arrigo, R.D. (1990) Teak (*Tectona grandis*), a tropical species of large-scale dendroclimatic potential. *Dendrochronologia* 8, 83–98.

Kozlowski, T.T. (1971) Cambial growth. In: Kozlowski, T.T. (ed.) *Growth and Development of Trees*. Academic Press, New York, pp. 1–62.

Larson, P.R. (1994) *The Vascular Cambium – Development and Structure*. Springer-Verlag, Berlin, 725pp.

Mello, H.A. (1963) Alguns aspectos da introdução da teca no Brasil. *Anuário Brasileiro de Economia Florestal* 15, 113–119.

Ometto, J.C. (1989) *Registros e estimativas de parâmetros metereológicos da região de Piracicaba-SP*. FEALQ, Piracicaba-SP, 76pp.

Pumijumnong, N. (1997) Dendrochronology with teak (*Tectona grandis*) in Thailand. *Proceedings of IUFRO Conference Forest Products for Sustainable Forestry*. Washington State University, Pullman, p. 245.

Pumijumnong, N., Eckstein, D. and Sass, U. (1995) Tree-ring research on *Tectona grandis* in northern Thailand. *IAWA Journal* 16, 385–392.

Ranzani, G. (1966) *Carta de solos do município de Piracicaba*. Centro de Estudos de Solos, Universidade de São Paulo, Piracicaba-SP, 85pp.

Rao, K.S. and Dave, Y.S. (1981) Seasonal variation in the cambial anatomy of *Tectona grandis*, Verbenaceae. *Nordic Journal of Botany* 1, 535–542.

Waisel, Y. and Fahn, A. (1965) The effects of environment on wood formation and cambial activity in *Robinia pseudoacacia*. *New Phytologist* 64, 436–442.

10

Climate–Growth Relationships of Teak (*Tectona grandis* L.) from Northern Thailand

Nathsuda Pumijumnong

INTRODUCTION

The increasing demand for palaeoclimatic information has stimulated dendrochronologists to extend their study areas from the southern and northern temperate zone towards the equator (Baas and Vetter, 1989). The main problem is that the equatorial zone has only a few suitable tree species that form distinct tree rings. However, regions such as in northern Thailand display remarkable dry and wet seasons, influenced by the Asian monsoon. Consequently, tree species growing in that region form distinguishable growth rings during wet and dry periods which might be a source for palaeoclimatic information.

The Asian monsoon climate operates supraregionally, and long-term climatic data at high precision are important for the assessment of this climatic system, especially in relation to the various dynamic patterns of the Asian monsoon. Basic knowledge about the relationships between the climate and the growth of trees is crucial for the prediction of future growth responses as a result of increasing temperature or effects that have been discussed as part of the 'global change' issue.

Dendrochronological investigations have been conducted in South East Asia by authors such as Buckley *et al.* (1995) and D'Arrigo *et al.* (1997) who studied the individuals of the species *Pinaceae* and *Podocarpaceae* in Thailand. Murphy and Whetton (1989), Jacoby and D'Arrigo (1990) and Palmer and Murphy (1993) investigated teak (*Tectona grandis* L.) trees from Java (Indonesia). Bhattacharyya *et al.* (1992) and Wood (1996) studied teak trees in India and found that previous October rainfall correlated with tree-ring width.

(eds R. Wimmer and R.E. Vetter)

Twenty-six teak chronologies in northern Thailand were compiled by Pumijumnong (1995) and grouped into three regional chronologies (west, east, and south) according to their correlation with seasonal rainfall as well as topographic aspects. The western regional chronology is a compilation of 13 sites. The eastern regional chronologies include seven chronologies, one from the Phrae Province and one from the Lampang Province. Finally, the southern regional chronology is based on four site chronologies from the Tak, Maehongson and Lamphun Provinces.

The results of these studies demonstrate that teak tree-ring chronologies in northern Thailand are strongly positively correlated with the rainfall in April, May and June of the current year as well as June and July of the preceding year. This means that radial growth of teak is mainly driven by precipitation that occurs at the end of the dry season and beginning of the rainy season. In a recent study, Pumijumnong (1997) showed that the earlywood vessels of teak have completed their formation between mid-April and the end of May, when the cambium is most active. The growing period continues for 5 more months and shuts down at the end of October.

MATERIAL AND METHODS

In South East Asia, teak is the best known wood species, a widely used and highly valuable timber resource known for centuries. The tree's native distribution is restricted to South East Asia, and in Thailand, generally, in its natural habitat, teak is found all over the north and extends to the northeast. The primary teak areas are confined to the hilly or mountainous regions and the altitudinal range is between 100 and 900 m a.s.l. Teak prefers an annual rainfall between 1000 and 2000 mm, and the temperature may range between 18 and 43°C. Teak grows best on well-drained soils in deciduous forests that occupy alluvial flats and moist slopes along river streams (Banijbhatana, 1957).

This study presents data from a site in the Phrae Province (Fig. 10.1) and compares them with the previous compiled chronologies. The Phrae Province is located in northern Thailand and is influenced by the monsoon with expressed seasonality (Figs 10.2 and 10.3). The study site is at 300–400 m a.s.l. and is close to hilltribe villages. The slope angles of the site are between 10 and 15%. The area is mostly dominated by naturally grown teak, since the climatic conditions and the topography are well suited for teak growth. The wood quality of teak from this area is considered to be very high.

Other species typically associated with teak in the upper canopy are *Xylia kerrii*, *Lagerstroemia calyculata*, *Afzelia xylocarpa* and *Pterocarpus macrocarpus*. Common species found in the lower layer are *Gmelia arborea* and *Vitex penduncularis*. Some bamboo species are also very common in the undergrowth of these forests.

Climate data were made available from a station in Muang District, Phrae Province, about 60 km south of the sample site. Data cover the period of

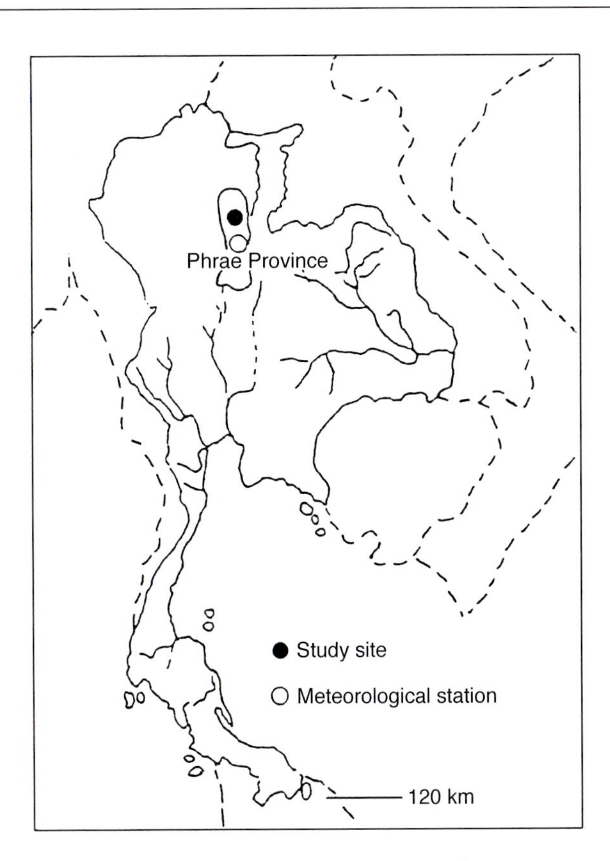

Fig. 10.1. The location of the new sample site (Phrae 5), 60 km north of the meteorological station located at the town of Phrae.

1911–1996 for rainfall and 1951–1996 for temperature. The sample site (Phrae 5) was visited in May 1997 where ten teak trees showing good cylindrical and straight boles with no visible damage were selected. Two cores per tree were taken using a regular increment corer. Each core was dried and glued in a grooved wooden stick. The surfaces of the samples were cut with a sharp knife to make the anatomical structure visible under the stereo-microscope (Fig. 10.4). Tree-ring widths were measured to the nearest 0.001 mm under a binocular microscope attached to a linear measuring stage connected to a computer. The raw data series were plotted and the plots inspected on the light table to identify false and missing rings. The accuracy of crossdating was subsequently checked with the COFECHA program (Holmes, 1983, 1994). Each individual series was detrended, firstly by fitting a negative exponential or alternatively a straight line and, secondly, by fitting a cubic spline (50% response period = 66 year) to the data. Autoregressive modelling was applied to the detrended series using ARSTAN (Cook, 1985) and the series were

Fig. 10.2. Photographs illustrating the dramatic difference between wet and dry seasons.

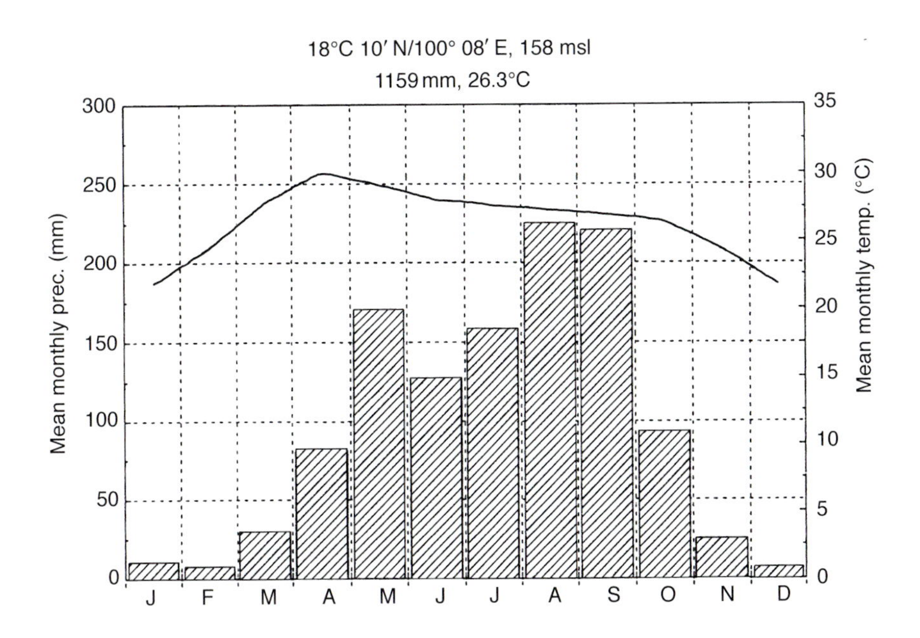

Fig. 10.3. Monthly rainfall in mm (bars) of the Phrae Province averaged over 86 years (1911–1996), and monthly temperature (°C) averaged over 46 years (1951–1996).

averaged using the robust mean to obtain residual chronologies with white noise.

RESULTS AND DISCUSSION

After successful crossdating of the ten tree-ring width series the fairly high mean sensitivity of 0.47 was calculated, revealing the high-frequency signal in these data (Fig. 10.5). Autocorrelation was used as a measure of persistence – inverse to the sensitivity – and it was moderately low at 0.50. Using detrended data, the correlation between the trees was 0.32 and the signal-to-noise ratio was 0.21. The variation held in common among the series may also be assessed by the percentage of variation explained through the first principal component (first eigenvector) of the correlation matrix of the tree-ring series over the common period (Fig. 10.8). The variance explained by the first eigenvector is 42%. Table 10.1 lists statistical indicators of Phrae 5 in comparison with previously studied sites in the same province (Pumijumnong *et al.*, 1995a,b).

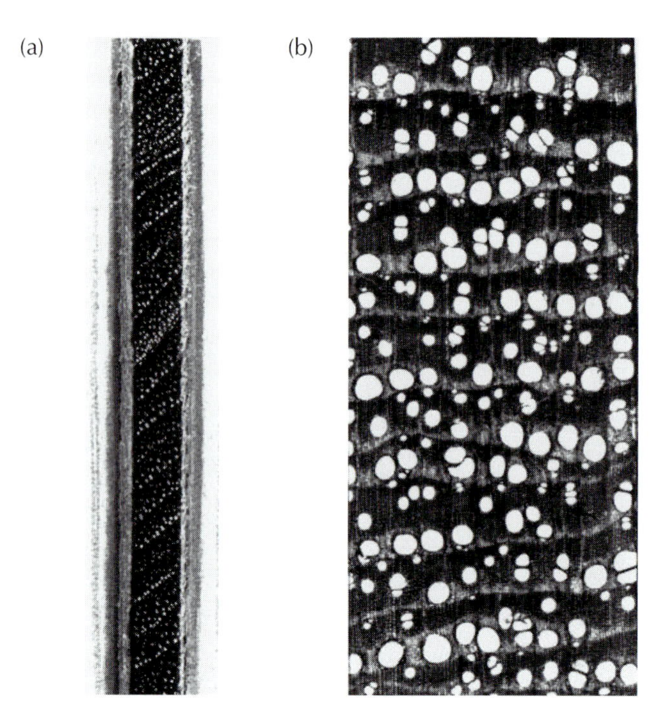

Fig. 10.4. (a) Surface of the core sample, (b) microscopic photography of teak.

Correlation Analysis

To investigate the growth–climate relationships, the meteorological data (1911–1996 for rainfall, 1951–1996 for temperature) were correlated with the standardized tree-ring chronology (Fig. 10.6). Seventeen months of rainfall data, from current November to previous July were correlated with the teak growth over the period of 1911–1996. To address possible effect between cambium age and relationships with climate, a juvenile part (1911–1950) as well as a more mature part (1951–1996) was correlated with monthly precipitation sums. Figure 10.7(a) shows that September precipitation is positively correlated with teak growth. Significant correlations were also found with previous August (positive) as well as January (negative). The shortened period 1911–1950 (Fig. 10.7b) shows a significant positive effect of current as well as preceding September on radial growth of teak. The second period 1951–1996 offers significant correlations between growth and precipitation of current year October as well as January (Fig. 10.7c). The correlation analysis

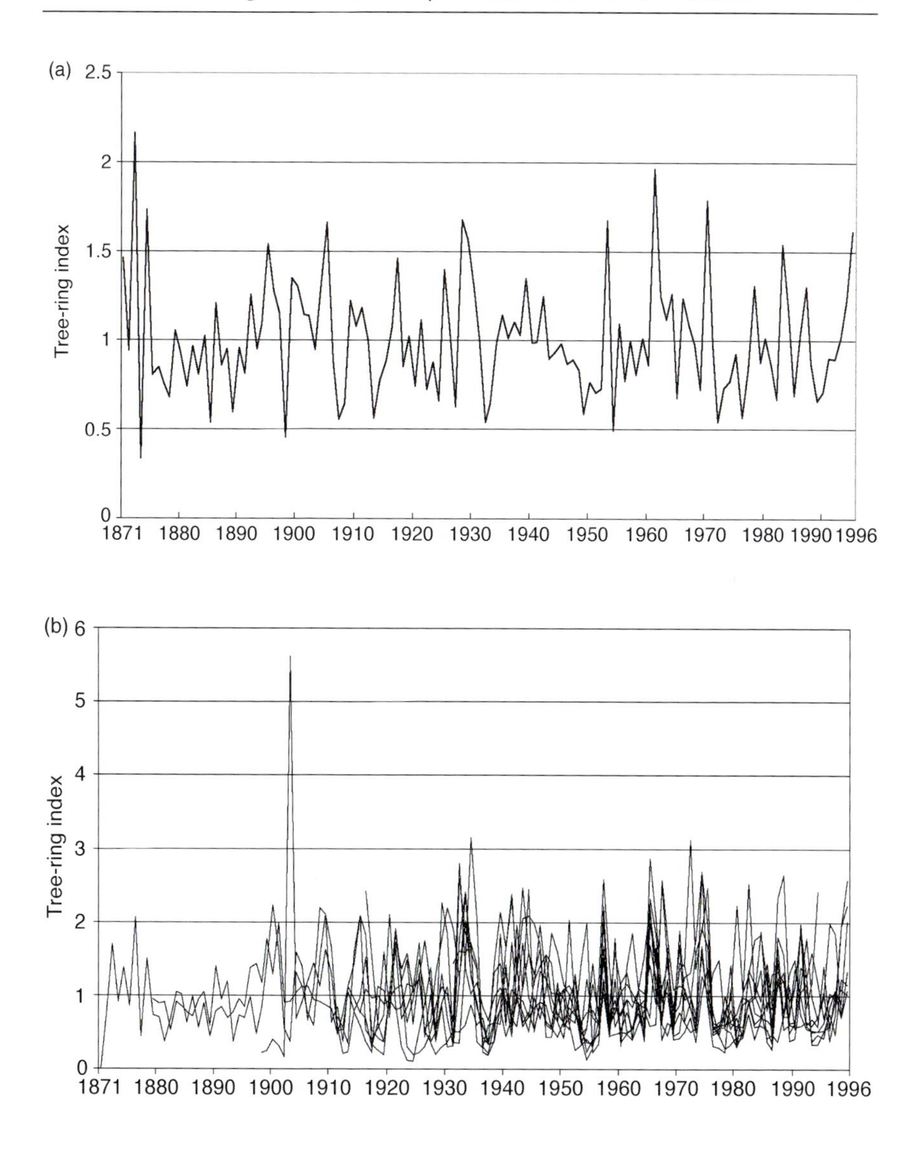

Fig. 10.5. Tree-ring indices of site Phrae 5 in Phrae Province. (a) Master chronology and (b) the ten individual series (one core from each of ten trees).

between growth and monthly mean temperature does not give significant coefficients (Fig. 10.7d). As a conclusion, tree growth of teak on the newly investigated site (Phrae 5) is mainly driven by rainfall during the late rain season. The previously investigated sites of the Phrae Province respond more to rain that falls earlier, during April through June (Pumijumnong *et al.*, 1995a). These differences are possibly related to the fact that human disturbances are

Table 10.1. Statistics of the raw data and detrended tree-ring series.

	Raw data									Detrended data	
Study area	Tree (n)	Radii (n)	Age (years)	Range in ages (years)	Time span	Ring width (mm)	Standard deviation	Mean sensitivity	Auto correlation	Correlation between trees	Variance 1st eigenvector
Phrae 1	12	22	102	54–142	1850–1991	1.8	1.37	0.48	0.51	0.3	36
Phrae 2	9	17	89	71–121	1871–1991	2.47	1.73	0.5	0.48	0.38	45
Phrae 3	13	18	96	58–133	1859–1991	1.84	1.41	0.48	0.56	0.28	35
Phrae 4	12	19	95	75–134	1858–1991	1.78	1.43	0.47	0.6	0.37	42
Phaing 1	14	16	89	49–123	1869–1991	2.29	1.67	0.44	0.61	0.45	50
Phaing 2	15	16	116	74–137	1855–1991	2.15	1.57	0.46	0.6	0.51	55
Phrae 5	10	10	93	60–129	1937–1996	3.63	3.03	0.47	0.5	0.32	42

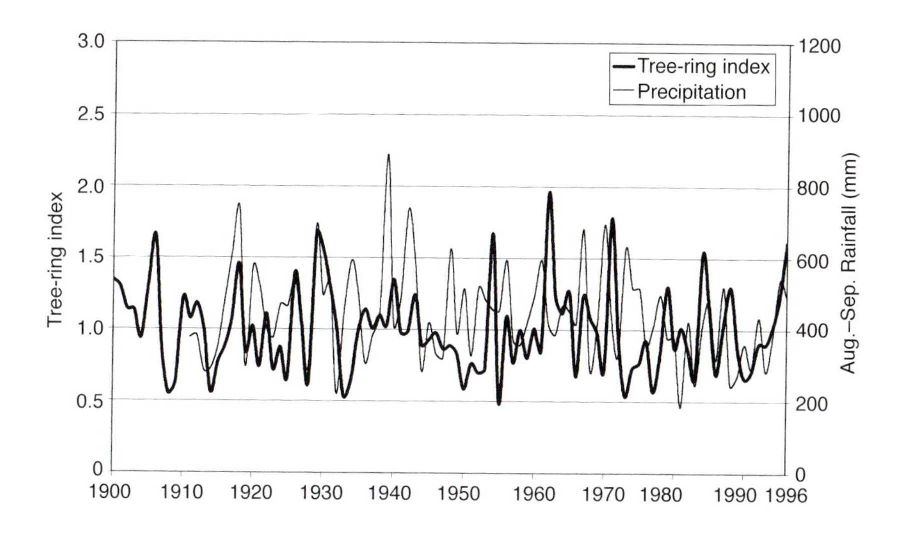

Fig. 10.6. A comparison of tree-ring chronology and rainfall data (August–September).

more relevant on the site Phrae 5 than on the other – earlier investigated – sites. As mentioned, the Phrae 5 site is located close to a village.

Eigenvector Scattergram

Together with the previously sampled chronologies, 27 series stood available for a correlation matrix over the common period of 1937–1991. The extracted first and second eigenvectors represent the biggest portions of the total variance held in common among the series and the eigenvector values of all sites are plotted in Fig. 10.8. Sites are enclosed in dashed lines which refer to provinces. The solid lines enclose the two major regional groups. The newly measured site Phrae 5 does not show similarity with the other Phrae sites, particularly along the first eigenvector.

So far, teak has been studied only in Indonesia (Java), India and Thailand. Vessel arrangement of teak can be classified as ring-porous and semi ring-porous (Fig. 10.4) according to Geiger (1915) and Coster (1927, 1928). The first teak chronology was produced by Berlage (1931). This author compared growth rings of a 400-year-old teak with climatic data and found a strong correlation between teak tree-ring width and November–May rainfall (rainy season in Java) but no correlation with June–October rainfall (dry season). These correlations are similar to those found for teak in Thailand by Pumijumnong *et al.* (1995a,b). DeBoer (1951), Jacoby and D'Arrigo (1990) and D'Arrigo *et al.* (1994) re-examined Berlage's teak chronology.

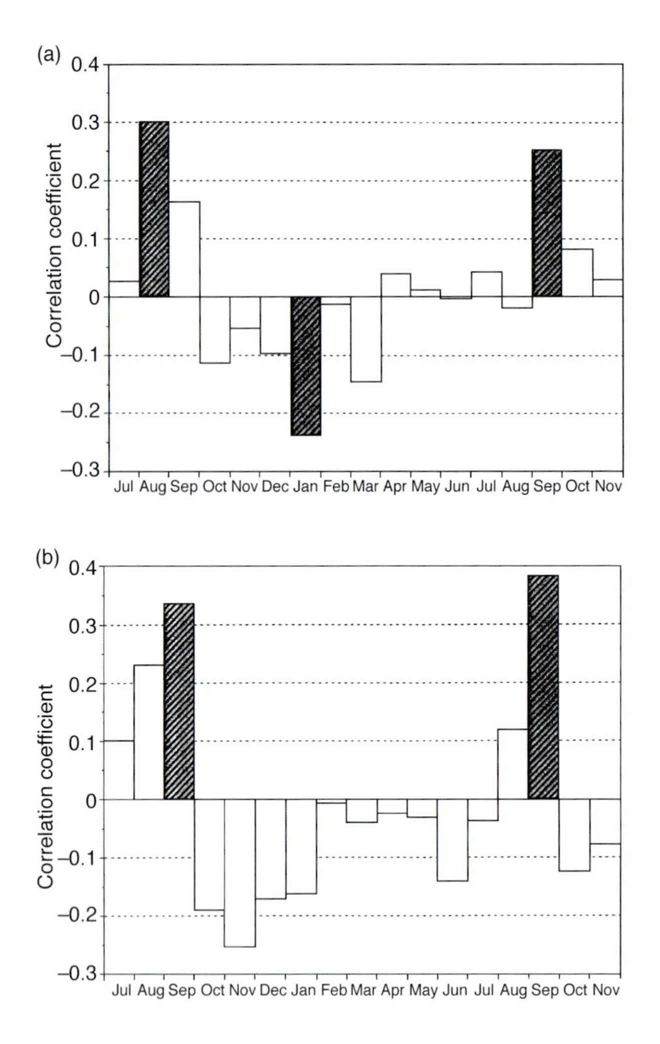

Fig. 10.7. (*and opposite*) Meteorological data from Muang District, Phrae Province (shaded bars, $P \leq 0.05$). (a) Precipitation 1911–1996; (b) precipitation 1911–1950; (c) precipitation 1951–1996; (d) temperature.

DeBoer (1951) found that tree-ring widths are positively correlated with the number of rain days. Correlations are also found with sunspots and the El Niño phenomenon. Murphy and Whetton (1989) studied Java teak by using Berlage's first teak chronology. They found a significant correlation between short-term growth fluctuations and the El Niño–Southern Oscillation phenomenon. Jacoby and D'Arrigo (1990) showed a strong correlation between tree-ring width and rainfall at the beginning of the rain season. In addition,

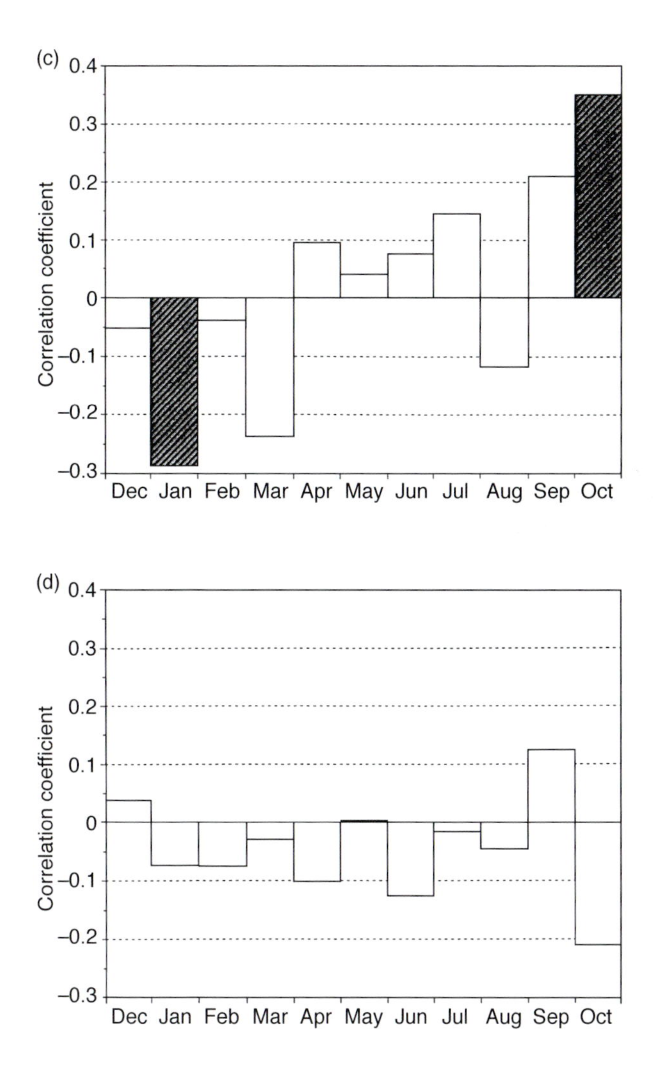

Fig. 10.7. *Continued.*

Jacoby (1989) demonstrated that amount of rainfall during the transition period from the dry season to the wet season critically influences growth teak in Java. Pant and Borgaonkar (1983) and Bhattacharyya *et al.* (1992) examined Indian teak, and they concluded that radial growth has a positive correlation with previous October rainfall. Wood (1996) studied synoptic dendroclimatology in the upper Narmada River Basin in India using stumps teak, and found that teak growth at the MNRF site is particularly well related with previous June–September–October rainfall. The climate–growth relationships

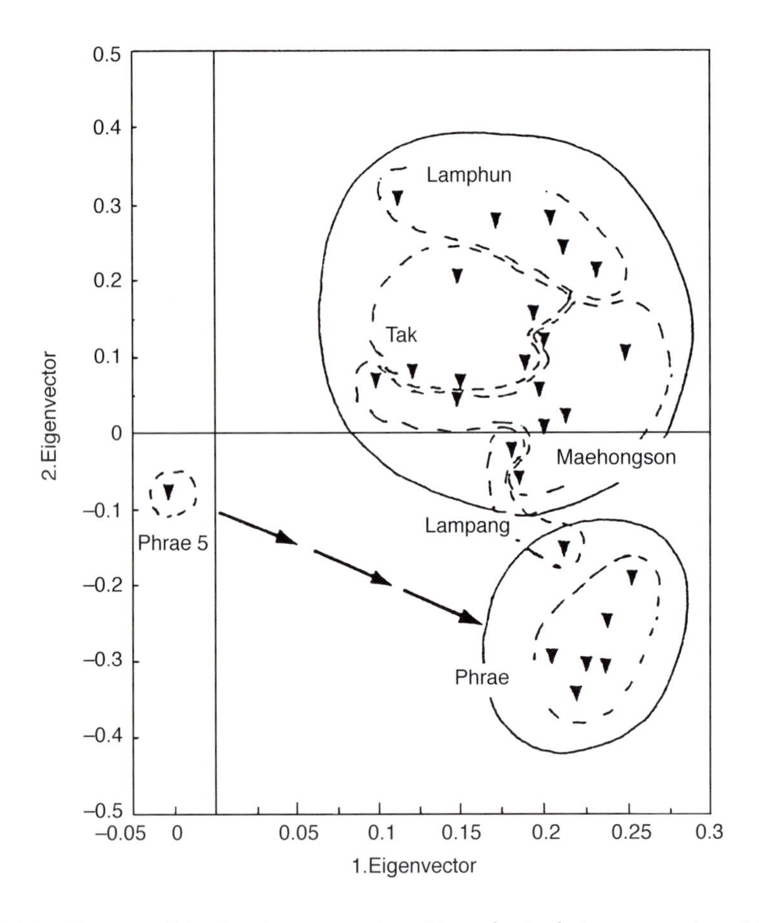

Fig. 10.8. Diagram of the 1st eigenvector (x-axis) vs. the 2nd eigenvector (y-axis) of 27 site chronologies; dotted lines encircle the sites by province, solid lines enclose the two clusters of sites on the basis of similarity of the tree-ring patterns.

from the teak trees growing in Phrae 5 show similarity with climate response of Indian teak.

CONCLUSIONS

A new teak tree-ring chronology from Phrae Province in northern Thailand correlates with September–October rainfall of the current year. This result differs from previous ones obtained in Thailand, even from the same province. However, to some extent a similarity was found with results from India. Rainfall is required for the tree growth throughout the entire growing season. No effects were found between monthly temperature and teak growth. Future work needs to concentrate on the wood anatomy of teak using image analysis. Certain anatomical variables might help to explain better the complex relationships between Asian monsoon climate and tree growth.

SUMMARY

Teak (*Tectona grandis* L.) is one of the few subtropical tree species which form clear annual growth rings. First teak tree-ring studies were documented about 150 years ago. These studies show that teak may reach an age up to 400 years. The wood is durable with excellent mechanical properties and has long been used as building material. A new teak chronology was built by sampling ten teak trees from the Phrae Province in northern Thailand. A correlation analysis revealed that the tree growth is mainly driven by September–October precipitation of the current year.

ACKNOWLEDGEMENTS

The author would like to thank Dieter Eckstein from Hamburg University for supervising his earlier work. Equipment was provided by the NWG Macintosh Centre, University of Sydney, and established at Mahidol University as part of a collaborative research project that included also the Lamont-Doherty Earth Observatory, USA. Thanks are due to Mike Barbetti, Manas Watanasak, Brendan Buckley, Rosanne D'Arrigo and Paul Krusic for their cooperation. Mike Barbetti, NWG Macintosh Centre for Quaternary Dating, University of Sydney, Australia, is acknowledged for comments on the manuscript. Special thanks go to the head of the Mae Yom National Park and his staff for their kind support during fieldwork. Funding for this project was provided by the Mahidol University, Salaya, Phutthamonthon, Nakhonpathom 73170, Thailand.

REFERENCES

Baas, P. and Vetter, R.E. (eds) (1989) Growth rings in tropical woods. *IAWA Bulletin n.s.* 10, 95–174.

Banijbhatana, D. (1957) Teak forests in Thailand. In: *Tropical Silviculture*. Vol. II, FAO, Rome, pp. 193–205.

Berlage, H.P. (1931) Over het verband tusschen de dikte der jaarringen van Djatiboomen (*Tectona grandis* L.f.) en den regenval op Java. (About the relationship between annual ring width of Djati trees (*Tectona grandis* L.f.) and rainfall at Java.) *Tectona* 24, 939–953.

Bhattacharyya, A., Yadav, R.R., Borgaonkar, H.P. and Pant, G.B. (1992) Growth-ring analysis of Indian tropical trees: dendroclimatic potential. *Current Science* 62, 736–741.

Buckley, B.M., Barbetti, M., Watanasak, M., D'Arrigo, R., Boonchirdchoo, S. and Sarutanon, S. (1995) Dendrochronological investigations in Thailand. *IAWA Journal* 16, 393–409.

Cook, E.R. (1985) A time-series analysis approach to tree-ring standardization. Thesis, Department of Geosciences, University of Arizona, Tucson.

Coster, C. (1927) Zur Anatomie und Physiologie der Zuwachszonen- und Jahrringbildung in den Tropen. *Annales du Jardin Botanique de Buitenzorg* 37, 49–161.

Coster, C. (1928) Einiges über das Dickenwachstum und die Inhaltsstoffe des Djatistammes, *Tectona grandis* L.f. *Tectona* 17, 1056–1057.

D'Arrigo, R., Jacoby, G.C. and Krusic, P.J. (1994) Progress in dendroclimatic studies in Indonesia. *Terrestrial, Atmospheric and Oceanic Sciences* 5, 349–363.

D'Arrigo, R., Barbetti, M., Watanasak, M., Buckley, B., Krusic, P., Boonchirdchoo, S. and Suratanon, S. (1997) Progress in dendroclimatic studies of mountain pine in northern Thailand. *IAWA Journal* 18, 433–444.

DeBoer, H.J. (1951) Tree-ring measurement and weather fluctuations in Java from A.D. 1514. *Koninklijke Neterlandse Akademie von Wetenschappen* 54, 194–209.

Geiger, F. (1915) Anatomische Untersuchungen über die Jahrringbildung von *Tectona grandis*. *Jahrbuch für Wissenschaftliche Botanik* 55, 522–607.

Holmes, R.L. (1983) Computer-assisted quality control in tree-ring dating and measurement. *Tree-Ring Bulletin* 43, 69–78.

Holmes, R.L. (1994) *Dendrochronology Program Library User Manual*. Laboratory of Tree-Ring Research, University of Arizona, Tucson.

Jacoby, G.C. (1989) Overview of tree-ring analysis in tropical regions. *IAWA Bulletin n.s.* 10, 99–108.

Jacoby, G.C.and D'Arrigo, R.D. (1990) Teak (*Tectona grandis* L.), a tropical species of large-scale dendroclimatic potential. *Dendrochronologia* 8, 83–98.

Murphy, J.O. and Whetton, P.H. (1989) A re-analysis of tree-ring chronology (teak) from Java. *Koninklijke Neterlandse Akademie von Wetenschappen* B92, 241–257.

Palmer, J.G. and Murphy, J.O. (1993) An extended tree-ring chronology (teak) from Java. *Koninklijke Neterlandse Akademie von Wetenschappen* 96, 27–41.

Pant, G.B. and Borgaonkar, H.P. (1983) Growth ring of teak trees and regional climatology. In: Singh, L.R., Singh, S., Tiwari, R.C. and Srivastava, R.P. (eds) *Environmental Management*. The Allahabad Geographical Society, Department of Geography, University of Allahabad, India, pp. 154–158.

Pumijumnong, N. (1995) Dendrochronology with teak (*Tectona grandis* L.) in north Thailand. Thesis, Hamburg University.

Pumijumnong, N. (1997) Cambium development of teak (*Tectona grandis* L.) in Thailand and its relationship to climate. In: *International Symposium on Wood Science and Technology*, Wood–Human–Environment, 23–24 October 1997. Seoul, Korea 61–72.

Pumijumnong, N., Eckstein, D. and Sass, U. (1995a) Tree-ring research on *Tectona grandis* in northern Thailand. *IAWA Journal* 16(4), 385–392.

Pumijumnong, N., Eckstein, D. and Sass, U. (1995b) A network of *Tectona grandis* chronologies in northern Thailand. *Proceedings International Workshop on Asia and Pacific Dendrochronology*, 4–9 March 1995, Tsukuba, Japan, pp. 35–41.

Wood, M.L. (1996) Synoptic dendroclimatology in the upper Narmada River Basin: An exploratory study in Central India. Master of Science Thesis, University of Arizona.

11

Growth Periodicity in Relation to the Xylem Development in Three *Shorea* spp. (*Dipterocarpaceae*) Growing in Sarawak

Tomoyuki Fujii, Andrew Tukau Salang and Takeshi Fujiwara

INTRODUCTION

Various methods to measure the growth dynamics of tropical trees were reviewed by Worbes (1995) in relation to the climatic conditions which affect tree-ring formation, particularly the formation of annual rings. He concluded that the exact proof of the nature of growth periodicity of trees in regions with short dry seasons (less than 2 months) or indistinct dry seasons (monthly precipitation more than 60 mm) is still required in dendrochronological investigations, and the application of the methods is a great need in investigation in tropical forests. Among the various methods, wood anatomical investigations are required for the analysis of growth boundaries.

Shimaji and Nagatsuka (1971) applied the pinning method devised by Wolter (1968) and showed an accurate xylem growth pattern of the limited portion of a stem of *Abies firma* with very little dispersion of data. This method has been well investigated (Kuroda and Shimaji, 1984a,b; Kuroda, 1986) and was recently improved by Nobuchi *et al.* (1993) who adopted the epoxy-embedding and thin-sectioning method. The site of each stage of cell wall formation was observed to estimate accurately the time course of the formation on the concept that the cells destroyed by pinning keep the original cell wall organization at the time of the pinning. Shiokura (1989) applied the pinning method to tropical trees using a nail, and concluded this method is simple and effective to measure radial increment in trees, and recently Nobuchi *et al.* (1995) successfully applied the same method to *Hopea odorata* growing in a seasonal tropical forest, but the amount of radial growth was not always proportional to the rainfall.

As discussed by Sass *et al.* (1995), the so-called 'window method of Mariaux' has been successfully applied to tropical fast-grown trees providing

(eds R. Wimmer and R.E. Vetter)

exact information on the growth rhythm, but considering problems occurring due to the low growth rate, the strong wound reaction and the inconsistency of cambial activity around the stem circumference it was concluded that this method was not optimal to determine a possible periodicity of wood formation of slow-growing species.

In the earlier studies using the 'sampling method', Bannan (1955) and Imagawa and Ishida (1970) sampled small blocks containing cambium and adjoining tissues at intervals from living trees for microscopic analysis and depicted growth curves in tracheid number on *Thuja occidentalis*, *Pinus densiflora* and *Larix kaempferi*, respectively. Using intensive microscopic studies on the cambium zones and the differentiating xylem the pinning method has been able to be successfully applied to the periodical growth analyses of temperate trees. In this sense, microscopic observations of the zone of differentiating xylem in tropical trees, especially from a weakly seasonal climate region, by the sampling method are desired for the better understanding of the results by cambial wounding methods, such as the pinning and the window methods. Seasonal changes of the zone of differentiating xylem of three *Shorea* species (*Dipterocarpaceae*) growing in Sarawak were observed microscopically in detail using the thin-sectioning method in relation to growth boundaries developed during the sampling periods.

MATERIAL AND METHODS

Sampling Site and the Climate

The sampling forest is a logged-over natural hill forest in Sabal, Sarawak, east Malaysia, located in the southeast of Kuching and near to the boundary with Kalimantan, Indonesia, in Borneo Island. It is a natural tropical rain forest once harvested around 1980 and classified as mixed hill dipterocarp forest.

Sarawak is a rainy tropical region with indistinct dry seasons with a mean annual sum of precipitation of about 4000 mm. Monthly average temperature in Sarawak is almost constant all the year round with a range from 25 to 27°C. According to the climatic data offered by Perkhidmatan Kajicuaca Malaysia (Cawangan Sarawak), mean monthly precipitation sum during the sampling period was 348 mm (Fig. 11.1). The rainy season is usually from December to February and the relatively dry season is from May to September. Daily sunshine averaged 4.6 h day^{-1} and was shortest in a rainy season, 3.0 h day^{-1}, and longest in a rather dry season, 7.1 h day^{-1}.

One tree each of three *Shorea* species were selected for the experiment (Table 11.1). Two species were from Section *Rubroshorea*, and one species from Section *Richetioides*. Fresh sample blocks including the cambium and the zone of developing xylem were collected at breast height or just above the buttresses using a steel knife and a chisel, and immediately immersed in 1–2% glutaraldehyde in phosphate buffer solution. Sample blocks were about 3 cm in

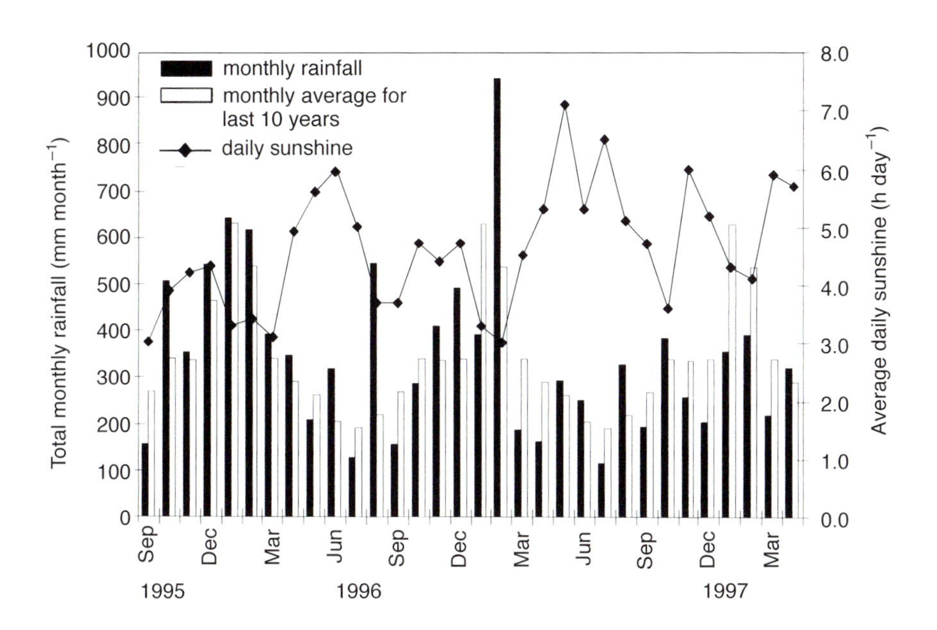

Fig. 11.1. Climate recorded at the airport in Kuching, Sarawak. Monthly rainfall is indicated in comparison with average of recent 10 years.

width and 5 cm in length, and wounds had openings a little wider. A distance between two subsequent samplings was about 20 cm tangentially to minimize the effects of a previous wounding as examined by Sass *et al.* (1995). On the second sampling, a block was cut out from about 20 cm above the first sampling. Sampling was repeated seasonally nine times from September 1994 to March 1997. The samples collected on 12 September and 9 October in 1994, 20 February, 26 June and 1 November in 1995, and 16 February in 1996 were analysed with the thin-sectioning method, but those collected on the last three occasions have not yet been analysed.

On 19 March 1997, fresh wood blocks (about 20 cm in tangential width) including the first wounds were cut out with a chain saw. They were kept wet with diluted glutaraldehyde during the transportation to Japan and then kept in a mixture of ethanol and glycerin.

Sample Preparation

The sample blocks were dissected carefully into small blocks (*c.*3 mm square in cross section and 5 mm in length), fixed with 1–2% glutaraldehyde diluted in phosphate buffer solution, and embedded in epoxy resin through graded

Table 11.1. Samples collected at Sabal, Sarawak, Malaysia.

Sample no. SARF-	Botanical name	Section	Diameter (cm)	Radius increase (mm)	Specific gravity*		
					Ave.	Low	High
TR06585	*Shorea patoiensis* Ash.	Richetioides	31	26	0.72	0.65	0.79
TR06586	*Shorea pinanga* Scheff.	Rubroshorea	34	18	0.54	0.47	0.62
WA1	*Shorea dasyphylla* Foxw.	Rubroshorea	66	20	0.51	0.42	0.61

*The values for specific gravity are of standard wood samples from the xylarium.

ethanol series and propylene oxide. Thin cross sections (*c.*3 μm thick) were cut from the epoxy-embedded blocks with a rotary microtome equipped with a glass knife, were picked up on a slide glass, and were stained with safranin and crystal violet. They were mounted with Canada balsam and observed under a polarized light microscope with and without an analyser. For further detailed investigations, semi-ultrathin cross sections (0.2–1.0 μm thick) of trimmed area from the same blocks were cut with an ultramicrotome (LKB 2128 Ultrotome) equipped with a diamond knife.

Measurements of the Cambial Activity

Fibres with developing secondary walls were usually stained darker than mature ones, but the stainability of the developing fibre walls was not constant. For the analyses of the xylem development, the determination of fusiform initials and the fibres at the early stages of the deposition of the outer layer of the secondary wall (S1) and the inner layer of secondary wall (S3) were needed for the measurement of the cell number and the width of each developing zone. Cambial initials were determined on microphotographs according to the following criteria as suggested by Larson (1994): (i) fusiform initial adjoins the shortest ray cell in a cambial zone, (ii) fusiform initial is one of the undifferentiated cells with the narrowest radial diameter and is usually radially a little wider among them. According to the criteria described above, fusiform initials were determined and marked by black dots on microphotographs of semi-ultrathin sections (Fig. 11.5).

Fibres at the early stage of the deposition of S1 and also those at the early stage of the deposition of S3 were determined on polarized light micrographs as the first fibres with one layer with conspicuous birefringence and as those with two layers with conspicuous birefringence, respectively. Then, on corresponding ordinary light micrographs, the first fibres with S1 and S3 layers in each radial file were marked by small stars and black squares, respectively (Fig. 11.5). Numbers of cells were counted in several radial files not including vessels from the cambial initials to the first fibres with S1 layer and also to the first fibres with S3 layer. The distance between centres of cambial initials and those marked fibres were also measured in the same radial files.

Measurements of the Optical Density

Cross surfaces of wood blocks including the first wounds were finished with a disposable steel knife for observation under a dissecting microscope. Traumatic tissues caused by the first sampling were traced in tangential direction to the edge of the blocks, and small blocks including the traumatic tissues and the cambial zone were cut out.

Cross sections were cut with a sliding microtome equipped with a disposable knife blade, stained with safranin and gentian violet, and mounted with Canada balsam. They were observed microscopically, and then the optical density from the traumatic tissue to the cambial zone was scanned with a computer-aided microdensitometer (Dendro 2003). The measuring condition was as follows: objective lens × 25, measuring slit width 20 and 840 μm in radial and tangential directions, respectively. The densitometer was calibrated using an empty area of a microscopic slide and dark conditions, so that optical density obtained was a relative value.

RESULTS AND DISCUSSION

Seasonal Cambial Activity

Thin sections of the zones of the cambium and the developing xylem are shown in Figs 11.2–11.4. Because the tangential bands of thick-walled phloem fibres and mature xylem fibres were darkly stained, the primary wall zone including the cambial zone in the middle was observed as a conspicuously transparent zone. The width of the primary wall zone was narrow in some sections and very wide in some others depending on the sampling seasons (Fig. 11.5). The width of the zones of xylem in the primary wall stage and in the secondary wall thickening stage measured on the marked microphotographs was shown in the histograms expressed by the distance (Figs 11.6–11.8). Within the limitation of microscopical investigation of the samples collected within one and a half years, two rainy and one indistinct dry seasons, no dormant cambium, which is characterized by more or less flat and thin-walled cells, was observed (Figs 11.2–11.5). The cambial initials were always accompanied by xylem elements in primary wall stage, and the radial diameter of fibres increased gradually inward and finally reached the polygonal to round shape in cross sections (Fig. 11.5). These results suggest that the cambial activities such as cell division and surface expansion of the derivatives proceeded continuously independently of the changes of the width of cambial and primary wall zone of the samples.

The width of the primary wall zone varied from 170 to 360 μm in *S. pinanga*, from 120 to 280 μm in *S. dasyphylla* and from 180 to 380 μm in *S. patoiensis*. In contrast, the width of the zone of fibres developing S1 and S2 layers, which refers to the width between the fibres first with S1 and S3 layers in each radial file as shown in Fig. 11.4, varied widely from 0 in the sample of June 1995 to about 800 μm in the sample of November 1995 in *S. patoiensis* (Fig. 11.6). Those in the other samples had ranges of 30–360 μm in *S. pinanga* (Fig. 11.7) and 60–500 μm in *S. dasyphylla* (Fig. 11.8). Considering the possibility that the rate of xylem formation may be different in the direction in the stem, the activity of xylem development was evaluated by the ratio of the width of the zone of developing secondary wall to primary wall zone (S/P ratio). The S/P ratio showed ranges of 0.1–1.8 in *S. pinanga* (Fig. 11.7), 0.5–1.8 in *S. dasyphylla* (Fig. 11.8) and 0–2.3 in *S. patoiensis* (Fig. 11.6). Based on the

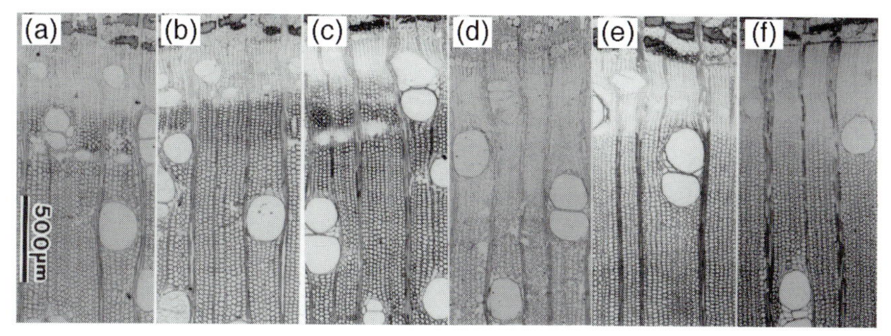

Fig. 11.2. *Shorea patoiensis* (TR6585)

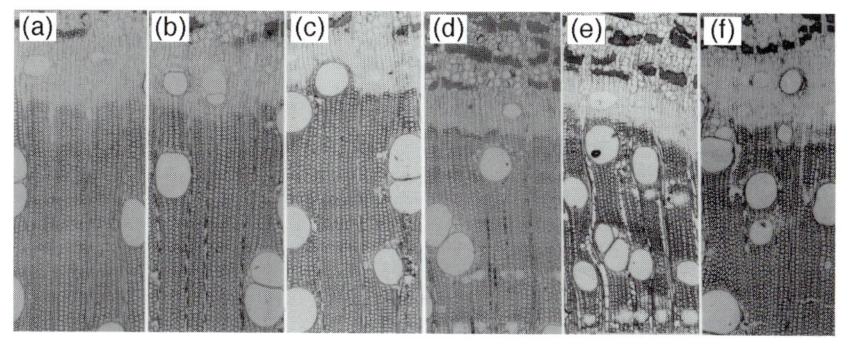

Fig. 11.3. *Shorea pinanga* (TR6586)

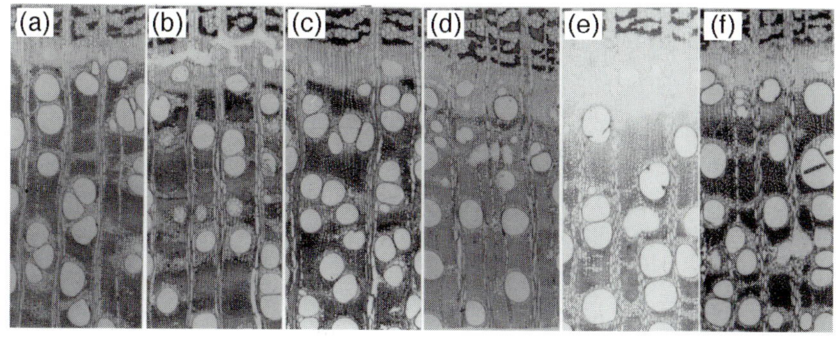

Fig. 11.4. *Shorea dasyphylla* (WA1)

Figs 11.2–11.4. Thin sections of cambial zone and the zone of developing xylem. a, September 1994; b, October 1994; c, February 1995; d, June 1995; e, November 1995; f, February 1996.

results that very low values of the S/P ratio and more or less wide primary wall zones were shown in the same sections, it is presumed that the rate of cell division in cambium and of cell expansion of the derivatives may probably slow down seasonally, but a certain width of thin-walled xylem elements around the

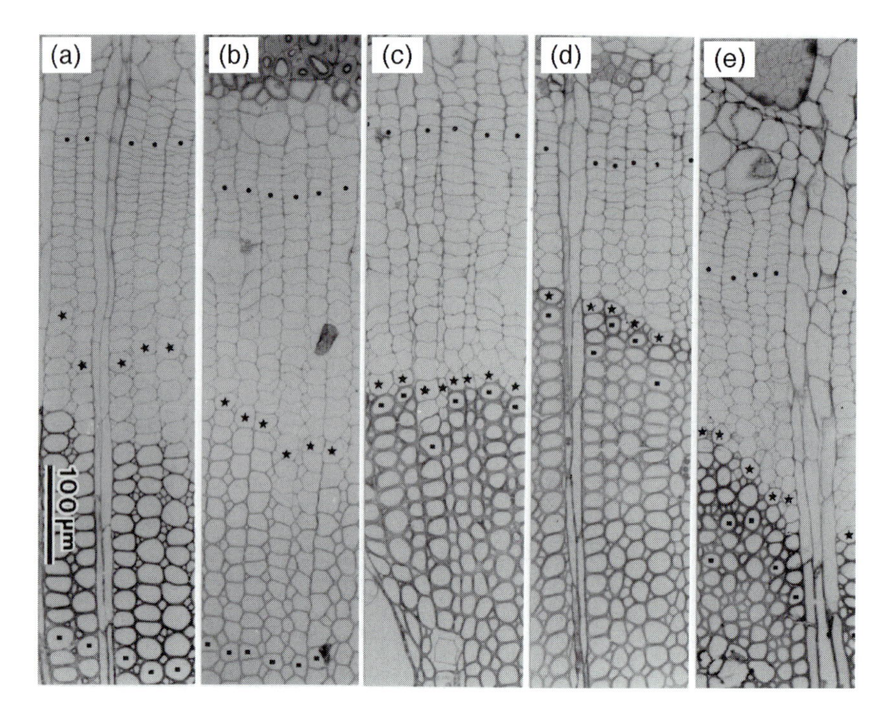

Fig. 11.5. Semi-ultrathin section of *Shorea patoiensis* (TR6585). a–e, The same as in Figs 11.2–11.4. Black dots, fusiform initials determined according to the criteria. Stars and black squares, fibres initiating S1 and S3 layers, respectively.

cambial zone is kept, and the secondary wall development slows down so remarkably as to reduce the population of the fibres under developing secondary walls. As the width of primary wall zone is determined by the frequency of cell division and the differentiating time of the xylem derivatives (Kennedy and Farrar, 1965) the seasonal changes of this width suggest that the frequencies of cell division and the duration for the xylem elements to fix their shape in the sample trees were not remarkably decreased. However, the conspicuous low activity in secondary wall formation that was recorded in some samples, such as *S. dasyphylla* in February 1995 and *S. pinanga* in February and June 1995, suggests that cells in the primary wall stage were inactive and transferred at very low frequency to the secondary wall deposition stage just before the sampling time.

All of the three sample trees showed the first decrease of the S/P ratio from September to October 1994 and an increase from June to November 1995, but they did not show a common tendency in other seasons. *S. pinanga* and *S. dasyphylla* showed the lowest ratio in February 1995, in the middle of the rainy season, in contrast to the relatively high value of 1.1 of *S. patoiensis* at the same

Fig. 11.6

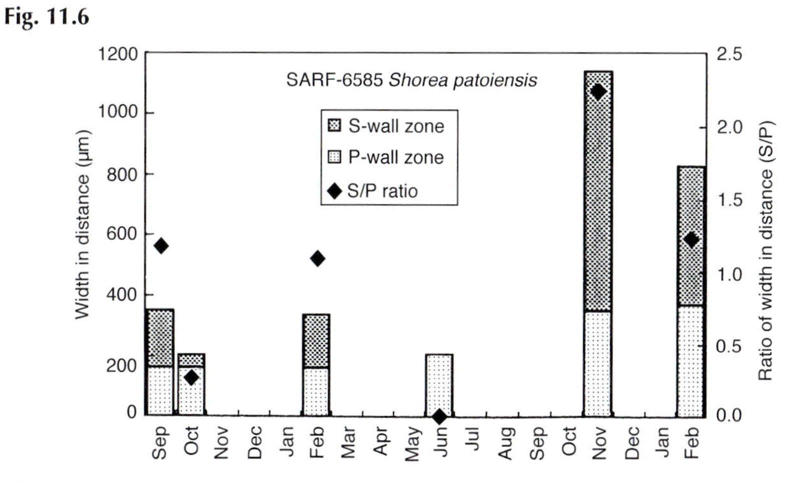

Fig. 11.7

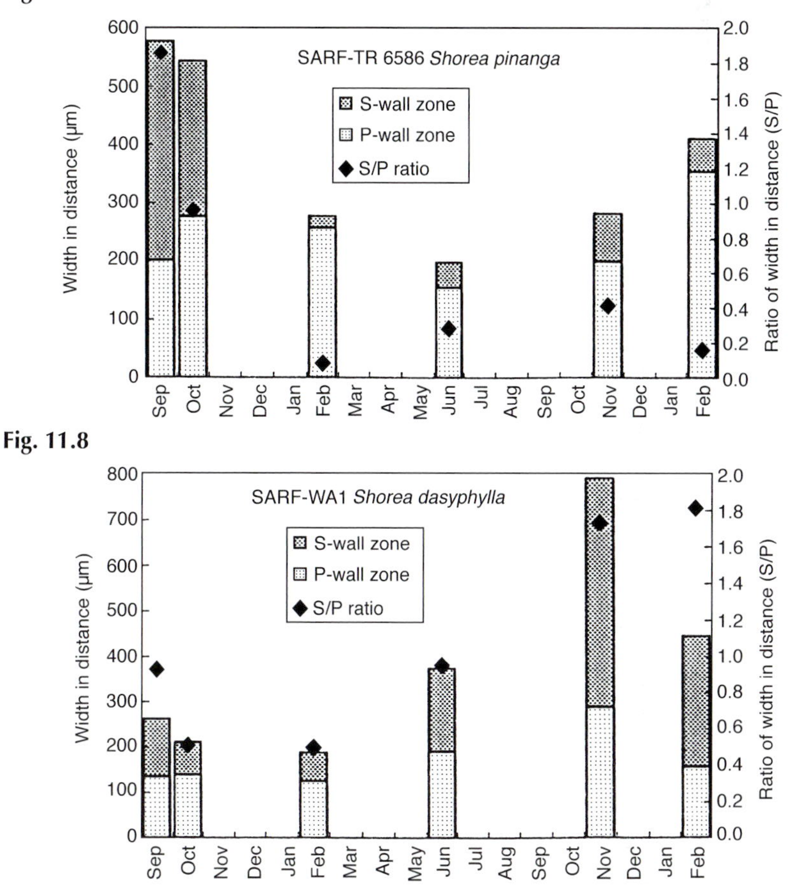

Figs 11.6–11.8. Width of the developing xylem zone in primary (P) and secondary (S) wall deposition stage and their ratios.

season. It should be noted here that the highest S/P ratios were recorded in November 1995 in *S. dasyphylla* and *S. patoiensis* (1.8 and 2.3, respectively), but it was 0.4 in *S. pinanga*. The S/P ratios in February 1995 and 1996 were almost at the same level in *S. pinanga* and *S. patoiensis*, but were quite different in *S. dasyphylla*. As a result, seasonal changes of the xylem development were clearly shown in the width of secondary wall zone rather than the width of primary wall zone. But, they were not simply related to the alternation of rainy and indistinct dry seasons. Although factors other than precipitation and sunshine, such as soil condition, water stress, flowering, fruiting or flushing during the sampling period, may be different among the sample trees in spite of the restriction of the site area, it becomes clear that the seasonal variation of the cambial activity is not common in the sample trees. These inconsistencies in the activity of the secondary wall formation in fibres among the samples suggest that the cambial activity of these sample trees is not as obviously related to the climatic condition (compare Figs 11.6–11.8 to Fig. 11.1) as has been reported for trees growing in the temperate zone and also in strongly seasonal tropics as reviewed by Worbes (1995).

Growth Boundary Formation

Various types of growth boundaries were visible in cross sections from the sample blocks collected in March 1997 which include the traumatic tissues reacting to the first sampling injuries. The traumatic tissues were traced tangentially in the region of the direct injuries from the first sampling and thick-walled fibre bands with gradual transition are seen (indicated by arrowheads in Figs 11.9–11.11). Because traumatic tissues have not been analysed in detail in these trees yet, it is not possible to exclude completely the possibility that the effects of seven samplings between October 1994 and November 1996 may be recorded as reaction woods in these cross sections. But, reaction tissues to the second sampling injuries located both about 20 cm above and about 20 cm in lateral were not conspicuously observed in these cross sections. Because of the distance of about 20 cm between subsequent sampling portions, the reaction tissues were supposed to be not so conspicuous in the sections as the first traumatic tissues, and also because it was hard to analyse them in detail, they were not taken care of in this study.

Several indistinct growth boundaries were observed in the xylem formed in the previous 2.5 years, during three rainy and two indistinct dry seasons, and were marked by thick-walled fibre bands, axial parenchyma bands including resin canals and occasionally vessels in tangential arrangement (Figs 11.9–11.11). In *Shorea patoiensis* (Fig. 11.9), six bands of axial resin canals in long tangential lines were remarkable and a further five tangential bands were observed in this section. Vessels in tangential arrangement in the band of thicker-walled fibres were also observed in the outer part. In *Shorea dasyphylla* (Fig. 11.10), two axial parenchyma bands including resin canals associated

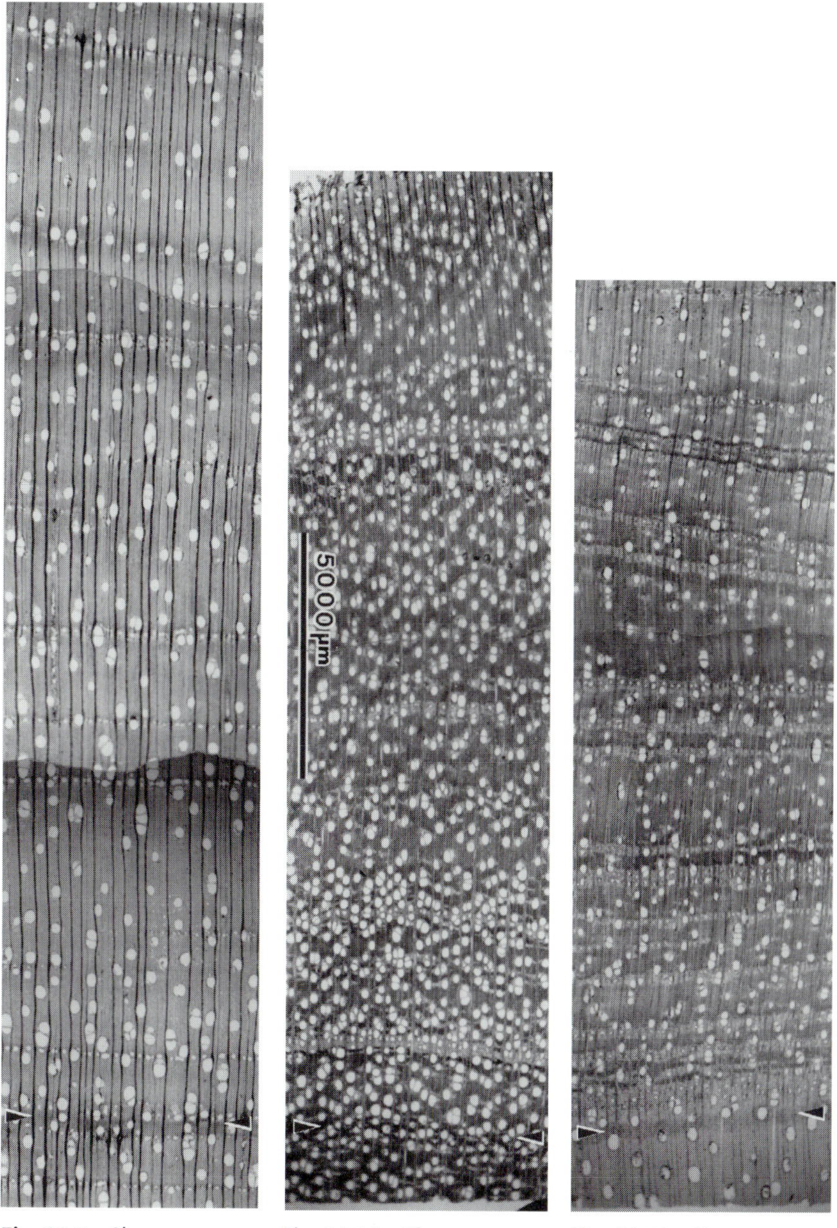

Fig. 11.9. *Shorea patoiensis* (TR6585)

Fig. 11.10. *Shorea dasyphylla* (WA1)

Fig. 11.11. *Shorea pinanga* (TR6586)

Figs 11.9–11.11. Cross-sections of xylem sampled in March, 1997 showing the xylem formed during 2.5 years near the first sampling injuries. Reaction wood caused by injuries during the first sampling is indicated by arrow heads.

with frequent narrower vessels outside were conspicuous and three others were between them. This type of growth ring boundary has been already reported in *Hopea odorata* to be usually annual (Nobuchi *et al.*, 1995). Although those conspicuous axial parenchyma bands may be assumed to be formed in response to the weak dry seasons, it cannot fit with the fact that the inner bands located only 150–200 μm from the first date-marking which was in early rainy season. In *Shorea pinanga* (Fig. 11.11), four abrupt changes of fibre wall thickness and 14–16 bands of axial resin canals in long tangential lines were shown in combination and independently. The bands of thick-walled fibres may be assumed to be reflecting growth in the weak dry season, but the abrupt changes from thinner to thicker-walled fibres and gradual transition from thicker to thinner-walled fibres are not easy to understand in the seasonal transition of the cambial activity.

Optical Densitometry

Optical densitograms of microscopic sections showed both long and short-term waves (Fig. 11.12). In comparison to the microphotographs (Figs 11.9–11.11), it is presumed that a short wave derives from vessels and/or axial parenchyma bands and a long one mainly from the changes in fibre wall thickness. The optical densitograms were not coincident among three species. The density decreased gradually at the beginning and showed later two weak maximum peaks in *S. patoiensis* and *S. dasyphylla*. In contrast, *S. pinanga* increased at first without any obvious peak and decreased in the last portion. The bands of axial parenchyma and thick-walled fibres were not so sharply expressed in these densitograms. However, the growth boundaries were detectable by comparing micrographs and densitograms. For example, axial parenchyma bands with/without resin canals were recorded as distinct minimum peaks in each tree. These results suggest that wood structural markers will help to analyse periodical growth of tropical trees growing in a rain forest using densitometry together with corresponding microphotographs.

CONCLUSIONS

The width of cambial and primary wall zone did not decrease remarkably. No dormancy of the cambium was observed. A very narrow zone of developing xylem under secondary wall deposition was observed, but not in the same season among samples. Changes of cambial activity were observed, but seasonal changes of xylem development were not simply related to the alternation of rainy and indistinct dry seasons. Climate factors, such as increase in the precipitation and sunshine, may be one of the causes affecting the growth rhythm, but inconsistent growth rhythm among the three trees could not be explained simply by climatic factors.

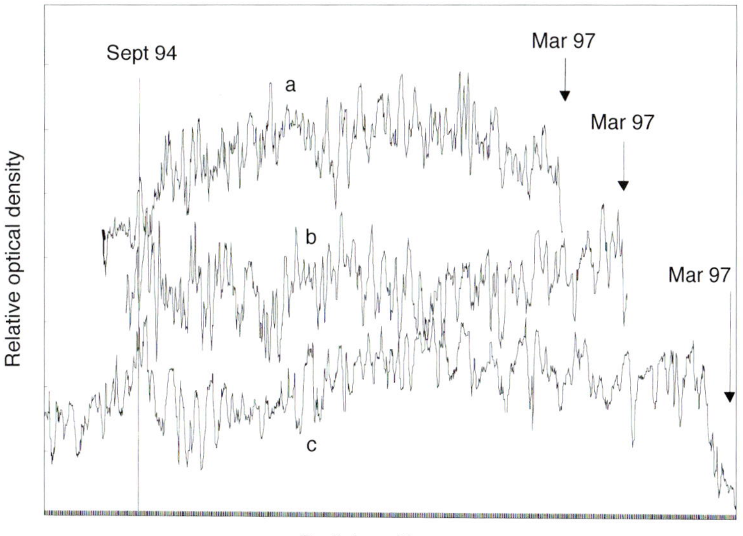

Fig. 11.12. Optical density distribution in the past 2.5 years. (a) *Shorea pinanga,* (b) *Shorea dasyphylla,* (c) *Shorea patoiensis.*

For future studies on the periodical growth of woody plants in aseasonal tropical rain forests, the following combination of methods is recommended.

1. Monitor the radial growth using, for example, dendrometers over an entire growing season (see review by Worbes, 1995) and the phenological phenomena such as flushing, flowering and fruiting.
2. Pin-mark bi-weekly or monthly.
3. Pin-mark all trees on same day as a reference line (Nobuchi *et al.*, 1995).
4. Sample fresh cambium tissues bi-weekly or monthly from pin-marked portions at first.
5. Analyse with the densitometer in combination with microscopy and relate to the radial growth rate.

SUMMARY

The zone of differentiating xylem of three *Shorea* spp. (*Dipterocarpaceae*) growing in Sarawak was analysed to investigate the growth boundaries in relation to the climate. Fresh sample blocks including the cambium and the zone of developing xylem were collected six times between September 1994 and February 1996. The width of the zone of developing xylem, both zones of expanding cell size and of thickening secondary walls, was measured. Seasonal changes were clearly shown in the width of secondary wall zone, but were not simply related to the alternation of rainy and indistinct dry seasons. The width of the primary wall zone did not decrease remarkably, suggesting the absence of cambial dormancy. Several indistinct growth boundaries were observed in the xylem formed in the previous 2.5 years, during three rainy and two indistinct dry seasons. Wood anatomical investigation on three *Shorea* species revealed that the growth boundaries were marked by thick-walled fibre bands and/or axial parenchyma bands with axial resin canals, and occasionally by vessels in tangential arrangement. Climate factors, such as an increase in precipitation and sunshine, may be one of the causes of boundary formation, but the inconsistent growth rhythm among the three trees could not be understood simply in the light of these climatic factors.

ACKNOWLEDGEMENTS

The authors thank the members of the Wood Anatomy Laboratory, Timber Research and Technical Training Centre, Sarawak Forest Department, Malaysia, for collecting fresh materials and embedding the samples. This work was partly supported by the Japan International Corporation Agency.

REFERENCES

Bannan, M.W. (1955) The vascular cambium and radial growth in *Thuja occidentalis* L. *Canadian Journal of Botany* 33, 113–138.

Imagawa, H. and Ishida, S. (1970) Study on the wood formation in trees. Report 1. Seasonal development of the xylem ring of Japanese larch stem, *Larix leptolepis* Gordon. *Research Bulletin of the College Experimental Forests*, Hokkaido University 27, 373–394.

Kennedy, R.W. and Farrar, J.L. (1965) Tracheid development in tilted seedlings. In: Cote, W.A. (ed.) *Cellular Ultrastructure of Woody Plants*. Syracuse University Press, pp. 419–453.

Kuroda, K. (1986) Wound effects on cytodifferentiation in the secondary xylem of woody plants. *Wood Research* 72, 67–118.

Kuroda, K. and Shimaji, K. (1984a) Wound effects on xylem cell differentiation in a conifer. *IAWA Bulletin n.s.* 5, 295–305.

Kuroda, K. and Shimaji, K. (1984b) The pinning method for marking growth in hardwood species. *Forest Science* 30, 548–554.

Larson, P.R. (1994) *The Vascular Cambium: Development and Structure*. Springer-Verlag, Berlin, Chapter 4, pp. 33–97.

Nobuchi, T., Fujisawa, T. and Saiki, H. (1993) An application of the pinning method to the marking of the differentiating zone and to the estimation of the time course of annual ring formation in sugi (*Cryptomeria japonica*). *Mokuzai Gakkaishi* 39, 716–723.

Nobuchi, T., Ogata, Y. and Siripatanadilok, S. (1995) Seasonal characteristics of wood formation in *Hopea odorata* and *Shorea henryana*. *IAWA Journal* 16, 361–369.

Sass, U., Killmann, W. and Eckstein, D. (1995) Wood formation in two species of *Dipterocarpaceae* in peninsular Malaysia. *IAWA Journal* 16, 371–384.

Shimaji, K. and Nagatsuka, Y. (1971) Pursuit of the time sequence of annual ring formation in Japanese fir (*Abies firma* Sieb. et Zucc.). *Journal of the Japan Wood Research Society* 17, 122–128.

Shiokura, T. (1989) A method to measure radial increment in tropical trees. *IAWA Bulletin n.s.* 10, 147–154.

Wolter, K.E. (1968) A new method for marking xylem growth. *Forest Science* 14, 102–104.

Worbes, M. (1995) How to measure growth dynamics in tropical trees. A review. *IAWA Journal* 16, 337–351.

12

Pinus tropicalis Growth Responses to Seasonal Precipitation Changes in Western Cuba

Margarita M. Chernavskaya, Henri D. Grissino-Mayer, Alexander N. Krenke and Andrey V. Pushin

INTRODUCTION

The total tree-ring width, earlywood width and latewood width, as well as wood density, depends mostly on environmental and climatic conditions of the current and previous years. It is well known that the sensitivity of trees to climatic variability increases when limiting factors occur. Many different chronologies have been used for revealing past climatic trends that have included reconstructions of air temperature, precipitation and pressure. Results obtained by Fritts (1976), Briffa *et al.* (1983, 1988), Shiyatov (1986), Schweingruber (1988), Wu *et al.* (1990) and others demonstrate the usefulness of dendrochronological methods for analysing climate in extratropical latitudes where the growth of trees can be limited by thermal conditions. Our study deals with the variability of radial growth of trees growing in tropical climates where precipitation is the most dynamic temporal factor.

MATERIAL AND METHODS

Cuba is characterized by a tropical trade climate with a mean January temperature of 22.5°C and a mean August temperature of 28.0°C. Total annual precipitation is 1000–1200 mm, with some years experiencing up to 2200 mm. The rainy (May to October) and dry (November to April) seasons comprise equal portions of the annual total. Meteorological data between 1906 and 1986 from Las Vegas, located in the Havana province, and two short time series from two sites nearest to our study site – La Palma (1967–1986) and Pinar del Rio (1967–1984) – were used to analyse the effects of precipitation on pine growth.

Between eight and ten trees of *Pinus tropicalis* were sampled from each of the three sites, all within the Sierra de los Organos, in the Pinar del Rio province of western Cuba. Two cores were taken from each tree using a regular increment borer. Site 3 is situated on the north-facing macroslope of Sierra de Rossario at a height of about 300–350 m a.s.l. Site 4 is situated on the south-facing macroslope of the Sierra de los Organos at the same elevation as the previous one. This site constitutes a slope of northwest exposure with slope inclination <25°. Site 5 is situated on a pass in the Los Jasminos Mountains at a height of 200 m a.s.l.

Because of the extremely individual response of tree-ring growth to environmental changes it was a problem to synchronize the samples not only from the same site, but sometimes also from one tree. So, further analysis was mostly carried out on the base of individual cores. Lightwood and darkwood widths (the terms 'earlywood' and 'latewood' may not be appropriate in the traditional sense) were measured with a model MBC-10 microscope, while wood density measurements were made with an RDK-1D X-ray densitometric instrument. The latter allowed us to determine more precisely the boundary between annual layers. The biological growth trends were removed from the tree-ring time series by standard dendrochronological methods.

RESULTS

Our analyses revealed typical features of the annual ring formation under tropical climatic conditions with a well-expressed course of annual precipitation. Under these conditions, growth occurs throughout the year and trees have rather wide rings visible to the unaided eye. The formation of lightwood occurred during the wet summer season, while darkwood formed during the dry winter season. With violation of the normal annual precipitation course (decrease of monthly precipitation within the wet season and/or increase of precipitation in one of the months during the dry season) layers of darkwood form within a layer of lightwood and vice versa (Fig. 12.1). This testifies to high sensitiveness of *Pinus tropicalis* to intra-annual precipitation variability, but it causes the formation of intra-annual ('false') rings that make further revealing of annual tree rings, as well as crossdating, hard. Density diagrams were made for all samples to overcome this impediment. The false rings were identified as variability of density of low amplitude. Abrupt changes in wood density were specified as tree-ring boundaries.

Comparison of inter-annual precipitation variability with tree-ring growth showed that a decrease of at least 150 mm average monthly precipitation was necessary for formation of darker and denser wood after the lightwood. In contrast, an increase of average monthly precipitation of 70–100 mm was sufficient to induce formation of lightwood (Table 12.1). The trees appeared to be more sensitive to an increase in precipitation during a dry season than to a reduction of precipitation during a rainy period. We found that the formation

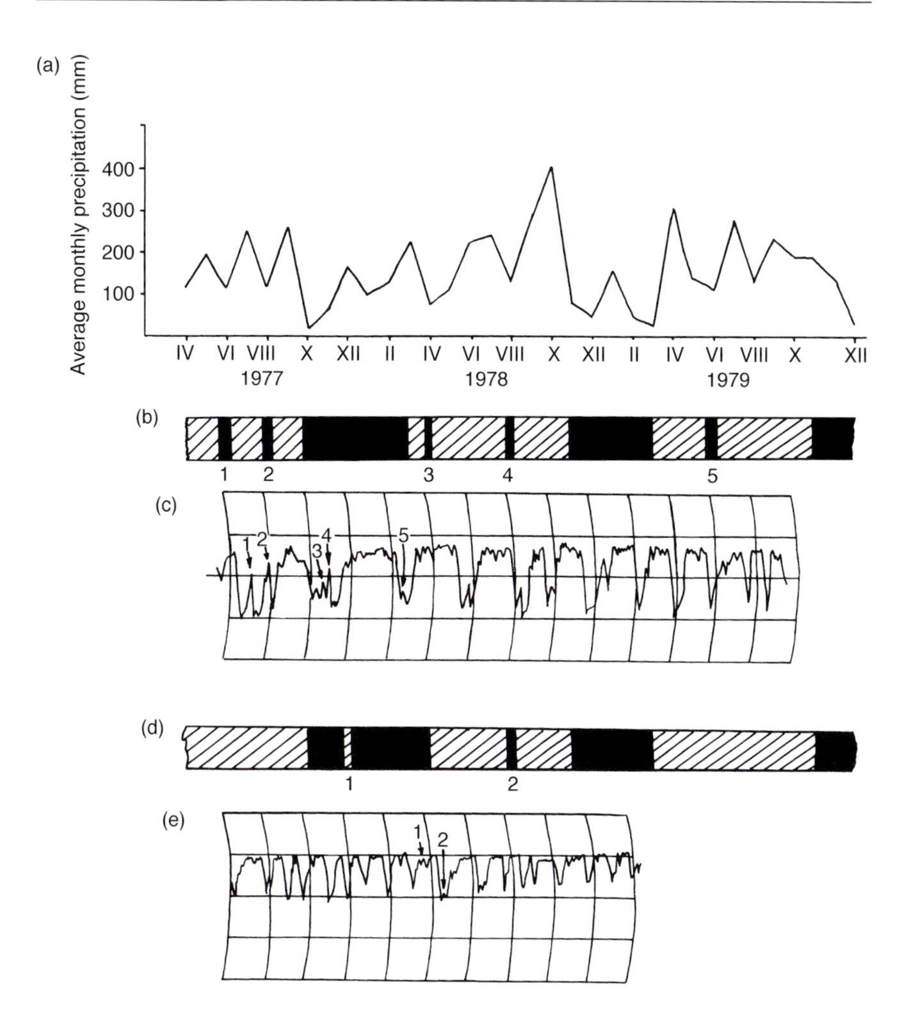

Fig. 12.1. Tree-ring growth response to intra-annual precipitation variability in western Cuba. (a) Average monthly precipitation in La Palma (April 1977 to December 1979); (b) alternation of lightwood (inclined lines) and darkwood, sample 44, site 4; (c) density diagram of sample 44 (1, 2, ... 5 – 'false' rings); (d) alternation of lightwood and darkwood, sample 23, site 3; (e) density diagram of sample 23 (1,2 – 'false' rings).

of lightwood did not occur when average monthly precipitation was less than 200 mm.

We also found a positive correlation (all correlation coefficients exceeding 0.5) between growth of darkwood and dry season precipitation at all of our study sites (Fig. 12.2). A negative correlation between growth of lightwood and wet season precipitation was typical at sites 3 and 4. A positive correlation

Table 12.1. Abrupt change in the amount of precipitation followed by the formation of layers with different density (1 – number of cases, 2 – precipitation change in mm, 3 – standard deviation).

Meteorological station	Seasons					
	Wet			Dry		
	1	2	3	1	2	3
La Palma	13	−215	138	16	104	60
Las Vegas	33	−167	119	23	73	42

between growth of lightwood and wet season precipitation was found only at site 5, obviously due to good drainage and lasting insolation.

CONCLUSIONS

Pinus tropicalis sensitiveness thresholds to average monthly precipitation change were estimated for dry and wet seasons, respectively. We conclude that the association between tree-ring growth of *Pinus tropicalis* and inter-annual precipitation demonstrates the possibility of climate reconstructions being conducted on dendrochronological tree-ring data from tropical regions.

SUMMARY

We investigated the tree-ring growth variability of *Pinus tropicalis* growing under tropical climate conditions at three sites in western Cuba. We found that the amount and temporal dynamics of precipitation was the dominant factor that determined the formation of tree-ring structure under stable temperature conditions.

ACKNOWLEDGEMENTS

The authors are sincerely grateful for the comments and constructive criticisms of an anonymous reviewer of this manuscript.

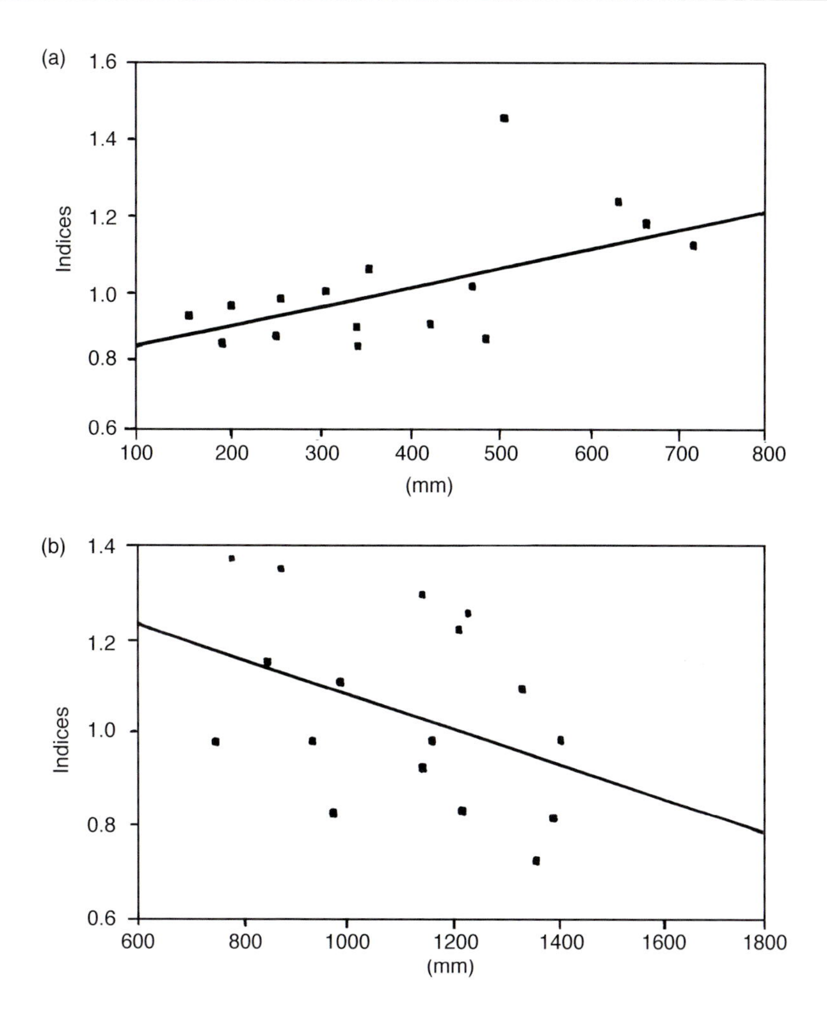

Fig. 12.2. Relationship between darkwood (a) and lightwood (b) growth and amount of precipitation for dry and wet seasons, respectively (sample 14, site 3, La Palma).

REFERENCES

Briffa, K.R., Jones, P.D., Wigley, T.M.L., Pilcher, J.R. and Baillie, M.G.L. (1983) Climate reconstruction from tree rings. Part I. Basic methodology and preliminary results from England. *Journal of Climatology* 3, 233–242.

Briffa, K.R., Jones, P.D., Pilcher, J.R. and Hughes, M.K. (1988) Reconstructing summer temperature in northern Fennoscandinavia back to A.D. 1700 using tree-ring data from Scots pine. *Arctic and Alpine Research* 20, 385–394.

Fritts, H.C. (1976) *Tree Rings and Climate.* Academic Press, New York, 567pp.

Schweingruber, F.H. (1988) *Tree Rings: Basics and Applications of Dendrochronology*. Reidel Publishing Company, Dordrecht, 276pp.

Shiyatov, S.G. (1986) *Dendrochronology of the Upper Forest Boundary in the Urals*. Nauka, Moscow, 186pp.

Wu, X., Zhan, X., Sun, L. and Cheng, Z. (1990) Reconstructing Middle-Tibet climate during the last 600 years by the dendroclimatological method. *Acta Meteorologica Sinica*, 4, 294–304.

Tree Rings and Historical Aspects

D

13

Occurrence of Moon Rings in Oak from Poland during the Holocene

Marek Krąpiec

INTRODUCTION

A type of discoloration in the heartwood of European oak (*Quercus robur* L. and *Quercus petraea* Lieb.) called 'included sapwood' or 'moon rings' was first described by Duhamel du Monceau (1758). Moon rings appear on cross-sections as bright zones within darker coloured heartwood (Fig. 13.1). Moon rings have also been reported for other European species such as *Robinia pseudoacacia* L. (Erteld *et al.*, 1963), for *Larix* and *Thuja* (Trendelenburg, 1955), and for a few tropical species (Scheiber, 1965). Moon rings occur in standing trees and they are caused by broken-off living branches during long-lasting frost periods. The subsequently formed tree ring has some similarity with wound tissue and the wood properties differ from regular wood. The low phenol content as well as the poor formation of tyloses in the earlywood disturbs the heartwood formation which is the reason why moon rings have a light colour and do not convert into dark heartwood.

Wood anatomy does not differ significantly from regular heartwood, and the same is true for the overall chemical composition. However, moon ring zones differ from heartwood not only in the colour but also in their reduced mechanical properties such as wood density, elasticity and hardness (Dujesiefken and Liese, 1986; Dzbeński and Krutul, 1994; Charrier *et al.*, 1995). In addition, moon ring wood is less durable and consequently more susceptible to biodegradation (Krzysik, 1974; Dujesiefken and Bauch, 1987). The formation of moon rings in oak wood has to be considered as a severe defect because a reduction in value of up to 80% for veneer logs may occur (Dujesiefken *et al.*, 1984). As an example, Polish wood industries have lost quite a bit of money when they started to import apparently cheap oak wood from Russian Bashkirie.

(eds R. Wimmer and R.E. Vetter)

Fig. 13.1. Specimen of subfossil oak wood from Bogaczewo pile dwelling with multiple moon rings (MR).

Moon rings can be best seen at 10–12 m above the ground in zones where broken-off branches are most likely. At lower parts of the stem moon ring zones may include less tree rings than higher up in the tree (Dujesiefken and Liese, 1986).

Moon rings are observed in 1–2% of Central European oak trees (Paclt, 1954, 1989; Bolychevtsev, 1970; Dujesiefken and Liese, 1986). Single enclosed moon rings occur most frequently in England (Tapper *et al.*, 1978), France (Henry, 1896; Charrier *et al.*, 1995), Central Europe (Liese, 1942; Dujesiefken and Liese, 1986; Dzbeński and Krutul, 1994) and in the Asian part of Russia (beyond the Ural – Bolychevtsev, 1970), whereas double and triple moon rings are less frequent and were noted only in those European regions with continental climate. In particular, oak trees from Pskow Oblast which are growing at the northern ecological limit form up to five zones of moon rings (Alekseev (cited in Bolychevtsev, 1970)).

The idea for this investigation arose during a dendrochronological study of oak from the Hallstatt period (*c.* 2500 years ago), Mazury (NE Poland). Moon rings occurred especially frequently in these samples. The analysis of moon ring frequency presented here was done on oak samples from southern Poland. The study includes samples from living trees as well as wood samples going back to the early Holocene. The focus is to demonstrate whether and how moon rings could serve as dendrochronological reference points, useful for dating and palaeoclimate reconstructions.

MATERIAL AND METHODS

Oak samples with moon rings were identified in a set of around 3000 oak slices collected in Holocene alluvial deposits, in archaeological excavations as well as in constructions and buildings. With a few gaps, the samples cover almost the entire postglacial period (the last 10,000 years). About 700 samples of subfossil oak trees, collected since the mid-1980s by the Department of Stratigraphy and Regional Geology of the University of Mining and Metallurgy in Cracow, were of primary importance for the investigation of moon ring frequency during the Holocene.

Oak trees from riparian forests grew in fertile and constantly watered habitats. After flooding over the river bank, the oak trees growing on these sites were overthrown and transported only a short distance before they accumulated and were buried in the river deposits. Therefore, the wood samples can be considered of local origin.

Using these samples, it was possible to establish a standard chronology for the period 474 BC–AD 1555 (Krąpiec, 1996). With the remaining samples, floating chronologies were built and dated applying the curve fitting 'wiggle matching' method according to Pearson (1986). The method is based on reconstructed short-term changes of ^{14}C concentration in the past, represented by the standard calibration curve. ^{14}C is analysed in selected tree rings, of which the relative ages are exactly known from at least three samples. The obtained floating calibration curve is dated by matching it with the standard calibration curve. Age determination with this method delivers results to the nearest 20–30 years (Krąpiec, 1992, 1994).

In addition to the historical samples, living oak trees were sampled by cutting stem discs or taking cores by means of a regular Pressler increment borer. Trees from southern Poland that grew on river plains, similar to the environment of the subfossil oaks, were chosen for investigation. The samples were taken from the forest districts: Pińczów, Niepołomice (two localities), Sieniawa, Radymno and Krasiczyn (Table 13.1 and Fig. 13.2).

TREE-RING DATING

Moon ring zones were identified on sanded cross-sectional surfaces of discs or on cores using a stereo-microscope. For all samples taken from living trees ring widths were measured to the nearest 0.01 mm using a self-made measuring stage which was hooked up to a standard personal computer (Krąpiec, 1992). Correlation and crossdating of the obtained sequences was done with a set of self-written computer programs (Krawczyk and Krąpiec, 1995). For each locality a correlation diagram with marked moon ring zones was computed.

Control analyses of annual growth rings of subfossil wood were carried out only in such cases when limits of moon ring zones had not been exactly determined during earlier datings. As most of the analysed samples were taken

Table 13.1. Description of the investigated sites in southern Poland localities with moon rings in living oak trees.

No.	Site	Tree age (years)	Date of sampling	Number of samples	Type of samples	Moon ring frequency
1	Forest district Pińczów, Michałów area, section 96	117–126	1994 1995/96	16	cores 2 slices 14	4/16
2	Forest district Niepołomice, Lipówka Reservation	102–186	1996/97	12	cores	6/12
3	Forest district Niepołomice, Chobot area, section 461	86–110	1996/97	17	cores	8/17
4	Forest district Sieniawa, Głażyna area, section 250	90–95	1996/97	16	slices	15/16
5	Forest district Radymno, Krzywa Pałka area, section 58f	75–86	1996/97	15	slices	6/15
6	Forest district Krasiczyn, Łęownia area, section 21	70–85	1996/97	10	slices	10/10

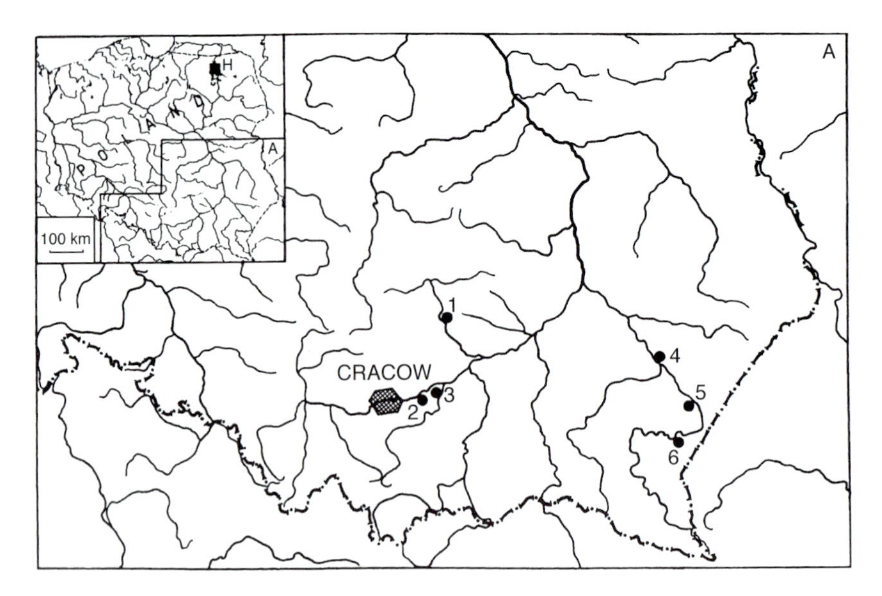

Fig. 13.2. Sample localities of oak wood with moon rings. Living tree stands: (1) Pińczów; (2) Niepołomice Forest – Lipówka Reservation; (3) Niepołomice Forest – Chobot; (4) Sieniawa; (5) Radymno; (6) Krasiczyn; H – pile dwellings from Hallstatt period, Mazury.

from lowermost parts of the stems, where the widths of moon ring zones are smaller, datings of the last tree rings from those zones can differ by about 2–5 years from real dates of moon ring zone formation.

RESULTS AND DISCUSSION

Analysis of Living Oak Trees

Moon rings were found on all investigated sites (Table 13.1). The fewest moon rings were observed at site 1 (Pińczów district). Moon rings were found on this site in four out of 16 samples analysed (Fig. 13.3). In the trees of site 1 the moon

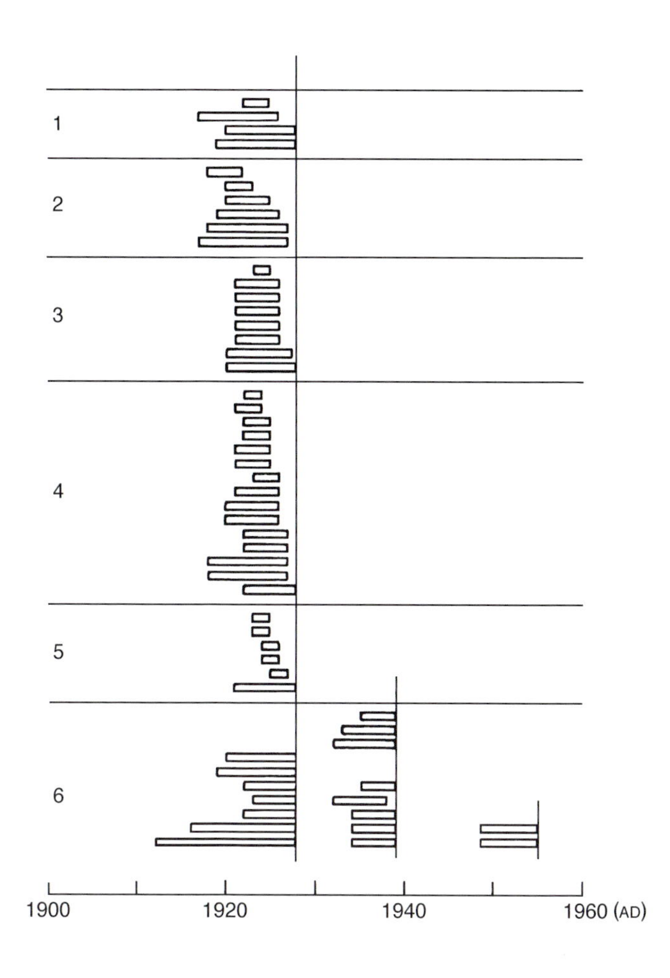

Fig. 13.3. Moon ring zones in samples taken from living oak trees from Southern Poland. Locality numbers are described in Table 13.1 and Fig. 13.2.

ring zones were formed after the severe winter in 1928/29. Moon rings formed in that winter were also found on sites 2 and 3. About half of the samples show moon rings of that severe winter and they encompass several annual rings (Fig. 13.3). A similar situation can be reported for site 5 (Radymno district): moon rings on that site were present in about 40% of the investigated samples and their formation also dates to the winter of 1928/29. The observed moon ring zones on site 5 are relatively narrow and include only a few annual rings (Fig. 13.3). However, the investigated samples of site 5 were taken from the bottom part of the stems. At site 4 (Sieniawa district) moon rings were observed in 94% of the examined samples and they were also formed in response to the severe winter in 1928/29.

Comparing all investigated sites, site 6 (Krasiczyn district) significantly differs from the others through the presence of double or even triple moon ring zones. The moon rings formed in 1928/29 and 1939/40 were observed in over 70% of the analysed samples, while the severe winter in 1955/56 caused moon rings in 20% of the trees.

Comparing the investigated sites it can clearly be seen that the winters of 1928/29 and 1939/40 caused most moon rings in the investigated oak trees. This finding is in agreement with results from other European sites. It is also interesting that the severe winter 1955/56 caused moon rings only in southern Poland but not in Germany and Russia. As a general conclusion, moon ring frequency in oak seems to increase from west to east in Europe, reaching its maximum at the natural ecological limits as well as on sites where oak has been introduced by humans.

Analysis of Subfossil and Historical Wood

In the sample set of subfossil oakwood and timber coming from various constructions and buildings in southern Poland 53 zones with moon rings have been identified. The percentage of samples showing moon rings is similar to the one observed in living trees (about 2%). However, it should be noted that this material coming from different time periods is much more diverse. For the past 1000 years about 2500 samples were available for this investigation.

The high frequency of moon rings for sub-periods of the 'Little Ice Age' is particularly remarkable. During that period and later, moon rings frequently formed in southern Poland, namely in the years 1422, 1460, 1555, 1736, 1794, 1827 and 1890. For Western Europe, Dujesiefken and Liese (1986) reported frequent formation of moon rings in the years 1736, 1794, 1827 and 1890 (Table 13.2). The 'winters of the century' of 1423/24 in Poland, and 1459/60 in Poland and the Baltic Sea basin, as well as of 1556/57 with frost from October through to end of March (Rojecki *et al.*, 1965), coincide with high moon ring frequencies.

Earlier, in medieval times, moon rings occur less frequently; only the years 1332 (NE Poland), 1186 (SE Poland) and 920 (NW Poland) were found.

Table 13.2. Years with moon rings in Poland (author's own results) and in other European countries (Dujesiefken and Liese, 1986).

Years of moon ring formation	Site
1955/56	Krasiczyn (SE Poland)
1939/40	Krasiczyn (SE Poland), Moscow (Russia), Baden-Württemberg, Bayern, Hessen (Germany)
1928/29	Pińczów, Niepołomice, Radymno, Sieniawa, Krasiczyn (S Poland); Wrocław (SE Poland), Niedersachsen (Germany)
1890/91 (1891/92?)	Niepołomice (S Poland), north Moscow (Russia)
1879/80	Blois, Festigny, Nancy, Vosges (France), Schleswig-Holstein (Germany)
1827/28 (1829/30?)	Niepołomice (S Poland)
1829/30	Blois, Festigny, Vosges (France), Bayern (Germany)
1798/99	Bayern (Germany)
1794/95	Bytom (S Poland), Blois, Festigny, Vosges (France),
1788/89	Blois, Festigny, Vosges (France),
1736/37 (1739/40?)	Krzyżanowice (S Poland)
1739/40	Bayern (Germany)
1708/09	Festigny, Nancy (France), Bayern (Germany)
1555/56 (1556/57?)	Żary (SW Poland)
1459/60	Wrocław (SW Poland)
1449/50	SE England
1438/39	SE England
1422/23 (1423/24?)	Branice (S Poland)
1332/33	Olsztyn (NE Poland)
1186/87	Czermno (SE Poland)
920/21	Wolin (NW Poland)
553/54	Wolica, Branice, Krzyżanowice (S Poland)
490/91	Wolica (S Poland)

However, available historical sources do not report extremely cold winters in those years.

Another maximum of moon ring zones was found in the 5th and 6th centuries. During that time, palaeogeographical changes occurred in most of the river valleys of southern Poland with intensive accumulation of oak stems in alluvial deposits. This indicates a phase of humidity and cooling in Poland and for the entire Central Europe (Kalicki and Krąpiec, 1996). Moon ring formations were dated to the time around AD 553 and also to around AD 490. AD 553 could be associated with climatic changes caused by a volcanic eruption as reported by Baillie (1994).

During the Hallstatt period (750–350 BC) another peak of frequent moon rings was observed. For that period, oak samples with moon rings came from southern Poland as well as from oaks from the Mazury area, NE Poland (see Fig. 13.2).

In the investigated historical constructions, logs from young oak trees were found, usually less than 100 years old and of 25–30 cm in diameter. Because the subsequent moon ring zones are often separated by only a few regular heartwood rings, they can be very helpful for dating of short sequences from those times (see Fig. 13.4). The moon ring zones in Mazury area (NE Poland) were dated to around 442 BC and 380 BC.

An earlier period marked by numerous oaks with moon ring zones was dated to around 1630 BC. Oak stems from those times were found in alluvial deposits of Vistula in Grabie near Cracow and accumulation of them in those

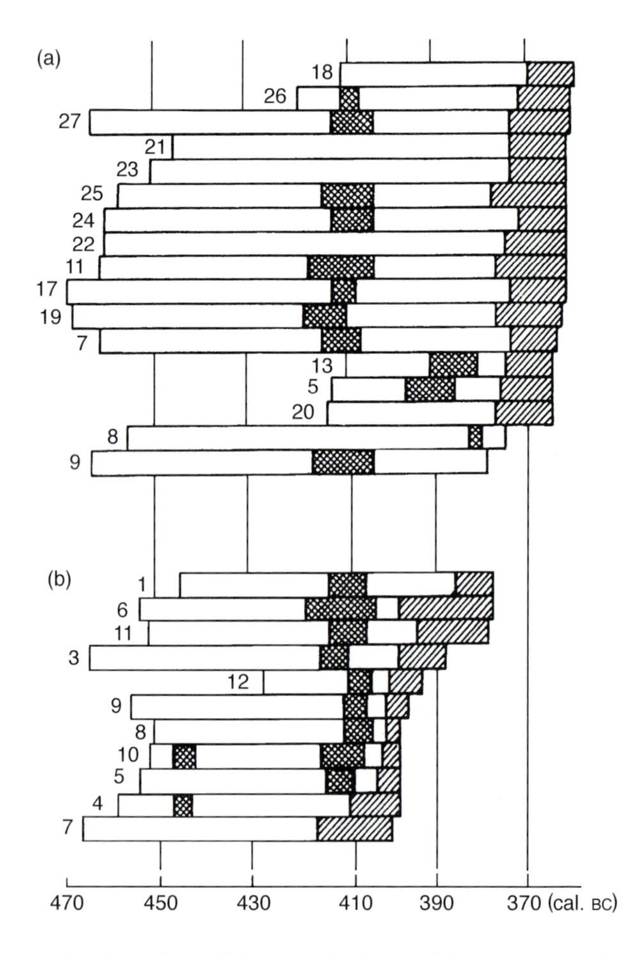

Fig. 13.4. Dendrochronological datings of oak wood from Ryn (a) and Bogaczewo (b) pile dwellings from Hallstatt period. Moon rings crosshatched, sapwood marked with diagonal lines.

sediments can be linked to a phase of cooling and humidity in Central Europe. The stems from Grabie exhibit traces of mechanical damage. This could be caused by ice rocks carried over by spring floods as a consequence of climatic anomalies after the huge Santorini volcano eruption (Baillie and Munro, 1988). No moon ring formations were observed in oak samples that were dated to the Holocene climatic optimum, the so-called Atlanticum. On the other hand, one moon ring zone was determined in a single sample representing the oldest oaks from southern Poland and dated to around 6700 BC. These oak trees migrated into the area after the regression of the glacier in response to climatic warming.

CONCLUSIONS

The presented study shows that moon rings can be found in oak wood representing the whole Holocene. Periods with increased moon ring frequencies are generally associated with periods of cool and humid climate, well known from palaeobotanic and palaeoclimatic studies. Extreme climatic events as caused by, for example, volcanic eruption could be another reason for moon ring formation. For the last millennium it is possible to identify exact dates of moon ring formations in oak wood from southern Poland. The findings generally coincide with results reported for other European areas (Germany, Russia, France). Therefore, the identification of moon rings may be useful for dendrochronological dating. The presented investigation is only a beginning of constructing moon ring chronologies for all of Europe during the whole Holocene. Collections of subfossil timber are available in various European laboratories and this resource could be used in the future.

SUMMARY

A type of discoloration in the heartwood of European oak is called ‘included sapwood’ or ‘moon rings’. These moon rings appear on cross-sections as bright zones within darker coloured heartwood. The presented study shows that moon rings can be found in oak wood representing the whole Holocene. Periods with increased moon ring frequencies are generally associated with cool and humid climate. The identification of moon rings might be a useful feature in dendrochronological dating. The results should encourage others to investigate moon rings in other regions.

ACKNOWLEDGEMENTS

The study presented here was financed from University of Mining and Metallurgy, Cracow, grant no. 10.140.79, and the chapter was finally written due to the convincing persuasion of Dr Rupert Wimmer, to whom the author is cordially thankful.

REFERENCES

Baillie, M.G.L. (1994) Dendrochronology raises questions about the nature of the AD 536 dust-veil event. *The Holocene* 4, 212–217.

Baillie, M.G.L. and Munro, M.A.R. (1988) Irish tree-rings, Santorini and volcanic dust veils. *Nature* 332, 344–346.

Bolychevtsev, V.G. (1970) Annual ring of oak as evidence of secular climatic cycles. *Lesovedenie* 1, 15–23 (in Russian).

Charrier, B., Janin, G., Haluk, J.P. and Mosedale, J.R. (1995) Colour and chemical characteristics of moon rings in oakwood. *Holzforschung* 49, 287–292.

Duhamel du Monceau, M. (1758) Des maladies des arbres. In: *La Physique des Arbres.* Guerin & Delatour, Paris, pp. 337–354.

Dujesiefken, D. and Bauch, J. (1987) Biologische Charakterisierung von Eichenholz mit Mondringen. *Holz als Roh- und Werkstoff* 45, 365–370.

Dujesiefken, D. and Liese, W. (1986) Vorkommen und Entstehung der Mondringe (*Quercus* spp.). *Fortwissenschaftliches Centralblatt* 105, 137–155.

Dujesiefken, D., Liese, W. and Bauch, J. (1984) Discoloration in the heartwood of oak-trees. *IAWA Bulletin n. s.* 5, 128–132.

Dzbeński, W. and Krutul, D. (1994) Fizyko-chemiczne właściwości drewna dębowego z wewnętrznym bielem. *Proceedings of the XVII Symposium 'Preservation of wood'*, PAN & SGGW, Rogów, pp. 127–134.

Erteld, W., Mette, H.J. and Achterberg, W. (1963) *Holzfehler in Wort und Bild.* VEB Fachbuchverlag, Leipzig, 78pp.

Henry, E. (1896) Sur la lunure ou double aubier du chene. *Bulletin de la Société des Sciences de Nancy* 14, 68–79.

Kalicki, T. and Krąpiec, M. (1996) Reconstruction of phases of the 'black oaks' accumulation and of flood phases. *Geographical Studies* (Special Issue) 9, 78–85.

Krąpiec, M. (1992) Skale dendrochronologiczne późnego holocenu południowej i centralnej Polski. *Kwartalnik AGH – Geologia* 18(3), 37–119.

Krąpiec, M. (1994) 'Czarne dęby' – dendrochronologia i fazy akumulacji pni w dolinie Wisły. In: Starkel, L. and Prokop, P. (eds) *Proceedings of the Conference 'Environmental Changes of the Carpathians and Subcarpathian Basins'*, Institute of Geography and Spatial Organization, Polish Academy of Science, Warszawa, pp. 57–68.

Krąpiec, M. (1996) Subfossil oak chronology (474 BC–AD 1529) from Southern Poland. In: Dean, J.S., Meko, D.M. and Swetnam T.W. (eds) *Proceedings of the International Conference 'Tree Rings, Environment and Humanity', Radiocarbon*, Tucson, pp. 813–819.

Krawczyk, A. and Krąpiec, M. (1995) Dendrochronologiczna baza danych. *Materials of 2nd Conference 'Computers in Scientific Research'*, Wrocław Scientific Society, Wrocław, pp. 247–249.

Krzysik, F. (1974) *Nauka o Drewnie.* PWN, Warszawa, 653pp.

Liese, J. (1942) Frostschäden an Eichen. *Mitteilungen der Deutschen Dendrologischen Geseltschsft* 55, 321–324.

Paclt, J. (1954) Über die Entstehung der Mondringigkeit der Eiche. *Phytopathologische Zeitschrift* 21, 210–213.

Paclt, J. (1989) Mondringe: für einheitliche Erfassung ihrer Entstehungsjahre. *Beiträge zur Biologie der Pflanzen* 64, 17–22.

Pearson, G.W. (1986) Precise calendrical dating of known growth – period samples using a 'curve fitting' technique. *Radiocarbon* 28, 2B, 911–934.

Rojecki, A., Girguś, R. and Strupczewski, W. (1965) *Wyjątki ze źródeł historycznych o nadzwyczajnych zjawiskach hydrologiczno-meteorologicznych na ziemiach polskich w wiekach od X do XVI.* Wyd. Komunikacji i Łączności, Warszawa, 214pp.

Scheiber, C. (1965) *Tropenhölzer*. VEB Fachbuchverlag, Leipzig, 398pp.

Tapper, M., Fletcher, J.M. and Walker, F. (1978) Abnormal small earlywood vessels in oak. Their use as chronological indicators and their relation to arrested heartwood formation (included sapwood) after certain cold winters. *British Archaeological Reports* 51, 339–342.

Trendelenburg, R. and Mayer-Wegelin, H. (1955) *Das Holz als Rohstoff*. Carl Hanser Verlag, München, 541pp.

14

Chronologies for Historical Dating in High Asia/Nepal

Burghart Schmidt, Thomasz Wazny, Kuber Malla, Elisabeth Höfs and Mitra Khalessi

INTRODUCTION

In the year 1985, during his first journey, the Tibetologist, D. Schuh from the University Bonn, Germany, explored the southern Mustang and the Muktinath Valley in search of historical records related to that area (Fig. 14.1). He was truly impressed by the numerous caves in the hills on the northern bank of the Dzong River. These caves, constructed by humans, can easily be seen by those who travel from the Kali Gandaki valley up to the sanctuary of Muktinath. In the following year, D. Schuh – accompanied by R. Bielmeier, C. Cueppers and B. Schmidt – started a preliminary survey of these cave-systems below the Dzong village with the goal of obtaining more information about these remnants of an old culture (Schuh, 1992–1993). One bigger cave had painted walls and wood that was suitable for tree-ring investigations. Therefore, this first survey also allowed a first trial of crossdating of historical timber from southern Mustang in Nepal.

In 1992, a research programme was started by the Nepalese Department of Archaeology and the German Research Foundation. This interdisciplinary project was initiated by Tibetologists and architectural historians. The project was designed by settlement archaeologists, historical settlement geographers and ethnologists, together with archeozoologists and researchers with backgrounds in applied photogrammetry and dendrochronology. One major goal of this 'Nepal–German Project on High-Mountain Archaeology' was the investigation of the settlement history (settlement processes), and the formation of states in the High Himalaya, characterized by Tibetan culture and tradition (Haffner and Pohle, 1993). For this reason the construction of a tree-ring chronology was performed by the Dendrochronological Laboratory at the University of Cologne.

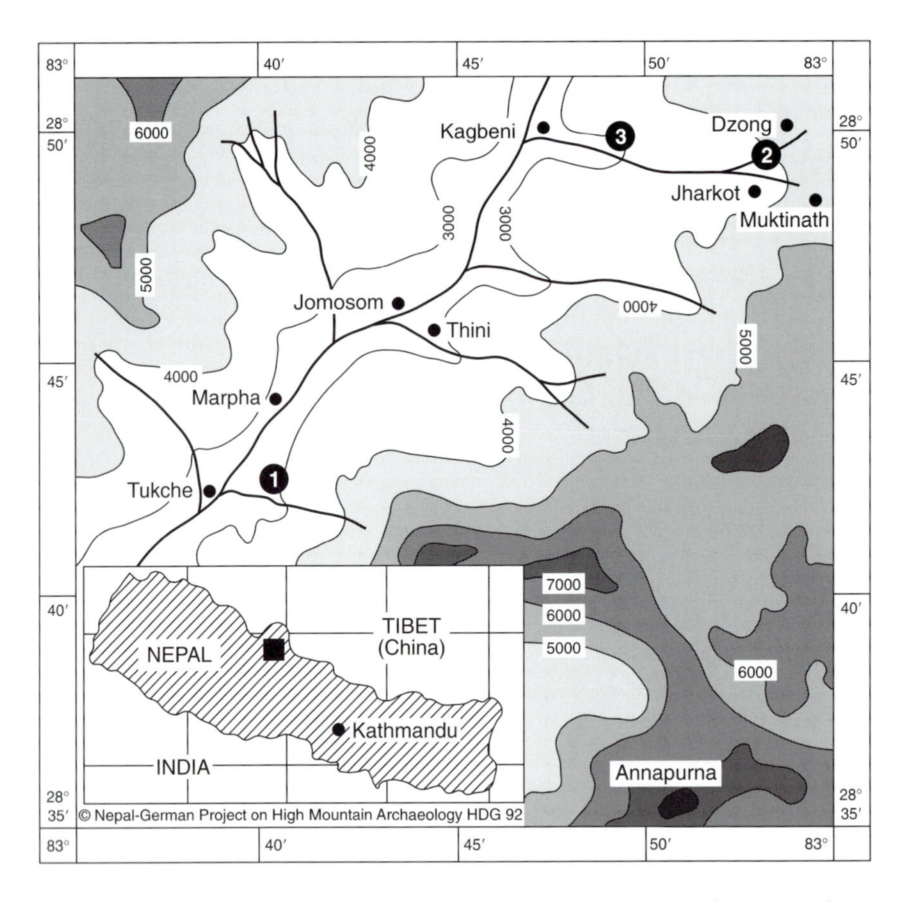

Fig. 14.1. Map of the southern part of Mustang district (numbers indicate sample locations).

MATERIAL AND METHODS

In previous trips to the area (1989 and 1992) about 400 wood samples were collected with the aim of constructing a tree-ring calendar for the Southern Mustang region. Samples were taken from living trees of nearby forests at Thini as well as from numerous houses and ruins between Marpha in the south and Dzakot (Jharkot) in the north. Because with the living pine tree samples from Thini we were only able to establish a chronology back to the year 1804, pillars from an old ruined monastery in Muktinath (felling year: 1906) were additionally sampled. The monastery samples extended our chronology back to 1768. With this basic chronology we were able to work further back, and a strong chronology back to 1455 has been established (Schmidt, 1992–1993).

Between 1993 and 1997, in cooperation with historians, another set of historical samples were taken from different structures. The objects and locations are listed in Table 14.1. During that time cores from about 350 living trees were taken from sites in the districts of North and South Mustang as well as from Manang and Khumbu.

The tree species used in this dendrochronological research were *Pinus wallichiana*, *Abies spectabilis* and *Picea smithiana*.

Investigations were carried out according to standard methods in dendrochronology. Details can be found elsewhere (e.g. Eckstein *et al.*, 1984; Schweingruber, 1988; Schmidt *et al.*, 1990). In order to improve working conditions in the field, a laboratory was established in Jomosom (2700 m). The preliminary results which we obtained on site helped us to continuously adjust our sampling strategies.

RESULTS AND DISCUSSION

A few hundred samples were taken from the ruins of the fortification Garab Dzong (Old Thini). The large number of dated timbers from the foundations of the ruin as well as from the other houses, monasteries and forts assure a well-replicated tree-ring chronology of Nepal that covers the period from AD 1324 to 1997.

The Manang district is located on the opposite side of the Annapurna range, at a distance of about 50 km. The longest site chronology from that area was obtained for Ngawal and spans 300 years from AD 1697 to 1996 (Table 14.2). The chronologies from Manang show high correlation with the

Table 14.1. Overview of the analysed historical samples from different objects and locations.

Objects	Location	Region	No. of samples
Castles	Djarkot	Mustang	30
	Kagbeni	Mustang	40
	Lupra	Mustang	10
	Ngawal	Manang	10
Monasteries	Djarkot	Mustang	10
	Kagbeni	Mustang	30
	Lupra	Mustang	15
	Braga	Manang	2
Houses	Djarkot	Mustang	90
	Khingar	Mustang	110
	Kagbeni	Mustang	120
Archaeological excavation	Garab Dzong	Mustang	900
	Muktinath valley	Mustang	35

Table 14.2. Chronologies of sites from the dry area of Mustang and the moister areas of Manang and Khumbu.

Region/site	Species	No. of trees	Period
North Mustang			
Tangbe	*Pinus wallichiana*	86	1850–1996
South Mustang			
Thini	*Pinus wallichiana*	40	1819–1993
Tukche	*Pinus wallichiana*	10	1890–1990
Manang			
Manang (north)	*Pinus wallichiana*	29	1738–1996
Manang (south)	*Pinus wallichiana*	13	1726–1996
Ngawal	*Pinus wallichana*	42	1697–1996
Pisang I	*Pinus wallichiana*	27	1738–1996
Khumbu			
Lamjura (north)	*Abies spectabilis*	25	1720–1997
Lamjura (south)	*Abies spectabilis*	18	1794–1997
Phakding	*Pinus wallichiana*	39	1919–1997
Monjo	*Pinus wallichiana*	22	1921–1997
Namche	*Pinus wallichiana*	30	1957–1997
Khumjung	*Abies spectabilis*	16	1901–1997
Tengpoche (north)	*Abies spectabilis*	16	1876–1997
Tengpoche (south)	*Abies spectabilis*	15	1911–1997
Thame	*Abies spectabilis*	22	1942–1997

chronologies from North and South Mustang. This let us conclude that historical timbers from Manang should also be datable with the standard chronology for Mustang (Fig. 14.2).

From the eastern part of Nepal, in the Khumbu area, we collected samples from Lamjura in the south up to Tengpoche in the north (Table 14.2). The growth patterns of these trees are less homogenous than in the drier area of Mustang (Fig. 14.3). The correlation of this site chronology with the one from Mustang/Manang (distance: about 200 km) is not significant.

Further comparisons with site chronologies from Nepal, established by Paul Krusic (Tree-Ring Laboratory, Columbia University) and Bhattacharyya *et al.* (1992) are planned to proof the dating-range of the Mustang calendar. In addition, dendroclimatic analysis of the South Mustang chronology is in progress.

From eastern Tibet, Bräuning (1994) has established chronologies for *Juniperus*, *Picea*, *Abies* and *Larix* for the purpose of dendroclimatological analysis (Bräuning and Lehmkuhl, 1996; Zimmermann *et al.*, 1997). More chronologies are available from Kashmir (Hughes and Davies,1987; Hughes, 1992) as well as Karakorum (Esper *et al.*, 1995).

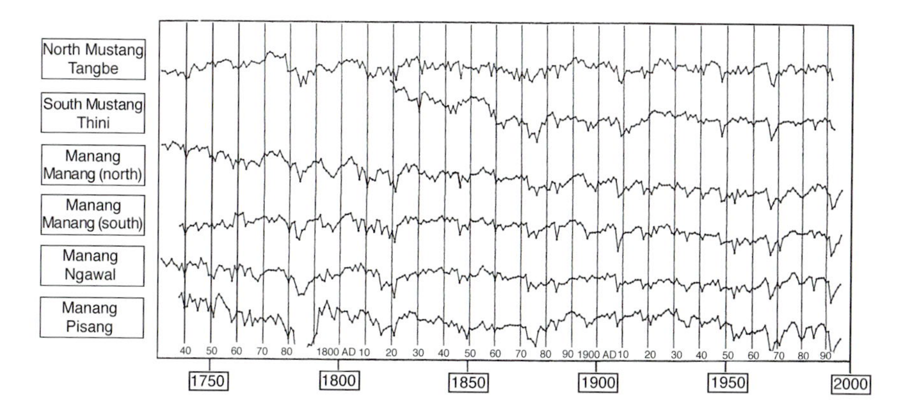

Fig. 14.2. Chronologies (ring widths) from North and South Mustang and from the Manang area. The growth patterns of both areas are highly correlated.

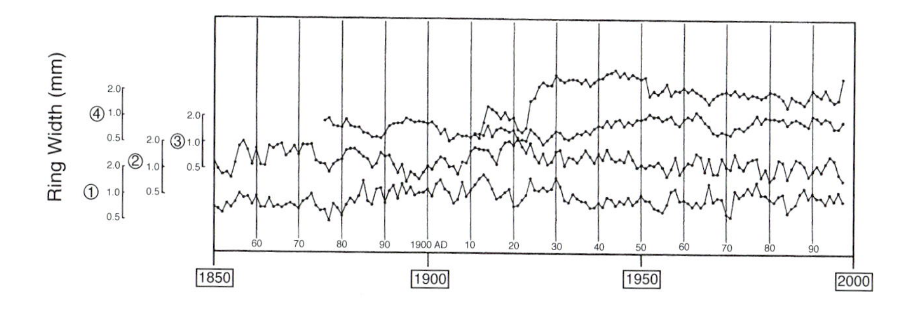

Fig. 14.3. Chronologies from the Khumbu area. 1: Lamjura (north), 2: Lamjura (south), 3: Tengpoche (north), 4: Tengpoche (south). The diagrams show the low similarity between site chronologies in this area.

CONCLUSIONS

In the closed area of South Mustang between Muktinath in the north and Tukche in the south of the Kali Gandaki valley, the dendrochronological results provide useful information about the history and dynamics of the local settlements as well as about the history of the local architecture, castles and monasteries along this old and famous trade route between Tibet and India.

In cooperation with the historians, more than 1700 samples were analysed which were taken from archaeological excavations, old houses, monasteries and castles. The large number of dated tree-ring series assure a good replication of this first tree-ring calendar of Nepal from AD 1324 to 1997.

With this tree-ring 'calendar' many historical objects from North Mustang and also from the Manang area can now be crossdated. Dendroclimatological studies are planned for the future to investigate the climate history and interactions with settlement dynamics of this high mountain area.

SUMMARY

In South Mustang, Nepal, dendrochronological results provide important information about the history and dynamics of the local settlements, local architecture, castles and monasteries along this old and famous trade route between Tibet and India. More than 1700 samples were analysed taken from archaeological excavations, old houses, monasteries and castles. A first master chronology was established for Nepal covering the time-span between AD 1324 to 1997.

REFERENCES

Bhattacharyya, A., LaMarche, V.C. and Hughes, M.K. (1992) Tree-ring chronologies from Nepal. *Tree-Ring Bulletin* 52, 59–66.

Bräuning, A. (1994) Dendrochronology for the last 1400 years in eastern Tibet. *GeoJournal* 34(1), 75–95.

Bräuning, A. and Lehmkuhl, F. (1996) Glazialmorphologische und dendrochronologische Untersuchungen neuzeitlicher Eisrandlagen Ost- und Südtibets. *Erdkunde* 50, pp. 341–359.

Eckstein, D., Baillie, M.G.L. and Egger, H. (1984) *Dendrochronological Dating*. Handbook for Archaeologists, No. 2. European Science Foundation, Strasbourg, 55pp.

Esper, J., Bosshard, A., Schweingruber, F. and Winiger, M. (1995) Tree-rings from the upper timberline in the Karakorum as climatic indicators for the last 1000 years. *Dendrochronologia* 13, pp. 79–88.

Haffner, W. and Pohle, P. (1993) *Settlement Processes and Formation of States in the High Himalayas Characterized by Tibetan Culture and Tradition*. Ancient Nepal, No. 134, Kathmandu.

Hughes, M.K. (1982) Global data base: Asia. In: Hughes, M.K., Kelly, P.M., Pilcher, J.R. and LaMarche V.C. Jr (eds) *Climate from Tree Rings*. Cambridge University Press, Cambridge, pp. 157–158.

Hughes, M.K. (1992) Dendroclimatic evidence from the Western Himalaya. In: Bradley, R.S. and Jones, P.D. (eds) *Climate Since AD 1500*. Routledge, London, pp. 415–431.

Hughes, M.K. and Davies, A.C. (1987) Dendroclimatology in Kashmir using tree ring width and densities in subalpine conifers. In: Kairiukstis, L., Bednarz, Z. and Feliksik, E. (eds) *Methods of Dendrochronology. East–West Approaches*. IIASA/Polish Academy of Sciences, pp. 163–176.

Schmidt, B. (1992–1993) *Dendrochronological Research in South Mustang*. Ancient Nepal, No. 130–133, Kathmandu.

Schmidt, B., Köhren-Jansen, H. and Freckmann, K. (1990) *Kleine Hausgeschichte der*

Mosellandschaft. Band 1 Dendrochronologie und Bauforschung. Schmidt, B. and Freckmann, K. (eds), Köln, 336pp.

Schuh, D. (1992–1993) *Introduction*. Ancient Nepal, No. 130–133, Kathmandu.

Schweingruber, F.H. (1988) *Tree Rings: Basics and Applications of Dendrochronology*. Reidel, Dordrecht, 276pp.

Zimmermann, B., Schleser, G. and Bräuning, A. (1997) Preliminary results of a Tibetan stable C-isotope chronology dating from 1200 to 1994. *Isotopes in Environmental and Health Studies* 33, 157–165.

E

Tree-Ring Analysis and Environmental Interactions

15

A Comparison between Repeated Timber Inventories and Dendrochronological Time Series for Forest Monitoring

Franco Biondi

INTRODUCTION

Forest monitoring relies on retrospective information to distinguish between 'normal' and 'abnormal' conditions in the status and trends of ecological parameters and processes (Franklin, 1989; Kolb *et al.*, 1994). Because of both its economic relevance and ease of measurement, growth is probably the parameter that national forest agencies, such as the USDA Forest Service, have monitored at the most detailed spatial scale for the longest time (e.g. Powell *et al.*, 1993). Identification of long-term patterns in forest growth is indeed a primary objective of basic and applied research, with direct implications in forest management and silviculture, as well as in ecological, palaeoclimatological and global change studies. Two ground-based sources of information on such patterns are (i) repeated forest inventories, targeted by forest mensurationists, and (ii) tree rings, targeted by dendrochronologists. Traditionally, mensurationists are interested in future, absolute, average values of stand growth, and have focused on quantifying and predicting timber resources in relation to site characteristics and management practices (Clutter *et al.*, 1983). On the other hand, dendrochronologists are interested in past, relative, year-to-year values of tree growth, and have concentrated on reconstructing environmental variables – precipitation, temperature, air pressure, streamflow, drought severity, sunshine, etc. – and dating past events – fire, insect outbreaks, volcanic eruptions, landslides, floods, windthrows, earthquakes, frosts, etc. – recorded in annual xylem layers of selected trees (Hughes *et al.*, 1982; Jacoby and Hornbeck, 1987; Bartholin *et al.*, 1992).

Relatively little research has focused on comparing and combining forest inventories and dendrochronological records. Differences in the mensurational and dendrochronological approaches have hindered communication between the two fields, even though the idea that wood growth provides 'absolute history' is so widespread and popular that it has been used for advertisement campaigns. As an example of methodological differences, mensurationists avoid computing the dimensionless ring indices used by dendrochronologists and prefer to convert ring widths into ring areas, or basal area increments, to quantify temporal trends (Hornbeck *et al.*, 1988). Still, Thammincha (1981) found good agreement between his annual ring-index series for southern Finland and the data from National Forest Inventories. He was also able to improve accuracy of stand growth predictions by including information on inter-annual variation derived from tree-ring data. Recently, chemical markers in tree rings have been targeted as potential indicators of forest health (Lewis, 1995). Even more encouraging, the French National Forest Office has adopted dendrochronological methods to provide information on the past history of forest stands included in its permanent plot network for forest ecosystem monitoring (Lebourgeois, 1997). In this chapter, I present a case study of how repeated forest inventories and tree-ring chronologies can augment and benefit each other for tackling scientific problems in the temporal domain.

MATERIAL AND METHODS

Study Area

The study was conducted on the Gus Pearson Natural Area, a 800 × 400 m permanent plot established in 1908 within unmanaged ponderosa pine (*Pinus ponderosa* Dougl. ex Laws. var. *scopulorum*) in north-central Arizona (Fig. 15.1). Ponderosa pine forests around Flagstaff occupy a crucial place in the history and development of modern dendrochronology. It was in these forests, in 1901, that A.E. Douglass began collecting wood specimens to test the hypothesis that annual tree growth, as measured by ring width, was influenced by climatic regime, and therefore could be used to reconstruct past climatic fluctuations (Douglass, 1909, 1914, 1919; Webb, 1983). At about the same time, in 1908, G.A. Pearson was sent to Arizona by the Forest Service to study natural regeneration of ponderosa pine (Pearson, 1942, 1950). In order to evaluate the impact of management practices on ecological processes, he established the permanent plot that now bears his name. Today, the Gus Pearson Natural Area is part of the Fort Valley Experimental Forest, within the Coconino National Forest.

Forest inventories in digital format were obtained from the US Forest Service. All pines with diameter at breast height (1.3–1.5 m; DBH) above 8.9 cm (3.5″) were first tagged and measured in 1920, and then remeasured every 5 years from 1920 to 1960, and every 10 years from 1960 to 1990. At every inventory, ingrowth (= pines whose DBH had exceeded the minimum value for tagging) was added to the database, and mortality (= pines that had

Fig. 15.1. Location of the study area. ★ = Gus Pearson Natural Area, Fort Valley Experimental Forest, northern Arizona, USA.

died) was recorded. The study area is divided into 29 100 × 100 m plots that were neither thinned nor burned during the period covered by the forest inventories. Spatial and temporal patterns of tree and stand growth derived from the repeated inventories have revealed a decline of individual tree growth over the 20th century (Biondi, 1994, 1996; Biondi *et al.*, 1994). Because of the rapid increase in stand density during the 20th century, large trees have experienced, on average, greater reduction in growth over time, and greater chances of dying, than small trees. During previous centuries, stand structure in southwestern ponderosa pine had been maintained at low density levels by frequent, low-intensity fire (Weaver, 1951; Cooper, 1960; Dieterich, 1980; White, 1985). Fire suppression programmes in the 1900s have allowed pine regeneration to grow into larger size classes, thereby occupying the space in and around clumps of much older trees, which experienced a relatively abrupt change in competition processes. The mechanisms hypothesized by Biondi (1994, 1996) for explaining growth and mortality patterns identified at the study area from 1920 to 1990 are summarized in Fig. 15.2.

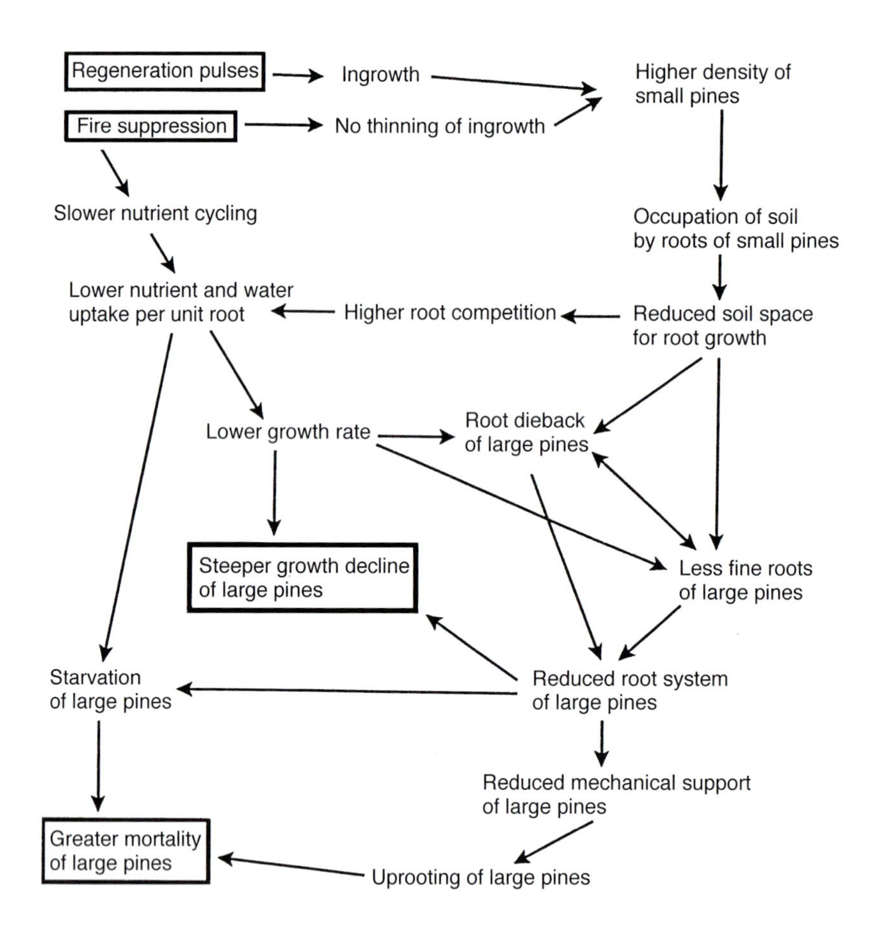

Fig. 15.2. Path diagram illustrating the most likely mechanisms behind the 20th century growth and mortality trends observed at the study area (for details, see Biondi *et al.*, 1994; Biondi, 1996).

Tree-Ring Data

Dendrochronological sampling consisted of two increment cores extracted from the stem of 58 tagged pines growing on the 29 plots. Two pines, one above and one below 50 cm DBH in 1990, were randomly selected on each plot in order to stratify the sample by two size classes. The 50 cm DBH cutoff was used to mimic the distinction between large ponderosa pines, called 'yellow pines' because of the orange and light brown colour of their outer bark, and small pines, called 'blackjacks' because of the dark grey, almost black colour of their outer bark (Schubert, 1974). Two increment cores were extracted from each tree about 1.0 m above ground level and 180° from each other whenever possible. Pines on plots 1–14, 16 and 27, i.e. those reported in Avery *et al.* (1976), were first sampled in 1987 (T.W. Swetnam, University of Arizona, 1990, personal communication), and later re-sampled to extend their tree-ring record until 1990. A total of 116 wood cores was collected between November 1990 and July 1991. Coordinates of each cored pine were obtained in the field using compass and measuring tape. At the laboratory, size of sampled pines was tested for spatial autocorrelation using geostatistical methods (Myers, 1991; Biondi *et al.*, 1994).

All increment cores were air dried, glued to wooden mounts after vertically aligning the xylem tracheids, mechanically sanded, then polished by hand with progressively finer sandpaper. Ring patterns were visually crossdated (Douglass, 1941; Stokes and Smiley, 1968) using a binocular microscope. Crossdating was independently verified by another researcher, then ring widths were measured to the nearest 0.01 mm by means of a sliding micrometer with a computer interface for data acquisition (Robinson and Evans, 1980). Dating accuracy was verified using the computer program COFECHA (Holmes, 1983). For comparison with forest inventories, every ring-width series was transformed into a ring-area series according to the procedure described by Phipps (1979). The two ring-area series from the same pine were averaged to quantify current annual basal area increment. The 10-year periodic basal area increment (PBAI) was computed as the 10-year sum of the current annual basal area increment.

Once crossdated rings are measured, it is necessary to reduce them into a manageable representation of short- and long-term historical patterns. Mean and variance of ring-width series are not stationary in time (Cook *et al.*, 1990a,b), and the removal of non-stationarity is called 'standardization' in the dendrochronological literature (Douglass, 1919; Schulman, 1956; Fritts, 1976; Hughes *et al.*, 1982; Cook and Kairiukstis, 1990). This procedure is intended to minimize growth variation due to phenomena acting at the individual tree or stand level. In the broadest sense, dendrochronological standardization could also be defined as the method used to combine all tree-ring samples into a single, average chronology. Four methods of producing an average tree-ring chronology were selected to represent the issues involved in growth trend detection. Separate tree-ring chronologies were developed for pines with 1990 DBH $\geq$ 50 cm and with 1990 DBH $<$ 50 cm. Standardization models were judged for their adherence to forest growth trends identified from forest inventories. Autoregressive modelling (Box and Jenkins, 1976; Biondi and Swetnam, 1987) was not considered in this chapter because emphasis was on interpreting, rather than removing, low-frequency variability. The methods can be summarized as follows:

$$\omega_t = [\oplus_i (w\, y_1^{-1}\, s^{-1})_{it}] + \alpha \qquad (15.1)$$

with ω_t = chronology value at year t; $\oplus_i$ = biweight robust mean (Mosteller and Tukey, 1977) of the i-values, $i = 1, \ldots, n_t$ (n_t is the number of measured specimens that included year t); w = crossdated ring width (mm); y_1 = modified negative exponential or straight line (Fritts *et al.*, 1969); s = cubic smoothing spline with 50% variance reduction at a 128-year period (Cook and Peters, 1981); α = difference between 1.000 and the arithmetic mean of the robust-mean chronology. These are the default options of the ARSTAN software program (Cook and Holmes, 1996).

$$\omega_t = n_t^{-1} \Sigma_i (w\, y_2^{-1})_{it} \qquad (15.2)$$

with Σ_i = summation of the i-values, i = 1, ... , n_t; y_2 = modified negative exponential or straight line with slope ≤0. This is the most widely known method, as first proposed by Douglass (1914), and later made popular by Fritts (1976).

$$\omega_t = n_t^{-1} \Sigma_i [\log (w_{it} + k) - y_{2\,it}] \qquad (15.3)$$

with k = constant added to avoid taking the logarithm (log) of zero. This method is based on an exponential model, as detailed by Biondi (1992). Taking the logarithm of ring widths is equivalent to plotting ring-width series on semi-logarithmic paper, a method commonly adopted by European dendrochronologists (Schweingruber, 1988, p. 51).

$$\omega_t = n_t^{-1} \Sigma_i a_{it} \qquad (15.4)$$

with a = ring area (cm^2) computed from w assuming a circular cross-section. This method was first proposed by Phipps (1979), and has been commonly applied in dendroecological studies (e.g. Peterson *et al.*, 1993).

RESULTS AND DISCUSSION

Sampled pines were not affected by spatial autocorrelation, because omni-directional sample variograms of stem DBH and of tree height fluctuated randomly with respect to distance. Increment cores were dated and measured up to ring-year 1990. Because most cores did not go back further than 1570, measured ring widths ranged from 1570 to 1990. The total number of measured rings was 20,197, of which 16,408 measured on large pines, and 4509 measured on small pines. No dating errors were identified after running the COFECHA program, even though some cores had low correlations with the rest of the sample. Running COFECHA on the combined data set (N = 116) or on each of the two data sets for large and small pines (N = 58 each) produced indistinguishable results in terms of dating.

The oldest sampled pine was a large dominant growing in a relatively open spot surrounded by thickets of small pines. DBH increased from 95.0 cm in 1920 to 105.4 cm in 1990. Rings formed earlier than 1600 were initially dated against the Flagstaff chronology (Douglass, 1940, 1947). For best results, some of the specimens used by Douglass himself to build his chronology were retrieved from the Laboratory of Tree-Ring Research collections. Dating was reliable back to 1487; I dated the first visible ring as AD 1380±5, hence pine 124 on plot 18 was older than 600 years in 1990. Since the first visible ring did not include the pith, and the two cores were not taken at ground level, the true age was at least 30–50 years greater. Maximum reported age of southwestern ponderosa pine is about 750 years (Swetnam and Brown, 1992).

Maximum ages of sampled ponderosa pines were greater than those reported by White (1985). White analysed a total of 236 pines, and found that only five of them were older than 400 years, with a maximum age of 406 years in 1980 (White, 1985). The pines he sampled were located on plots 1–3 and 6–11, and usually one core was taken from each tree. The year of the innermost, crossdated ring was used to define pine age, even though most cores did not include the stem pith (A.S. White, University of Maine, 1992, personal communication). In the present study, one out of 58 randomly selected pines was older than 600 years, and another 13 pines were older than 400 years. The old age of trees at the Gus Pearson Natural Area is even more remarkable if one considers the location and topography of the stand, on both sides of a major highway and on gently undulating terrain. Traditionally, the search for very old, climatically sensitive trees favours remote areas on steep, rocky slopes (Schulman, 1956; Swetnam and Brown, 1992).

Periodic basal area increment (PBAI) computed from increment cores showed trends parallel to those computed from forest inventory data (Fig. 15.3). Growth rates of large pines declined more than those of small pines. Based on forest inventories, large pines had larger PBAIs than small pines until the last decades, when the situation reversed. Based on tree rings, PBAIs of large pines remained larger than that of small pines throughout the 1900s (Fig. 15.3). Periodic basal area increment estimated from forest inventories was greater than that derived from tree rings, especially for pines with 1990 DBH < 50 cm. The identification of the cause for such systematic difference would require additional studies.

Different standardization options produced chronologies with different temporal trends (Figs 15.4 and 15.5). Tree-ring chronologies developed by adopting ARSTAN defaults did not show any decline over time; rather they displayed a rapid increase after about 1980. Since ring-width series ended in a low-growth period (Biondi, 1994), dividing ring width by a fitted curve value could artificially inflate the final part of the tree-ring chronology. A similar growth surge was evident in small pines when method (2) was used, but it was not present in the method (3) chronology (Fig. 15.5). For large pines, methods (2) and (3) produced extremely similar tree-ring chronologies, and replicated the growth decline seen in periodic basal area increment. Among standardization options that required fitting a growth curve to model the biological trend, fitting the curve to log-transformed ring widths and then computing ring indices as deviations from the growth curve was the most robust method with respect to end-series computational problems.

Temporal trends were correctly represented by the average of ring-area series, which highlighted the difference between large and small pines (Figs 15.4 and 15.5). The growth decline was steeper in large pines than in small pines, it levelled off after 1970, and, for small pines, began reversing in the last

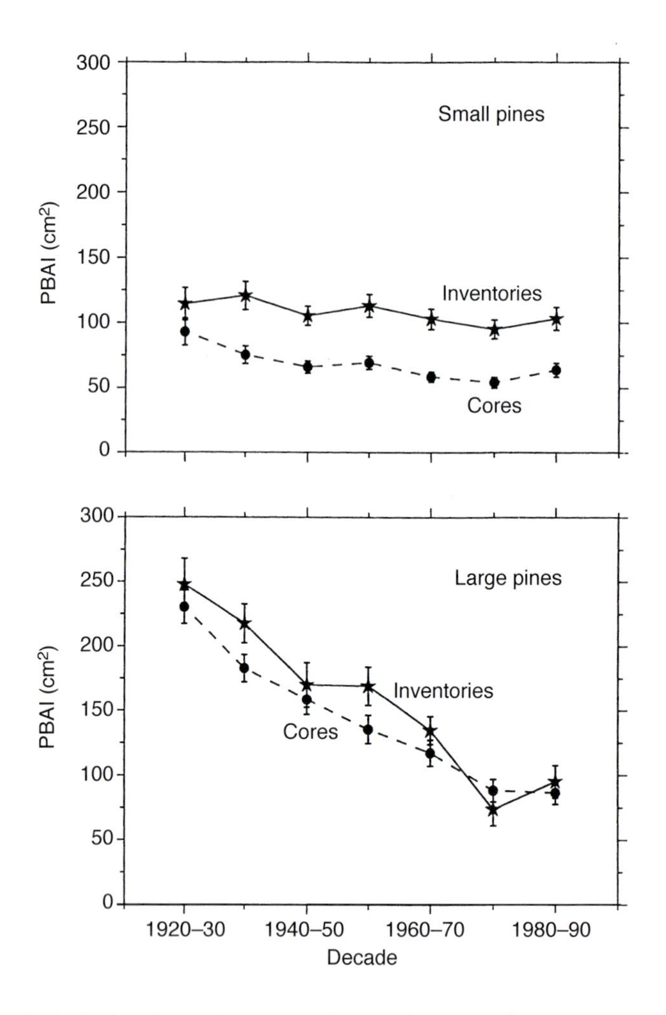

Fig. 15.3. Periodic basal area increment (PBAI) of pines with 1990 DBH above (large) and below (small) 50 cm. Standard error bars are also plotted to show variability and significance of parameter estimates. Values computed from repeated forest inventories (solid line) show similar fluctuations to, and usually exceed, those computed from increment cores (dashed line).

few years preceding 1990 (Fig. 15.5). These temporal trends were consistent with those identified from time-series plots of ring-width series (Biondi, 1994), and with trends detected graphically and statistically for periodic basal area increment of individual pines (Biondi, 1996). Therefore, inferences and comparisons of temporal trends for trees of different age/size classes should be based on ring-area series rather than ring-width series, presumably because ring-area series are less dependent on cambial age (Phipps, 1979). Considering

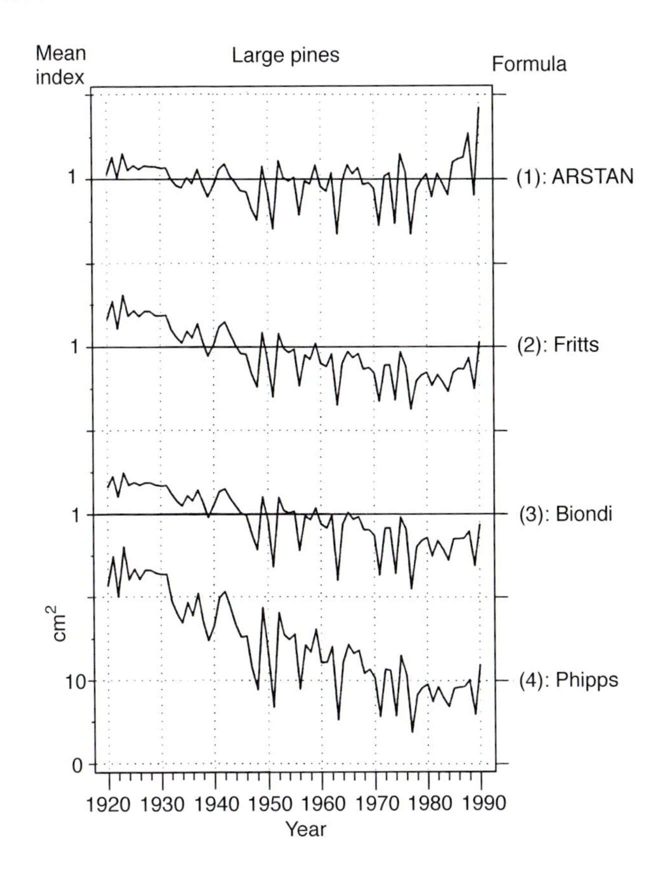

Fig. 15.4. Tree-ring chronologies from 1920 to 1990 based on tree-ring samples of pines with 1990 DBH above 50 cm (large pines). The four methods, from (1) to (4), used to compute the chronologies are detailed in the text.

the agreement between low-frequency growth patterns derived from forest inventories and from ring-area chronologies, the latter were chosen to represent tree growth at the study area over the past four centuries (Fig. 15.6). Large pines showed a unique growth surge in the early 1900s followed by a continued decline until about 1980. Small pines showed a less dramatic growth increase in the early 1900s followed by a less dramatic decline up to about 1980, when growth rates began increasing again.

CONCLUSIONS

Overall, the mensurational and dendrochronological approaches to forest monitoring have advantages and disadvantages, but the combination of

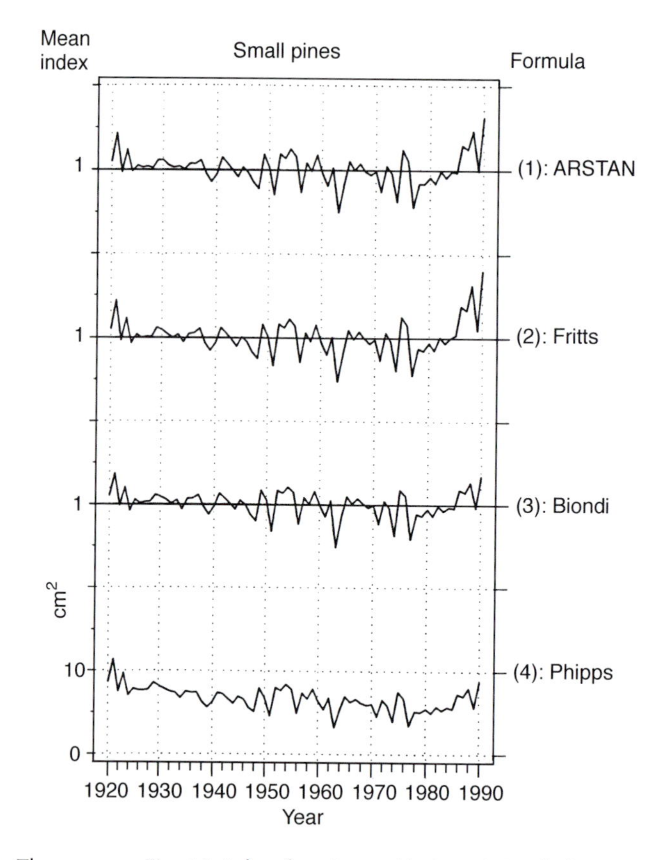

Fig. 15.5. The same as Fig. 15.4, but for pines with 1990 DBH below 50 cm (small pines).

tree-ring chronologies and repeated forest inventories is beneficial for detecting changes of forest growth at multiannual scales. Previous studies (Biondi, 1994, 1996) have shown that repeated forest inventories quantify growth of individual trees and of the entire stand, thus providing a complete picture, even in retrospective, of growth dynamics. It is then possible to ascertain whether there is a difference between growth trends of individual trees and of the forest stand they belong to. Ingrowth, the direct outcome of successful tree regeneration (Shifley *et al.*, 1993), and mortality, both of them highly variable processes, can be derived from the inventories. Furthermore, when combined with spatial information, forest inventories are a powerful tool to evaluate spatial processes, either alone or in combination with temporal ones (Biondi *et al.*, 1994). On the other hand, inventory data on permanent plots are usually collected at 5 to 10-year intervals. Hence, interpolation between successive inventories is needed to compute yearly growth values (Eriksson *et al.*, 1990), but, because of the

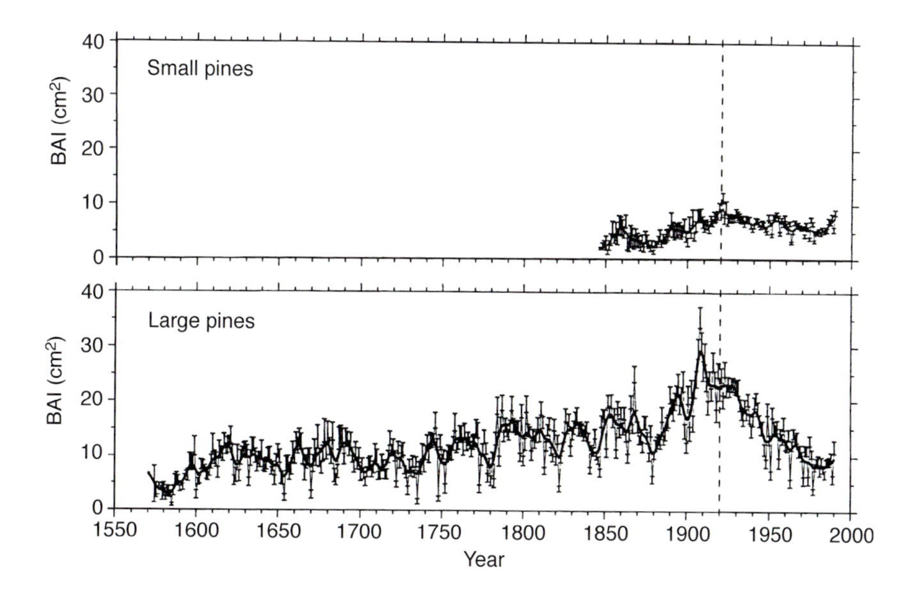

Fig. 15.6. Tree-ring chronologies for the whole period of available records, computed as the average of ring-area series. Standard error bars are plotted to show variability of annual values. It is evident that pines with 1990 DBH above 50 cm (large pines) were much older, and experienced a greater growth decline in the last century, than pines with 1990 DBH below 50 cm (small pines).

smooth pattern of interpolated annual growth rates, the effect of year-to-year environmental variation on tree growth remains undetected.

Compared to forest inventories, tree-ring records have limited spatial coverage, hence they cannot provide a complete picture of stand growth. On the other hand, forest inventories are restricted to the last century, but the longevity of tree species and the pace of stand dynamics in the mid-latitudes require a much longer record to disentangle variability at multiple time scales, from annual to decadal to centennial. For instance, the identification of 'abnormal' decline in tree radial growth has to be preceded by the definition of the 'normal', or expected, decline caused by the biological trend (Federer and Hornbeck, 1987; Fritts and Swetnam, 1989). Dendrochronological data can provide a long-term perspective on natural variability of individual tree growth, thereby giving managers additional information for evaluating recent trends. In this study, ring-area chronologies agreed with repeated forest inventories during their period of overlap, and then revealed that the establishment of the Gus Pearson Natural Area coincided with the largest peak of annual growth as well as with the beginning of the longest growth decline in the last four centuries (Fig. 15.6). When seen in historical perspective, the extreme growth decline experienced by large pines from 1920 to 1990 appears, in part, a return of growth rates to their long-term average preceding the growth surge of the early 1900s.

SUMMARY

Two ground-based sources of information on inter-annual to inter-decadal changes are (i) repeated timber inventories, and (ii) tree-ring chronologies. I present here a case study of how those two types of data can augment and benefit each other. At the Gus Pearson Natural Area, a ponderosa pine stand near Flagstaff (Arizona, USA), timber inventories were repeated by the US Forest Service from 1920 to 1990. The analysis of those data has revealed a decline of tree growth over the 20th century, as a reflection of increased stand density. Tree-ring data collected at the area after 1990 and spanning the last few centuries were compared to the inventory data for their ability to represent growth trends. Growth trends derived from ring-area chronologies agreed with repeated forest inventories, and revealed that decadal-scale growth rates in the 1900s have been anomalous. The mensurational and dendrochronological approaches to forest monitoring showed advantages (+) and disadvantages (−). Repeated forest inventories (+) quantified growth of individual trees and of the entire stand, thus providing a complete picture, even in retrospective; (−) had longer-than-annual resolution, and covered only the last decades. Dendrochronological data (+) quantified xylem growth of individual trees over their whole life span with annual resolution; (−) had limited spatial coverage, and led to different trends depending on the type of standardization option. Overall, the combination of tree-ring chronologies and repeated forest inventories is beneficial for detecting changes of forest growth at multi-annual scales.

ACKNOWLEDGEMENTS

I thank O.U. Bräker and two anonymous reviewers for their comments and suggestions.

REFERENCES

Avery, C.C., Larson, F.R. and Schubert, G.H. (1976) *Fifty-Year Records of Virgin Stand Development in Southwestern Ponderosa Pine*. USDA Forest Service General Technical Report RM-22.

Bartholin, T.S., Berglund, B.E., Eckstein, D. and Schweingruber, F.H. (eds) (1992) *Tree Rings and Environment*. LUNDQUA Report 34, Lund.

Biondi, F. (1992) Development of a tree-ring network for the Italian Peninsula. *Tree-Ring Bulletin* 52, 15–29.

Biondi, F. (1994) Spatial and temporal reconstruction of twentieth-century growth trends in a naturally-seeded pine forest. PhD dissertation, University of Arizona. University Microfilms International, Publication No. 9432850, Ann Arbor, Michigan.

Biondi, F. (1996) Decadal-scale dynamics at the Gus Pearson Natural Area: evidence for inverse (a)symmetric competition? *Canadian Journal of Forest Research* 26, 1397–1406.

Biondi, F. and Swetnam, T.W. (1987) Box–Jenkins models of forest interior tree-ring chronologies. *Tree-Ring Bulletin* 47, 71–95.

Biondi, F., Myers, D.E. and Avery, C.C. (1994) Geostatistically modeling stem size and increment in an old-growth forest. *Canadian Journal of Forest Research* 24, 1354–1368.

Box, G.E.P. and Jenkins, G.M. (1976) *Time Series Analysis: Forecasting and Control*, Revised Edn. Holden-Day, Oakland.

Clutter, J.L., Fortson, J.C., Pienaar, L.V., Brister, G.H. and Bailey, R.L. (1983) *Timber Management: A Quantitative Approach*. Wiley, New York.

Cook, E.R. and Holmes, R.L. (1996) ARSTAN: Chronology development. In: Grissino-Mayer, H.D., Holmes, R.L. and Fritts, H.C. (eds) *The International Tree-Ring Data Bank Program Library Version 2.0 User's Manual*. Laboratory of Tree-Ring Research, University of Arizona, Tucson, pp. 75–87.

Cook, E.R. and Kairiukstis, L.A. (eds) (1990) *Methods of Dendrochronology*. Kluwer, Academic Publishers, Dordrecht.

Cook, E.R. and Peters, K. (1981) The smoothing spline: a new approach to standardizing forest interior tree-ring width series for dendroclimatic studies. *Tree-Ring Bulletin* 41, 45–53.

Cook, E.R., Briffa, K., Shiyatov, S. and Mazepa, V. (1990a) Tree-ring standardization and growth-trend estimation. In: Cook, E.R. and Kairiukstis, L.A. (eds) *Methods of Dendrochronology*. Kluwer Academic Publishers, Dordrecht, pp. 104–123.

Cook, E.R., Shiyatov, S. and Mazepa, V. (1990b) Estimation of the mean chronology. In: Cook, E.R. and Kairiukstis, L.A. (eds) *Methods of Dendrochronology*. Kluwer Academic Publishers, Dordrecht, pp. 123–152.

Cooper, C.F. (1960) Changes in vegetation, structure, and growth of ponderosa pine forest since white settlement. *Ecological Monographs* 30, 129–164.

Dieterich, J.H. (1980) *Chimney Springs Forest Fire History*. USDA Forest Service Research Paper RM-220.

Douglass, A.E. (1909) Weather cycles in the growth of big trees. *Monthly Weather Review* 37, 225–237.

Douglass, A.E. (1914) A method of estimating rainfall by the growth of trees. In: Huntington, E. (ed.) *The Climatic Factor*. Carnegie Institution, Washington, pp. 101–121.

Douglass, A.E. (1919) *Climatic Cycles and Tree Growth*. Carnegie Institution, Washington.

Douglass, A.E. (1940) Estimated ring chronology, 150–1934 A.D. *Tree-Ring Bulletin* 6(4), 39 (insert).

Douglass, A.E. (1941) Crossdating in dendrochronology. *Journal of Forestry* 39, 825–831.

Douglass, A.E. (1947) Photographic tree-ring chronologies and the Flagstaff sequence. *Tree-Ring Bulletin* 14, 10–16.

Eriksson, M., Ek, A.R. and Burk, T.E. (1990) A framework for integrating forest growth and dendrochronological methodologies. *Forest Simulation Systems, Proceedings of the IUFRO Conference*. University of California, Division of Agriculture and Natural Resources, Bulletin 1927, pp. 165–173.

Federer, C.A. and Hornbeck, J.W. (1987) Expected decrease in diameter growth of even-aged red spruce. *Canadian Journal of Forest Research* 17, 266–269.

Franklin, J.F. (1989) Importance and justification of long-term studies in ecology. In: Likens, G.E. (ed.) *Long-Term Studies in Ecology: Approaches and Alternatives*. Springer-Verlag, New York, pp. 3–19.

Fritts, H.C. (1976) *Tree Rings and Climate*. Academic Press, London.

Fritts, H.C. and Swetnam, T.W. (1989) Dendroecology: A tool for evaluating variations in past and present forest environments. In: Begon, M., Fitter, A.H., Ford, E.D. and MacFadyen, A. (eds) *Advances in Ecological Research*, Vol. 19. Academic Press, London, pp. 111–188.

Fritts, H.C., Mosimann, J.E. and Bottorff, C.P. (1969) A revised computer program for standardizing tree-ring series. *Tree-Ring Bulletin* 29(1–2), 15–20.

Holmes, R.L. (1983) Computer-assisted quality control in tree-ring dating and measurement. *Tree-Ring Bulletin* 43, 69–78.

Hornbeck, J.W., Smith, R.B. and Federer, C.A. (1988) Growth trends in 10 species of trees in New England, 1950–1980. *Canadian Journal of Forest Research* 18, 1337–1340.

Hughes, M.K., Kelly, P.M., Pilcher, J.R. and LaMarche, V.C. Jr (eds) (1982) *Climate from Tree Rings*. Cambridge University Press, Cambridge.

Jacoby, G.C. Jr and Hornbeck, J.W. (compilers) (1987) *Proceedings of the International Symposium on Ecological Aspects of Tree-Ring Analysis*. US Department of Energy, CO_2 Conf-8608141, Washington, DC.

Kolb, T.E., Wagner, M.R. and Covington, W.W. (1994) Concepts of forest health: Utilitarian and ecosystem perspectives. *Journal of Forestry* 92, 10–15.

Lebourgeois, F. (1997) *RENECOFOR – Etude Dendrochronologique des 102 Peuplements du Réseau*. Office National des Forêts, Département des Recherches Techniques, Fontainebleau (France).

Lewis, T.E. (ed.) (1995) *Tree Rings as Indicators of Ecosystem Health*. CRC Press, Boca Raton.

Mosteller, F. and Tukey, J.W. (1977) *Data Analysis and Regression*. Addison-Wesley, Reading.

Myers, D.E. (1991) Interpolation and estimation with spatially located data. *Chemometrics and Intelligent Laboratory Systems* 11, 209–228.

Pearson, G.A. (1942) Herbaceous vegetation a factor in regeneration of ponderosa pine. *Ecological Monographs* 12, 315–338.

Pearson, G.A. (1950) *Management of Ponderosa Pine in the Southwest*. USDA Forest Service Agricultural Monograph No. 6.

Peterson, D.L., Arbaugh, M.J. and Robinson, L.J. (1993) Effects of ozone and climate on ponderosa pine (*Pinus ponderosa*) growth in the Colorado Rocky Mountains. *Canadian Journal of Forest Research* 23, 1750–1759.

Phipps, R.L. (1979) Simulation of wetlands forest vegetation dynamics. *Ecological Modelling* 7, 257–288.

Powell, D.S., Faulkner, J.L., Darr, D.R., Zhiliang, Z. and MacCleery, D.W. (1993) *Forest Resources of the United States, 1992*. USDA Forest Service General Technical Report RM-234. Fort Collins, CO.

Robinson, W.J. and Evans, R. (1980) A microcomputer-based tree-ring measuring system. *Tree-Ring Bulletin* 40, 59–64.

Schubert, G.H. (1974) *Silviculture of Southwestern Ponderosa Pine: The Status of Our Knowledge*. USDA Forest Service Research Paper RM-123.

Schulman, E. (1956) *Dendroclimatic Changes in Semiarid America*. University of Arizona Press, Tucson.

Schweingruber, F.H. (1988) *Tree Rings: Basics and Applications of Dendrochronology*. Reidel, Dordrecht.

Shifley, S.R., Ek, A.R. and Burk, T.E. (1993) A generalized methodology for estimating forest ingrowth at multiple threshold diameters. *Forest Science* 39, 776–798.

Stokes, M.A. and Smiley, T.L. (1968) *An Introduction to Tree-Ring Dating*. University of Chicago Press, Chicago.

Swetnam, T.W. and Brown, P.M. (1992) Oldest known conifers in the southwestern United States: temporal and spatial patterns of maximum age. In: Kaufmann, M.R., Moir, W.H. and Bassett, R.L. (technical coordinators), *Old-Growth Forests in the Southwest and Rocky Mountain Regions*. USDA Forest Service General Technical Report RM-213, pp. 24–38.

Thammincha, S. (1981) Climatic variation in radial growth of Scots pine and Norway spruce and its importance in growth estimation. *Acta Forestalia Fennica* 171, 1–57.

Weaver, H. (1951) Fire as an ecological factor in the southwestern ponderosa pine forests. *Journal of Forestry* 49, 93–98.

Webb, G.E. (1983) *Tree Rings and Telescopes*. University of Arizona Press, Tucson.

White, A.S. (1985) Presettlement regeneration patterns in a southwestern ponderosa pine stand. *Ecology* 66, 589–594.

16

Tree-Ring Patterns in an Old-Growth, Subalpine Forest in Southern Interior British Columbia

Roberta Parish, Joseph A. Antos and Richard J. Hebda

INTRODUCTION

Variation in the width of annual tree rings provides insights into the forces affecting forest development. Tree-ring characteristics provide a sensitive record of the annual interaction of external forces, such as disturbance and climate, with stand factors, such as species and competition. Because many factors affect tree-ring width and these factors interact in complex ways, it may be difficult to decipher which primary factors control radial growth, thus hindering the use of ring patterns for interpreting past climate and stand history. Most studies that attempt to reconstruct past climate variation have used relatively isolated trees in extreme environments where the effects of stand dynamics are minimized (cf. Fritts, 1976). Recently, however, dendroecology has been used successfully to separate climate response from competition (McLaughlin *et al.*, 1987; Piutti and Cescatti, 1997, Chapter 17, this volume), to characterize natural disturbance regimes (e.g. Stewart, 1986; Frelich and Lorimer, 1991; Veblen *et al.*, 1991a, 1994), and to distinguish natural from anthropogenic disturbances (Savage, 1991; Cherubini *et al.*, 1996; Payette *et al.*, 1996). Tree-ring patterns contain a wealth of information, and dendroecological analyses of trees in closed-canopied stands offer the possibility of examining the effects of climate, disturbance and competition on tree growth and stand development, provided that the effects of these factors on tree growth can be disentangled.

In subalpine forests, the harsh climate and short growing seasons should result in strong climate signals. However, disturbance and competition can also have major impacts on tree growth. Within the widespread, subalpine spruce–fir forests of western North America, disturbance by both wildfire and bark beetles is common (Baker and Veblen, 1990; Veblen *et al.*, 1991a,b, 1994), although long periods lacking major disturbance can occur. For example, in climatically wet areas of the interior of British Columbia, some subalpine forests have been without stand-destroying disturbances for up to 500 years (R. Willis, Weyerhaeuser Canada, unpublished data). Climate has changed since these stands originated. Temperature variations during the 'Little Ice Age', *c.*1450 to 1890 (Crowley, 1996) are well documented. Tree establishment, survival and growth are tempered through a climatic filter. Forests that established 200 to 500 years ago did so under a climate regime different from the present, and these changes have combined with disturbance and among-tree interactions to determine tree growth patterns.

Understanding the natural dynamics of subalpine forests and the factors affecting tree growth within these forests is an important management issue because subalpine forests in the interior of British Columbia are currently being harvested at an unprecedented rate. Forest managers are being asked to design harvesting systems that mimic natural disturbance regimes without adequate knowledge of the dynamics of these stands. The application of dendrochronological procedures to examine factors affecting these forests is therefore timely. As a component of our overall study of the dynamics of subalpine forests in British Columbia, here we interpret tree-ring patterns in an old-growth spruce–fir forest to disentangle the effects of disturbance and climate on tree growth. We also use pollen analysis of forest humus to clarify stand history.

METHODS

Site Description

The study area (50°49′ N, 119°54′ W) is located at Sicamous Creek in the Hunters Range on the Caribou Plateau in southeastern British Columbia. The bedrock of the region is granitic gneiss and the soils are sandy loam. The site is on a north-facing slope between 1550 and 1750 m elevation. No high-elevation weather stations occur near the site. Glacier, about 115 km from the site, is the closest high-elevation station and Salmon Arm, about 25 km from the site, is the closest station (Table 16.1).

The forest has many old-growth features: numerous dead standing trees, gaps of different sizes, and fallen decaying logs. Subalpine fir (*Abies lasiocarpa* (Hook.) Nutt.) and Engelmann spruce (*Picea engelmannii* Parry ex Engelm.) are the only trees present. Fir is more abundant than spruce and comprises 87.5% of stems ≥ 1.3 m tall (2292 stems) and 75% of the live canopy basal area (34.4 $m^2 ha^{-1}$). The site has a deciduous shrub layer 1–2 m tall and a well-developed herb layer.

Table 16.1. Location and 1951–1980 normals for the meteorological stations. Coefficients of the time series models are not significantly different between the two stations but the constants are significantly different at $P < 0.001$.

Longitude	Latitude	Longitude	Elevation (m)	Daily normal temperature (°C)	Total annual precipitation (mm)	Time period	Time series model of temperature*
Salmon Arm	50°42′N	119°15′W	506	7.6	533.7	1911–1992	$5.164 + 0.119_{t_1} - 0.328_{t_6} + 0.542_{t_{12}}$
Glacier	51°14′N	117°29′W	1241	2.6	1320.8	1908–1952	$1.608 + 0.104_{t_1} - 0.351_{t_6} + 0.529_{t_{12}}$

* where t_1 is the temperature at lag 1 (i.e. preceding or following month), t_6 is temperature at lag 6 (months) and t_{12} is temperature at lag 12 (months).

Field Sampling

We used discs from a stand being logged instead of increment cores because (i) our overall objectives included examining age structure and spatial pattern, which requires obtaining ages from as many trees as possible, and (ii) discs can provide more information than cores. In 1994, a 50×50 m plot was located in each of six small cutblocks within the stand. Trees ≥ 1.3 m tall were tagged and species recorded. The blocks were logged in winter 1994 and discs were cut from the base of 1048 sound tagged trees in summer 1995.

Pollen Sampling

An undisturbed block of the topmost 30 cm of forest humus exposed in a road cut in an unharvested part of the forest was taken. High-resolution samples for pollen analysis were taken of the top 10 cm segment at 0.5 or 1 cm levels in the laboratory. Samples of about 1 cm^3 were treated according to standard palynological methods (Faegri and Iversen, 1975) and mounted in glycerine jelly. For each sample at least 300 pollen and spores were identified and counted with reference to the pollen and spore collection at the Royal British Columbia Museum. Pollen and spore data were graphed using TILIA and TILIA-GRAPH software packages (Grimm, 1991–1993, 1991).

Disc Preparation and Chronology Building

Discs were air-dried and the lower side planed, sanded and dated following Stokes and Smiley (1968). Annual ring widths along one to three radii were measured to the nearest 0.01 mm using a Measu-chron digital positiometer. The computer program COFECHA (Holmes, 1983) was run to detect measurement and crossdating errors.

Tree-ring chronologies were constructed for both species. We chose, where possible, trees that grew steadily rather than suppressed and released trees. In steadily growing trees, missing rings were rare except for those associated with scarring. Many old spruce had small linear scars from 1862 and 1863. Destructive sampling of some discs revealed pitch accumulations and, in some cases, galleries typical of spruce bark beetle (*Dendroctonus rufipennis*) (L. Safranyik, Canadian Forestry Service, personal communication). Crossdating with COFECHA and visual inspection showed some trees lacked rings for as many as 6 years along radii adjacent to scars. To accommodate these missing years, the tree-ring record was split at this point and two sequential portions were analysed from the same radius.

All of our old firs were suppressed before *c.* 1850 and the statistics calculated by COFECHA from annual ring measurements showed strong intra- and inter-tree correlations from around 1850 to the present but very poor correlation before 1800. Poor correlations are typical of trees whose growth is not affected by major climatic factors (cf. Fritts, 1976), such as subcanopy individuals that are suppressed by neighbouring trees. Thus, we truncated the chronology at 1845. In contrast we were able to extend the spruce chronology back 330 years to the mid-1600s.

The computer program ARSTAN was used to produce index chronologies for both species (Cook and Holmes, 1984). We were also interested in determining whether trees that established at different times, under different stand and climatic conditions, responded similarly to climate patterns. Thus, in addition to the chronology for each species, we produced separate chronologies for the oldest trees (>250 years old) and young canopy trees (100–150 years old) for the purpose of examining relationships with climate data.

Growth Release

Growth releases were used to identify past canopy disturbances. We examined one radius from each of the 1048 trees. We defined release as a doubling of mean ring width sustained for at least 10 years. Because trees may take from 1 to 3 years to realize the increase, the year before the first measurable increase was defined as the year of release.

Climatic Analyses

Mean monthly temperature, total monthly precipitation (snow and rainfall), and total monthly snowfall were obtained from the Glacier and Salmon Arm meteorological stations (Environment Canada, 1994). Time series models using lag 1 (one month), lag 6 (six months), and lag 12 (annual) as explanatory variables provided a good fit (adjusted $r^2 > 0.9$) for the temperature data at each station (Table 16.1). Because the fitted regression coefficients were normally distributed and their estimators based on a large data sample, we were able to use z-tests to compare the coefficients of the two models. The constant was significantly different ($P < 0.001$) between models, which is consistent with the expectation of an overall temperature difference between stations differing 700 m in elevation. However, there were no significant differences between stations for the coefficients of the model parameters, which suggests that the two stations had similar temperature patterns despite lower temperatures at Glacier than Salmon Arm. Thus we will report on temperature data only from Salmon Arm, which is not only closer to the site but has a longer record than Glacier. A similar approach was applied to the precipitation data. In this case, time series models provided a poor fit to precipitation data (adjusted $r^2 < 0.5$) and showed that precipitation patterns at the two stations were different. Because of this difference, we examined the relationship between annual growth and precipitation and snowfall separately for each station.

Two types of analyses were used to explore the relationship of annual growth and climate: time series and response function analyses (Fritts *et al.*, 1991). Time series analysis, using the SAS ARIMA procedure (SAS Institute Inc., 1993), was used to compare climatic variables to annual ring indices generated from the ARSTAN program. Monthly mean temperature, total monthly precipitation and snowfall, as well as cumulative winter snowfall (November to April), were first detrended and prewhitened to remove any autocorrelation. The resulting residuals were then cross-correlated with the residuals from the prewhitened ring-index from the standard ARSTAN chronology, using the separate chronologies for young and old trees of each species. Detrending and prewhitening simplify the interpretation of the cross-correlogram by eliminating the possibility of significant cross-correlation resulting from the autocorrelation of one or both series (Nemec, 1996). Cross-correlations were considered significant at an approximately 5% level of significance if more than two standard errors from zero (SAS Institute Inc., 1993).

Response function analysis in the PRECON program (version 5.14) (Fritts *et al.*, 1991) uses principal components of climatic data to determine the major axis of variation to relate monthly temperature, precipitation and tree-ring index. Statistical confidence intervals in PRECON are determined by the bootstrap method (Guiot, 1991). Response function analysis reduces the effects of interdependencies among climatic variables but the results often depend on the choice of growing period and the inclusion of lag effects. We used the standard ARSTAN chronology to test for lag effects and the residual chronology, which is comparable to the residual of the prewhitened chronology, to examine climatic response. A 17-month growing period (June of the previous year to October of the current year) using mean monthly temperature and total monthly precipitation (34 variables) was selected.

RESULTS

Pollen analysis showed that this site has been fir–spruce forest for several centuries (Fig. 16.1). The weakly developed charcoal layer at 4–5 cm depth indicates a fire within the stand. The high percentage of non-arboreal pollen (NAP) types such as *Poaceae* and *Tubuliflorae* (*Asteraceae*) associated with the charcoal and the reduced fern understorey suggest a major wildfire. Immediately following the fire, spruce became more abundant than fir (Fig. 16.1).

The oldest trees were 337 years old and both spruce and fir of this age were measured. The spruce population comprised two quite distinct cohorts, one *c.* 260–340 years old, from the time of stand initiation, and the second *c.* 100–140 years old, dating from the mid to late 1800s (Fig. 16.2). Fir regenerated throughout the history of the stand, although a major recruitment episode occurred in 1855–1900, concurrent with the second spruce cohort (Fig. 16.2).

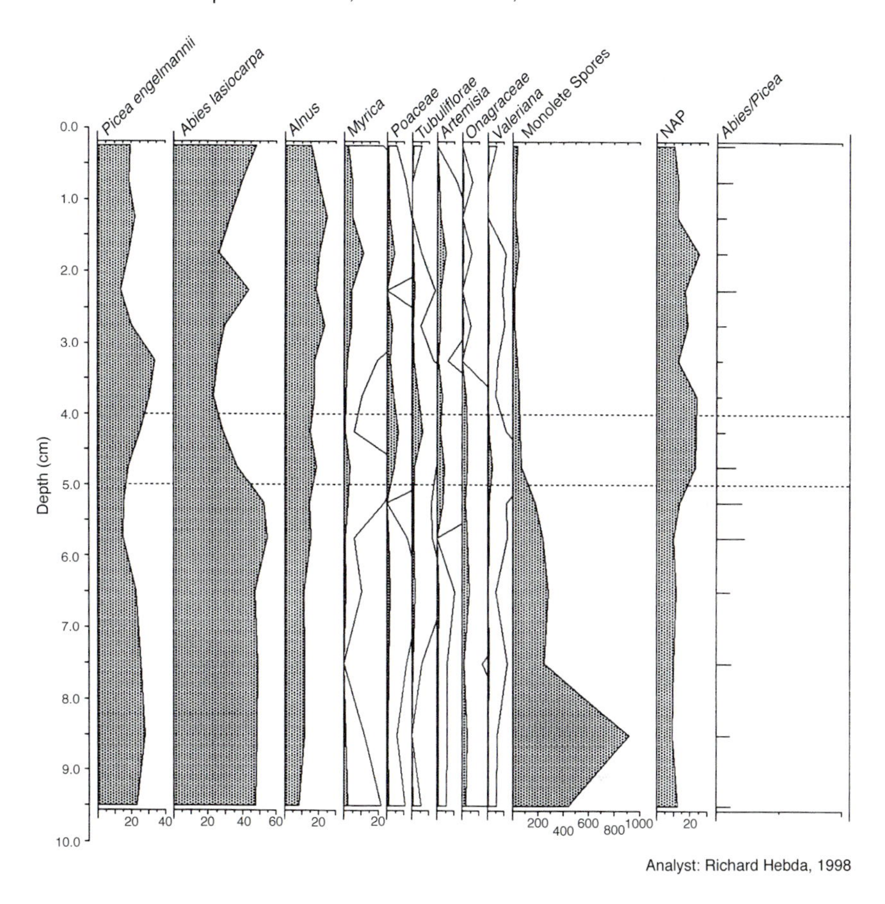

Fig. 16.1. Pollen and spore diagram for Sicamous Creek. A diffuse charcoal horizon occurred at 4–5 cm depth between the two dashed lines in the diagram. Monolete fern spores are shown but excluded from the sum of non-arboreal pollen (NAP) types.

Both species had suppressed individuals that were able to release. No release was recorded until *c.* 1840, after *c.* 200 years of stand development. The number of releases peaked *c.* 1860–70 and again *c.* 1930 (Fig. 16.3). A small number of trees (15) released twice: some twice within the first disturbance period and the remainder then and again in the 1930s.

COFECHA indicated an overall series correlation of 0.545 for spruce and 0.527 for fir. Both series had high autocorrelation values (Table 16.2), which suggests that growth in any given year is strongly influenced by growth in the preceding year. The PRECON results also indicate a significant lag effect of 3 years for spruce ($r^2 = 0.83$) and 2 years for fir ($r^2 = 0.65$).

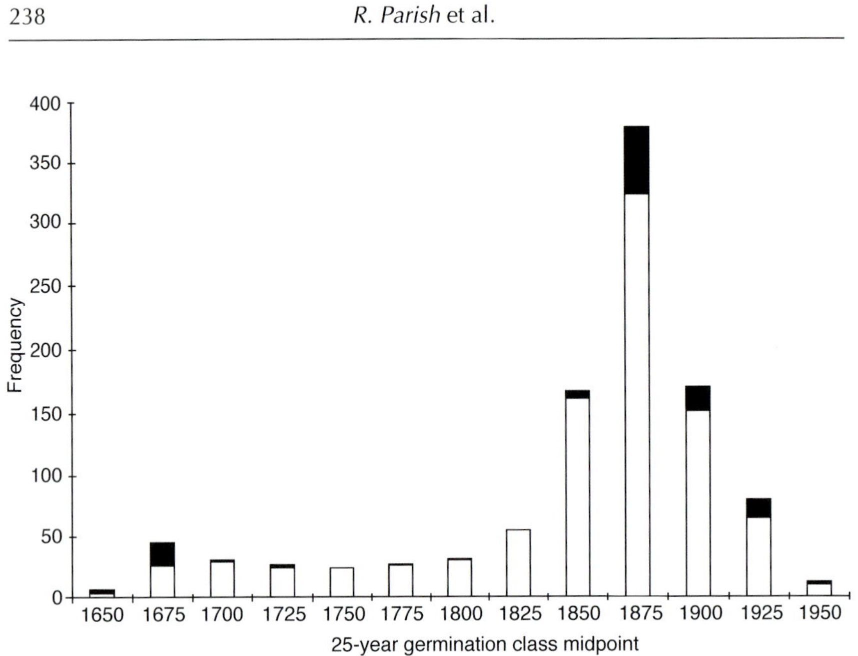

Fig. 16.2. Time of germination in 25-year classes of Engelmann spruce (black) and subalpine fir (white) ≥1.3 m tall.

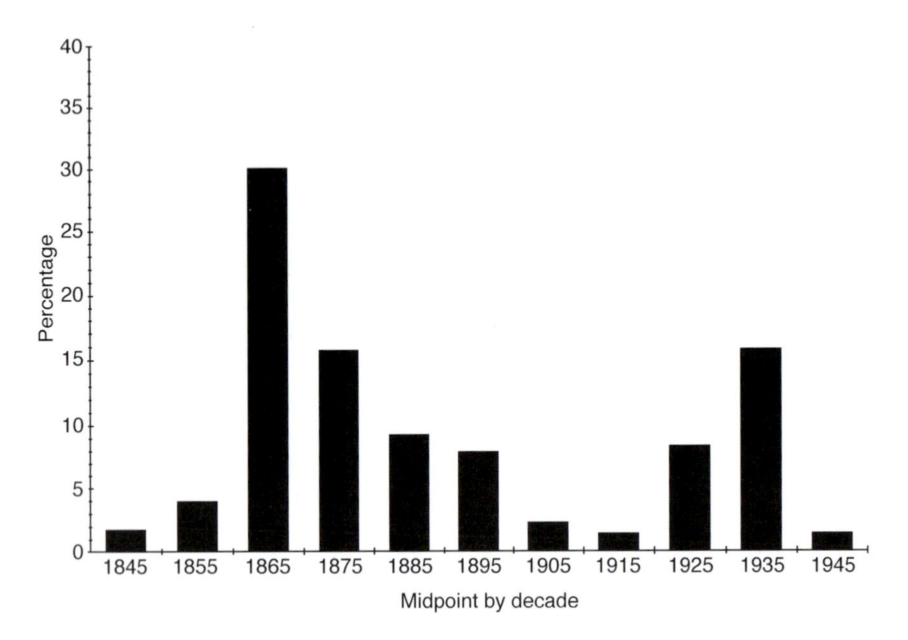

Fig. 16.3. Percentage of released Engelmann spruce and subalpine fir, by decade, in the period 1840 to 1950.

Table 16.2. Standard chronology statistics for Engelmann spruce and subalpine fir using the ARSTAN computer program.

	Engelmann spruce	Subalpine fir
Interval (years)	330	150
Number of trees/radii	25/41	45/63
Mean sensitivity	0.127	0.112
Standard deviation	0.184	0.197
First-order autocorrelation	0.618	0.744
Common interval (years)	98	94
Number of trees/radii	21/28	40/52
Signal to noise ratio	10.56	15.67
Variance first eigenvector	39.6	32.6
Inter-tree correlation	0.35	0.28
Intra-tree correlation	0.53	0.71

The climate–growth response of spruce and fir dating from time of stand initiation (*c*. 1650–1750) and from second cohort (*c*. 1850–1900) were very similar (Fig. 16.4). Time series analysis revealed no significant cross-correlation of either total monthly snowfall or winter snow accumulation with the growth of either species. Neither time series analysis nor PRECON showed a significant relationship between the annual growth indices of spruce or fir and total monthly precipitation data from either weather station. Temperature, however, had a significant effect on the annual growth of both species (Fig. 16.4). The most consistent relationship was a pronounced negative impact on current year growth from past summer temperatures. Both methods of analysis demonstrated significant relationships for the preceding July and August temperatures for both young and old trees of each species (Fig. 16.4). In contrast, temperatures during the current growing season and the preceding winter generally had a positive effect on growth, especially of spruce. For spruce, cross-correlations and response function coefficients were mostly significant for January, March, April, June and October; patterns were similar for fir but fewer cross-correlations and response function coefficients were significant (Fig. 16.4).

Because of the similarity of climate response of young and old trees, a single master chronology for each species was used to examine temporal patterns of growth. Spruce showed good growth in the latter half of the 1700s followed by slow growth through the early 1800s from *c*.1800–1840, with especially narrow rings at the end of that period in 1836–1838 (Fig. 16.5). In several spruce, the ring in 1802 was very narrow with deformed and crushed cells, suggesting frost damage. Both spruce and fir showed enhanced growth after 1870 and from *c*. 1933 to 1950. Slow growth generally occurred during the 1920s and from the 1950s to the 1970s.

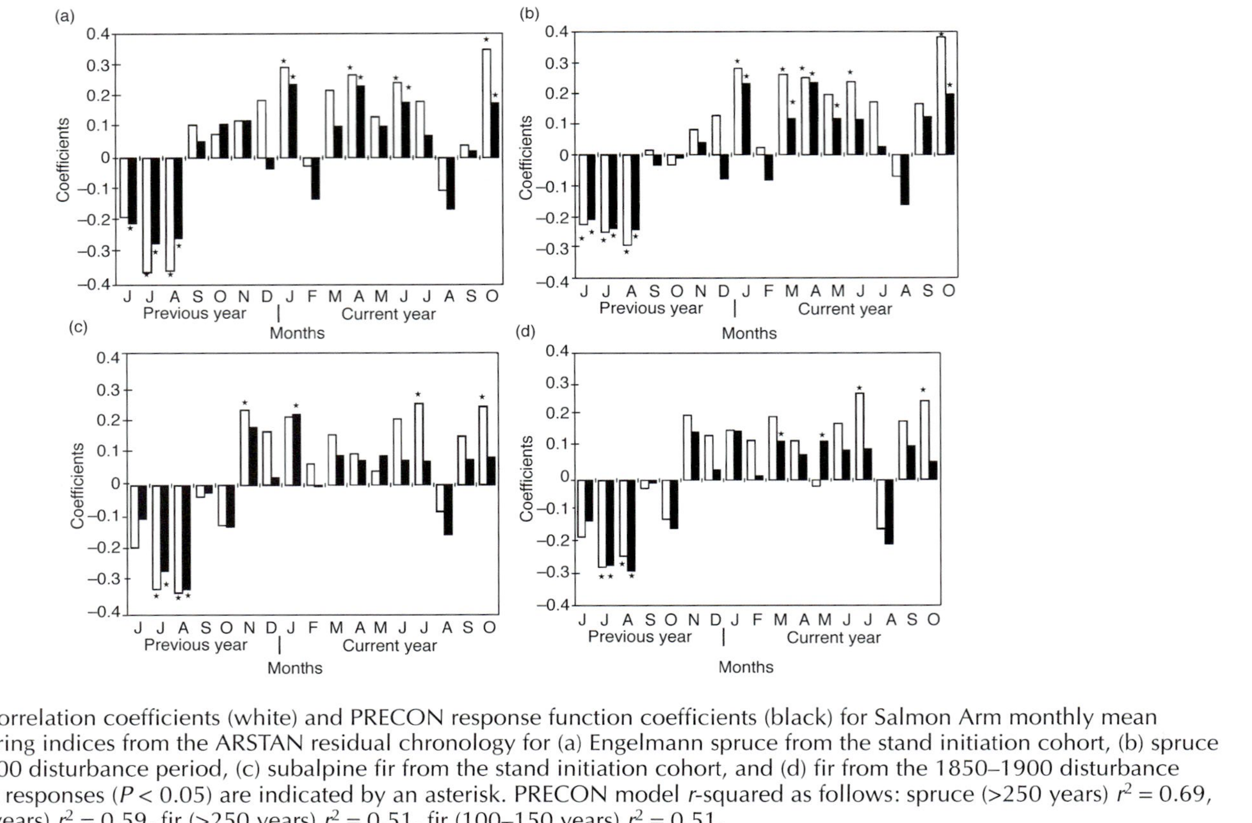

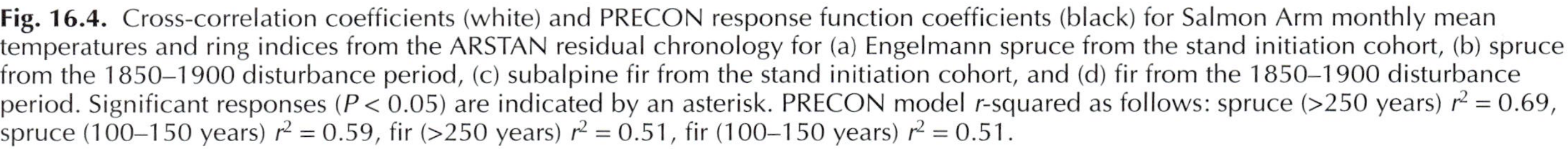
Fig. 16.4. Cross-correlation coefficients (white) and PRECON response function coefficients (black) for Salmon Arm monthly mean temperatures and ring indices from the ARSTAN residual chronology for (a) Engelmann spruce from the stand initiation cohort, (b) spruce from the 1850–1900 disturbance period, (c) subalpine fir from the stand initiation cohort, and (d) fir from the 1850–1900 disturbance period. Significant responses ($P < 0.05$) are indicated by an asterisk. PRECON model r-squared as follows: spruce (>250 years) $r^2 = 0.69$, spruce (100–150 years) $r^2 = 0.59$, fir (>250 years) $r^2 = 0.51$, fir (100–150 years) $r^2 = 0.51$.

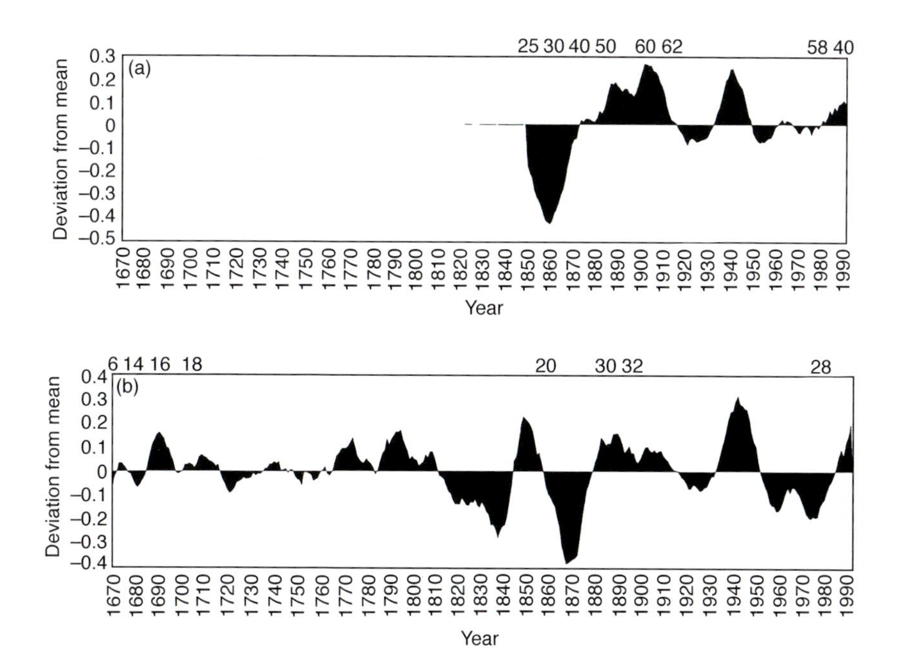

Fig. 16.5. Tree-ring chronologies smoothed with a 10-year moving average, shown as deviations from the mean for (a) subalpine fir and (b) Engelmann spruce. The number of radii used in the chronology are indicated at the top of the chart.

DISCUSSION

The tree-ring record showed clearly the influence of external forces on the development of this old-growth forest. A stand initiation date in the 1650s is supported by the consistent maximum age of 337 years found for both species in different parts of the forest. This age is well under the potential maximum age of 500 to 600 years for Engelmann spruce (Alexander and Shepperd, 1990). The presence of charcoal on the surface throughout the site confirms that the site was burned within the past few centuries. Age structure and pollen analysis suggest that the resulting forest remained sufficiently open for the first 80 years for spruce to establish freely.

No widespread disturbance was recorded in the tree-ring record for 200 years after stand initiation. After 1840, the canopy opened sufficiently to release existing trees. The many trees released and the rapid growth of a second cohort of both species indicates that canopy mortality was very high during the 1860s and 1870s; 75% of the present canopy trees released or established during that time. A light to moderate period of disturbance occurred *c.* 1927 to 1936 that released trees but failed to open the canopy sufficiently to allow new germinants to grow rapidly.

The presence of healed scars and beetle galleries in 1862 and 1863 suggests spruce bark beetle as the cause of canopy mortality and the subsequent release of spruce and fir. Spruce bark beetle has caused major damage to Engelmann spruce stands throughout much of the species range and can kill most canopy spruce in a stand (Alexander and Shepperd, 1990; Baker and Veblen, 1990; Veblen *et al.*, 1991b). The pollen record shows, however, that fir was abundant throughout the stand history so that it is probable that the balsam bark beetle (*Dryocoetes confusus*), which attacks and kills mature fir, also caused mortality. Heavy canopy removal probably resulted from the combined action of two host-specific insects. In contrast, the less intense disturbance of the 1930s was the result of balsam bark beetle, which was reported to be present at that time (Koot and Hodge, 1992). Spruce bark beetle was not recorded in the area in the 1930s (Erickson, 1987) and no spruce show scars from that period.

Bark beetle outbreaks are usually associated with large-scale windthrows or other accumulations of slash, although endemic populations infest downed, injured, diseased or otherwise stressed trees (Lane and Goheen, 1979; James and Goheen, 1981; Safranyik *et al.*, 1983). Blowdown may have been responsible for initiating the disturbances by bark beetles, but we have no data to address this issue. Climate warming in the late 1800s and in the 1930s could also have influenced bark beetle dynamics. Warm summers and mild winters are known to benefit beetle reproduction and survival and, in the presence of suitable host material, increase levels of attack (Massey and Wygant, 1954; Safranyik *et al.*, 1990). The present outbreak of spruce bark beetle in Alaska has been related to unprecedented climate warming (Holsten and Burnside, 1997).

Spruce and fir in the Hunters Range in southern British Columbia show growth patterns similar to those reported at other high-elevation sites (e.g. Colenutt and Luckman, 1991), at high latitudes (e.g. Jacoby and Cook, 1981; D'Arrigo *et al.*, 1992) and from the Pacific Northwest (e.g. Fritts and Lough, 1985; Wiles *et al.*, 1996). These chronologies generally indicate that growth was good during the 1700s but decreased severely during the early 1800s. The onset of enhanced growth at the end of the 1800s depends on location. Prolonged warming at high latitudes began in the 1840s through to the 1940s (Jacoby and D'Arrigo, 1989). A warming trend reconstructed in the 1870s for the Pacific coast (Wiles *et al.*, 1996) coincides with the period of enhanced spruce and fir growth *c.* 1877 to 1893 at our site. In contrast, reconstructions based on latewood density (Briffa *et al.*, 1992) report 1870 to 1900 as a cold period in British Columbia and the Pacific Northwest. The Fritts and Lough (1985) Pacific Northwest reconstruction, however, shows warming in the 1860s and 1870s followed by cooling in the early 1900s before warming again in the 1930s, a pattern more closely resembling tree growth at the Hunters Range site. Regional temperature data for the southern British Columbia mountains from 1895 to 1992 show trends of cooling from about 1900 to the late 1910s, warming to the 1940s, cooling to the 1970s and subsequent warming (Findlay *et al.*, 1994). The trends closely match the 20th century pattern of spruce and fir growth at our site.

Graumlich and Brubaker (1986) attributed slow growth on coastal high-elevation sites to cool, wet winters, which lead to high snow accumulations. Peterson and Peterson (1994) found that growth of spruce and subalpine fir at coastal locations was negatively correlated to the depth of the spring snowpack. Our lack of relationship could be a function of the absence of direct snow measurements from the Hunters Range; however, snow levels in the Interior ranges are less than in coastal mountains and may not limit growth.

The history of a tree significantly affects its growth, and tree-ring growth response to climate has been shown to relate to tree age (Szeicz and MacDonald, 1994). However, we found no evidence that old trees (>250 years) and young trees (100–150 years) responded differently to weather patterns during the 1900s. Once in the canopy, neither age nor past history appear to have major influences. All of the fir and many of the spruce in the old cohort were initially suppressed in a subcanopy position, for up to 200 years in some cases, until released by the disturbance in the mid-1800s. The 100–150-year-old trees, which were used in our chronology, had not gone through a period of suppression. Therefore, trees that had and had not been suppressed responded similarly to climate variation – a point worthy of consideration by forest managers.

Time series analysis and PRECON gave very similar results; cross-correlations and response function coefficients between growth and climate variables were generally similar, although significance sometimes differed (see also Fritts and Wu, 1986). This is encouraging and indicates that either procedure can be used with similar outcomes. However, prudence would indicate that use of both can reinforce interpretations. Perhaps with other data sets outcomes would be less similar; this warrants further investigation.

The most consistently significant relationship between ring width and temperature is the negative growth response to the previous August temperature (see also Peterson and Peterson, 1994; Villalba *et al.*, 1994; Ettl and Peterson, 1995). The basis for this response is most likely related to the reproductive cycle of spruce and fir. In spruce and true firs, both seed- and pollen-cone bud development occurs from mid-July until autumn (Owens and Molder, 1984, 1985). Cones remain dormant over the winter until late April to early May at high elevations. Pollination usually takes place in May and seeds are mature by September. Warm temperatures in mid-July through August, during the late stage of slow shoot elongation, promote cone differentiation (Ross, 1985). Woodward *et al.* (1994) found that the size of subalpine fir crops was positively related to good radial growth in the previous 2 years but growth rings were narrow in the cone crop year. Both spruce and fir are known for the variability of their annual seed production (Alexander and Shepperd,1990; Alexander *et al.*, 1990).

The strong negative relationship in both spruce and fir between current year growth and previous July and August temperatures, which signalled the allocation of resources to reproduction at the expense of vegetative growth, must be factored into models of tree growth. This supports the conclusion of Graumlich (1993) that differentiation in the timing of seasonal warmth is needed to predict species growth accurately. Another implication is the potential impact of high temperatures associated with global warming. In the 1917–1940 and 1965–1980 warming trends, April and May had the greatest changes in temperature (Jones and Kelly, 1983). These are the months with which spruce and fir have a positive temperature association but, if a shift occurs to warm July and Augusts, will this encourage reproduction at the expense of radial growth?

Tree-ring patterns in the forest we studied are strongly influenced both by climate, which has not only direct effects on diameter growth but also indirect effects via changes in resource allocation, and disturbance, which has major effects on tree growth via canopy tree mortality. Clearly, dendrochronology is a useful tool in examining the factors affecting tree growth and stand dynamics in subalpine forests. Understanding the history of these forests is essential to making the most informed management decisions possible. Tree rings provide a window into the history of a stand and their analysis provides a powerful tool for the elucidation of the importance of past events.

SUMMARY

Patterns of radial increment, timing of tree establishment and pollen analysis were used to elucidate the effects of two external forces, disturbance and climate, on the growth of Engelmann spruce and subalpine fir in an old-growth forest. The stand was initiated by a major disturbance, probably fire, in the 1650s. Abrupt increases in radial increment (release), indicate a major disturbance by bark beetles in the mid-1800s and a minor one in the 1930s. However, spruce and fir that established after the disturbance in the mid-1800s had a similar radial growth response to climate as did trees dating from stand origin.

Annual radial growth was significantly related to temperature, but not to precipitation or snow accumulation. Warm temperatures during the current year generally had a positive effect on growth of both spruce and fir, as would be expected for subalpine forests. However, July and August temperatures in the previous year had a significant negative effect on growth, probably because warm late summer temperatures promote reproductive bud initiation at the expense of vegetative growth. Thus the effects of temperature on resource allocation can greatly modify the direct relationship between growth and resource acquisition, an important consideration when examining the effects of weather patterns on growth. The results of this study clearly illustrate the utility of examining both disturbance history and climate patterns when studying tree-ring patterns and what they reveal about factors affecting tree growth.

ACKNOWLEDGEMENTS

Funding for this work was provided by Forest Renewal BC. Special thanks to University of Victoria students, K. Brown, C. Davidson, D. Gillan, N. Goudie and B. Small for technical assistance and to V. Sit, Research Branch, BC Ministry of Forests, for statistical advice. Dr I. Walker, Okanagan University College, helped collect humus samples. The helpful and constructive comments of two anonymous reviewers improved this manuscript. We are grateful to Dr R. Alfaro of the Canadian Forest Service for the use of his laboratory to measure annual rings and to Dr L. Safranyik of the Canadian Forest Service for identifying beetle galleries.

REFERENCES

Alexander, R.R. and Shepperd, W.D. (1990) *Picea engelmannii* Parry ex Engelm. In: Burns, R.M. and Honkala, B.H. (technical coordinators) *Silvics of North America: I Conifers*. USDA Forest Service, Agricultural Handbook 654, pp. 187–203.

Alexander, R.R., Shearer, R.C. and Shepperd, W.D. (1990) *Abies lasiocarpa* (Hook.) Nutt. In: Burns, R.M. and Honkala, B.H. (technical coordinators) *Silvics of North America: I Conifers*. USDA Forest Service, Agricultural Handbook 654, pp. 60–70.

Baker, W.L. and Veblen, T.T. (1990) Spruce beetles and fires in the nineteenth-century subalpine forests of western Colorado, USA. *Arctic and Alpine Research* 22, 65–80.

Briffa, K.R., Jones, P.D. and Schweingruber, F.H. (1992) Tree-ring density reconstructions of summer temperature patterns across western North America since 1600. *Journal of Climate* 5, 735–754.

Cherubini, P., Piussi, P. and Schweingruber, F.H. (1996) Spatiotemporal growth dynamics and disturbances in a subalpine spruce forest in the Alps: a dendroecological reconstruction. *Canadian Journal of Forest Research* 26, 991–1001.

Colenutt, M.E. and Luckman, B.H. (1991) Dendrochronological investigation of *Larix lyallii* at Larch Valley, Alberta. *Canadian Journal of Forest Research* 21, 1222–1233.

Cook, E. and Holmes R.L. (1984) *User Manual for Program ARSTAN*. Laboratory of Tree-Ring Research, University of Arizona, Tucson.

Crowley, T.J. (1996) Remembrance of things past: greenhouse lessons from the geologic record. *Consequences* 2, 2–11.

D'Arrigo, R.D., Jacoby, G.C. and Free, R.M. (1992) Tree-ring width and maximum latewood density at the North American tree line: parameters of climatic change. *Canadian Journal of Forest Research* 22, 1290–1296.

Environment Canada (1994) *Canadian Monthly Climate Data and 1961–1990 Normals*. Ottawa (CD-ROM).

Erickson, R.D. (1987) *Maps of the Major Forest Insect Infestations: Kamloops Forest Region 1912–1986*. FIDS Report 87–8, Canadian Forestry Service, Victoria, 68pp.

Ettl, G.J. and Peterson, D.L. (1995) Growth response of subalpine fir (*Abies lasiocarpa*) to climate in the Olympic Mountains, Washington, USA. *Global Change Biology* 1, 213–230.

Faegri, K. and Iversen, J. (1975) *Textbook of Pollen Analysis*, 2nd edn. Hafner Publishing Co., New York.

Findlay, B.F., Gullet, D.W., Malone, L., Reycraft, J., Skinner, W.R., Vincent, L. and Whitewood, R. (1994) Canadian national and regional annual temperature departures. In: Boden, T.A., Kaiser, D.P., Sepanski, R.J. and Stoss, F.W. (eds) *Trends '93: A Compendium of Data on Global Climate Change.* ORNL/CDIAC-65. Carbon Dioxide Information Analysis Center, Oak Ridge National Laboratory, TN, pp. 738–764.

Frelich, L.E. and Lorimer, C.G. (1991) Natural disturbance regimes in hemlock–hardwood forests of the upper Great Lakes Region. *Ecological Monographs* 6, 145–164.

Fritts, H.C. (1976) *Tree Rings and Climate.* Academic Press, New York, 567pp.

Fritts, H.C. and Lough, J.M. (1985) An estimate of average annual temperature variation for North America, 1602 to 1961. *Climatic Change* 7, 203–224.

Fritts, H.C and Wu, X. (1986) A comparison between response–function analysis and other regression techniques. *Tree-Ring Bulletin* 46, 31–46.

Fritts, H.C., Vaganov, E.A., Sviderskaya, I.V. and Shaskin, A.V. (1991) Climatic variation and tree-ring structure in conifers: empirical and mechanistic models of tree-ring width, number of cells, cell size, cell-wall thickness and wood density. *Climate Research* 1, 97–116.

Graumlich, L. (1993) Response of tree growth to climatic variation in the mixed conifer and deciduous forests of the upper Great Lakes region. *Canadian Journal of Forest Research* 23, 133–143.

Graumlich, L.J. and Brubaker, L.B. (1986) Reconstruction of annual temperature (1590–1979) for Longmire, Washington, derived from tree-rings. *Quaternary Research* 25, 223–234.

Grimm, E. (1991–1993) TILIA 2.0b (computer software). Illinois State Museum, Research and Collections Centre, Springfield.

Grimm, E. (1991) TILIA-GRAPH 1.25 (computer software). Illinois State Museum, Research and Collections Centre, Springfield.

Guiot, J. (1991) The bootstrapped response function. *Tree-Ring Bulletin* 51, 39–41.

Holmes, R.L. (1983) Computer-assisted quality control in tree-ring dating and measurement. *Tree-Ring Bulletin* 43, 69–78.

Holsten, E. and Burnside, R. (1997) Forest health in Alaska: An update. *Western Forester* 42, 8–9.

Jacoby, G.C. and Cook, E.R. (1981) Past temperature variations inferred from a 400-year tree-ring chronology from the Yukon Territory, Canada. *Arctic and Alpine Research* 13, 409–418.

Jacoby, G.C. and D'Arrigo, R.D. (1989) Reconstructed northern hemisphere annual temperature since 1671 based on high-latitude tree-ring data from North America. *Climatic Change* 14, 39–59.

James, R.L. and Goheen, D.J. (1981) Conifer mortality associated with root disease and insects in Colorado. *Plant Disease* 65, 506–507.

Jones, P.D. and Kelly, P.M. (1983) The spatial and temporal characteristics of northern hemisphere surface air temperature variations. *Journal of Climatology* 3, 243–252.

Koot, P. and Hodge, J. (1992) *History of Population Fluctuations and Infestations of Important Forest Insects in the Kamloops Forest Region.* Forestry Canada FIDS Report 92-11, Canadian Forestry Service, Victoria, 112pp.

Lane, B.B. and Goheen, D.J. (1979) Incidence of root disease in bark beetle-infested eastern Oregon and Washington true firs. *Plant Disease Reporter* 63, 262–266.

Massey, C.L. and Wygant, N.D. (1954) *Biology and Control of the Engelmann Spruce Beetle in Colorado*. USDA Circular 944, 35pp.

McLaughlin, S.B., Downing, D.J., Blasing, T.J., Cook, E.R. and Adams, H.S. (1987) An analysis of climate and competition as contributors to decline of red spruce in high elevation Appalachian forests of the Eastern United States. *Oecologia* 72, 487–501.

Nemec, A.F.L. (1996) *Analysis of Repeated Measures and Time Series: An Introduction with Forestry Examples.* Biometrics Information Handbook 6. BC Ministry of Forests, Victoria, 83pp.

Owens, J.N. and Molder, M. (1984) *The Reproductive Cycle of Interior Spruce*. BC Ministry of Forests, Victoria, 30pp.

Owens, J.N. and Molder, M. (1985) *The Reproductive Cycle of True Firs*. BC Ministry of Forests, Victoria, 35pp.

Payette, S., Fortin, M.J. and Morneau, C. (1996) The recent sugar maple decline in southern Quebec: probable causes deduced from tree rings. *Canadian Journal of Forest Research* 26, 1069–1078.

Peterson, D.W. and Peterson, D.L. (1994) Effects of climate on radial growth of subalpine conifers in the North Cascade Mountains. *Canadian Journal of Forest Research* 24, 1921–1932.

Piutti, E. and Cescatti, A. (1997) A quantitative analysis of the interactions between climatic response and intraspecific competition in European beech. *Canadian Journal of Forest Research* 27, 277–284.

Ross, S.D. (1985) Promotion of flowering in potted *Picea engelmannii* (Parry) grafts: effects of heat, drought and gibberellin A4/7, and their timing. *Canadian Journal of Forest Research* 15, 618–624.

Safranyik, L., Shrimpton, D.M. and Whitney, H.S. (1983) The role of insect–plant relationships in the population dynamics of forest pests. In: Isaev, A.S. (ed.) *Proceedings of International IUFRO/MAB Symposium*, 24–28 August 1981, Irkutsk, USSR, pp. 197–212.

Safranyik, L., Simmons, C. and Barclay, H.J. (1990) *A Conceptual Model of Spruce Beetle Population Dynamics*. Forestry Canada, Information Report BC-X-316, Victoria, BC, 13pp.

SAS Institute Inc. (1993) *SAS/ETS User's Guide*, Version 6, 2nd edn. SAS Institute Inc. Cary, NC, 1022pp.

Savage, M. (1991) Structural dynamics of a southwestern pine forest under chronic human influence. *Annals of the Association of American Geographers* 81, 271–289.

Stewart, G.H. (1986) Population dynamics of a montane conifer forest, western Cascade Range, Oregon, USA. *Ecology* 67, 534–544.

Stokes, M.A. and Smiley, T.L. (1968) *An Introduction to Tree-Ring Dating*. University of Chicago Press, Chicago, 67pp.

Szeicz, J.M. and MacDonald, G.M. (1994) Age-dependent tree-ring growth responses of subarctic white spruce to climate. *Canadian Journal of Forest Research* 24, 120–132.

Veblen, T.T., Hadley, K.S. and Reid, M.S. (1991a) Disturbance and stand development of a Colorado subalpine forest. *Journal of Biogeography* 18, 707–716.

Veblen, T.T., Hadley, K.S., Reid, M.S. and Rebertus, A.J. (1991b) The response of subalpine forests to spruce beetle outbreak in Colorado. *Ecology* 72, 213–231.

Veblen, T.T., Hadley, K.S., Nel, E.M., Kitzberger, T., Reid, M. and Villalba, R. (1994) Disturbance regime and disturbance interactions in a Rocky Mountain subalpine forest. *Journal of Ecology* 82, 125–135.

Villalba, R., Veblen, T.T. and Ogden, J. (1994) Climatic influences on the growth of subalpine trees in the Colorado Front Range. *Ecology* 75, 1450–1462.

Wiles, G.C., D'Arrigo, R.D. and Jacoby, G.C. (1996) Temperature changes along the Gulf of Alaska and the Pacific Northwest coast modeled from coastal tree rings. *Canadian Journal of Forest Research* 26, 474–481.

Woodward, A., Silsbee, D.G., Schreiner, E.G. and Means J.E. (1994) Influence of climate on radial growth and cone production in subalpine fir (*Abies lasiocarpa*) and mountain hemlock (*Tsuga mertensiana*). *Canadian Journal of Forest Research* 24, 1133–1143.

17

A New Detrending Method for the Analysis of the Climate–Competition Relations in Tree-Ring Sequences

Elena Piutti and Alessandro Cescatti

INTRODUCTION

Radial growth combines the effects of climate, site-specific factors, natural and human disturbances (Fritts and Swetnam, 1989; Cook and Kairiukstis, 1990). Variations in ring width may be used to date wood, to reconstruct past climates and to investigate single tree history and whole stand dynamics (Fritts, 1976; Schweingruber *et al.*, 1990). Due to the periodicity and rhythms of spring/summer wood, ring sequences look like bar codes and, analogously, embed information which can be extracted by statistical and numerical techniques (Fig. 17.1).

The variability in tree-ring series depends on internal and external factors such as climate, competition and their interactions (Tessier, 1989; Dhôte, 1994). The ring width of a single tree at time t can be mathematically related to these different factors with equation 17.1 (Graybill, 1982; Devall *et al.*, 1991), where R_t denotes the ring width or the basal area increment (BAI), C_t is the common climate component at time t, A_t is the age trend, $D1_t$ is a single tree disturbance signal, $D2_t$ is a disturbance signal induced by external factors common to all individuals, and ε_t is the random error.

$$R_t = \mathrm{f}(C_t, A_t, D1_t, D2_t) + \varepsilon_t \quad (17.1)$$

To analyse the relationship between climate and growth, several statistical techniques have been developed to remove the individual microsignals (A and $D1$) and isolate the common macrosignals (C and $D2$). Common standardization methods use monotone and polynomial functions, different 'smoothing' techniques, such as moving average, exponential weighted smoothing, smoothing spline and compound increment functions (Box and Jenkins, 1970;

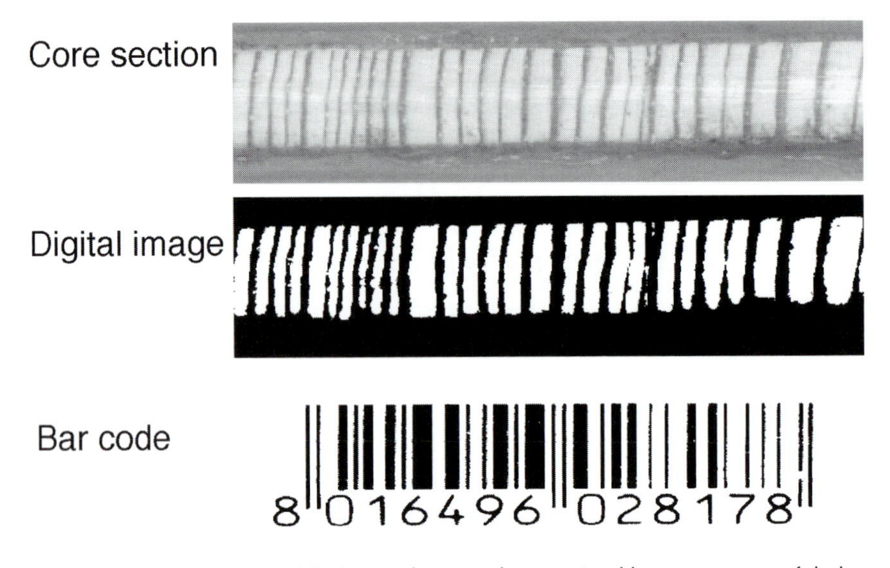

Fig. 17.1. Tree-ring series, like bar codes, are characterized by a sequence of dark and light stripes of different thickness. These sequences store information about tree history and forest dynamics.

Guiot *et al.*, 1982; Blasing *et al.*, 1983; Monserud, 1986; Visser and Molenaar, 1990). All these techniques emphasize the high-frequency signals (e.g. inter-annual variability induced by climate) by eliminating the low-frequency ones.

To reduce the effects of non-climatic factors on growth trends, the general sampling strategy adopted by dendrochronologists minimizes the effects of stand density on ring series through the selection of trees living at low competition levels (Nash III *et al.*, 1975; Schweingruber *et al.*, 1990). Due to limitations imposed by sampling criteria and statistical techniques for data standardization, the role of age trends (A_t), of individual microsignals ($D1_t$) and of the interactions between microsignals (competition) and macrosignals (climate) have been rarely documented (Blasing *et al.*, 1983). In the present chapter, these topics are faced by comparing the results of different detrending methods in the analysis of the interplay between tree growth, climate and competition.

MATERIAL AND METHODS

Study Site and Climate

The study site is located in the northeastern Italian Pre-Alps (46°02′ N, 12°25′ E; Fig. 17.2). The geological substrate is limestone, and soils are well-developed and drained brown earth, characterized by zoogenic humus (Marchisio *et al.*,

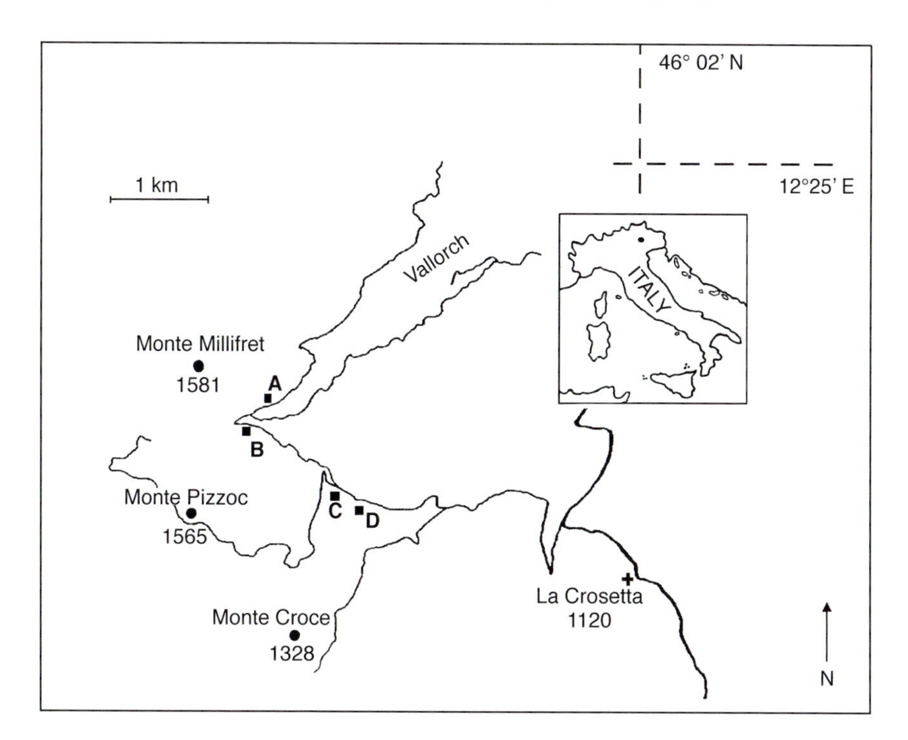

Fig. 17.2. Location of the four study sites (A–D), and of La Crosetta meteorological station (Cansiglio Plateau, northeastern Pre-Alps, Italy).

1994). The climate is considered optimum for beech stands in the Italian Pre-Alps.

The examined forest is located at 1250–1425 m a.s.l. in a pure beech (*Fagus sylvatica* L.) compartment, composed of even-aged stands from 80 to 140 years old. Four sample plots (1239–7488 m^2), unthinned in the last three decades, have been selected in stands with different densities, but characterized by homogeneous site conditions (parent rock, soil, ground vegetation, climate). The silvicultural regime is a shelterwood system with a 140-year rotation and a 20-year regeneration period.

Climatic data has been used from the meteorological station of La Crosetta (BL) located at 1120 m a.s.l, 3 km from the study site. Daily precipitation and temperature of the last two decades (1974–1993) are used as inputs; in this period environmental conditions were nearly unchanged in terms of stand density (almost no natural mortality and thinnings in the last 30 years). For the climatological analysis, monthly precipitation and mean monthly temperature were considered from September of the previous year to August of the current year. In order to estimate other environmental variables (potential and real evapotranspiration, water deficit and water surplus), the water balance was calculated using the Thornthwaite method (Thornthwaite and Mather, 1955) as modified by Newhall (1972), with the software developed by Cescatti (1992). The soil water capacity has been fixed at 60 mm for all the study sites, according to the soil analysis performed in the area by Marchisio *et al.* (1994).

Tree and Stand Variables

Topographical position, stem diameter at 1.30 m (dbh) and height were surveyed for each tree within the experimental plots (Fig. 17.3). Sample trees for coring were selected by random sampling (74 trees in total), stratified by diameter classes. Trees at the plot edges were considered competitors of the inner plot trees, but their growth data were not used in the analysis to avoid errors induced by edge effects (Cole and Newton, 1987).

For each sampled tree, stem circumference at breast height was measured and the mean diameter was calculated assuming a circular cross-section. With a forest calliper the direction of the mean diameter was located around the stem and two opposite cores were extracted along this direction. Cores were glued on wooden supports, sanded and smoothed, then tree-ring widths were measured with an electronic ring measuring machine (CCTRMD, Aniol, 1987). The synchronization of increment series was checked by visually identifying ring-width patterns.

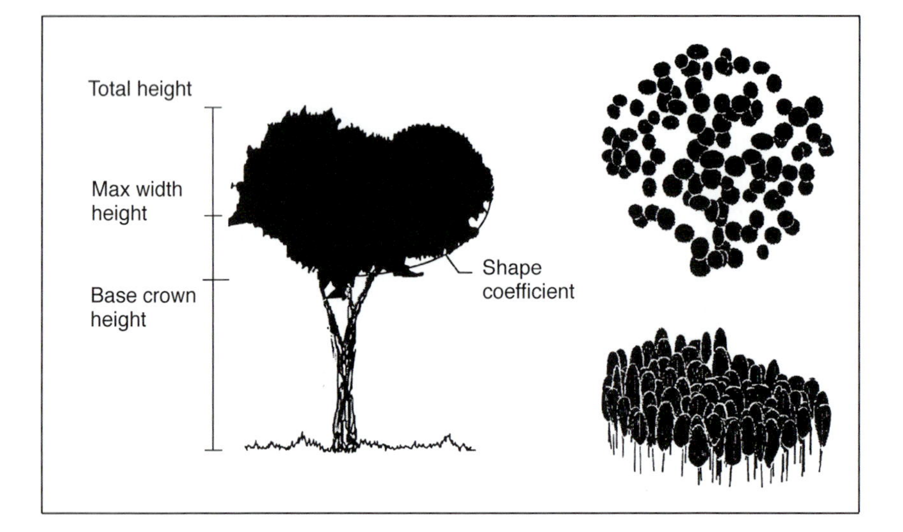

Fig. 17.3. To characterize the stand spatial structure, tree position and crown morphology have been surveyed in each of the four experimental plots (FOREST software, Cescatti, 1997).

Growth Trend Elimination

Data standardization is generally used to extract high-frequency signal from tree-ring series (Fritts, 1976). In this study, a third degree polynomial function has been compared with a more mechanistic method, based on a single-tree distance-dependent growth model.

The importance of growth models in forest management and in ecological investigations is widely documented (Mohren *et al.*, 1994). Among empirical growth models, good results are provided by non-linear empirical equations, defined as a potential growth function, multiplied by a modifier dependent on limiting factors (Wensel *et al.*, 1987; Daniels and Burkhart, 1988; Larsen, 1994). The potential growth trend has been analysed with the Chapman–Richards generalization of Von Bertalanffy's classical growth equation (Pienaar and Turnbull, 1973; Zeide, 1993); this equation quantifies the potential tree growth as the difference between anabolic and catabolic processes (Lorimer, 1983; Dale *et al.*, 1985). The competition term has been quantified using algorithms (competition indices) based on tree position and morphometrical variables (dbh, height, crown area, etc.) of subject trees and of their neighbours (Fig. 17.3). Seventeen different algorithms (Alemdag, 1978; Daniels *et al.*, 1986; Tomé and Burkhart, 1989; Biging and Dobbertin, 1992) have been tested in combination with four different algorithms for the selection of the competitors (constant distance, variable distance, overlapped area and angle count). All competition indices have been computed using the FOREST software developed by Cescatti (1997), which provides a framework for the estimation and the application of single-tree growth models (Fig. 17.4). The distance-dependent competition indices have been used as predictors of the reductive function Y (equation 17.2), characterized by a negative exponential shape, where k is a parameter larger than 1 and CI is the value of the competition index computed for single trees in each year of the simulation period (1974–1993).

$$Y = k^{-CI} \tag{17.2}$$

Fifty-four different versions of the growth model, obtained by combining the competition indices and the competitor selection algorithms, have been fitted by non-linear regression on the observed BAI, using the quasi-Newton method and considering the individual tree basal area and the competition index as predictors. Among the 54 tested growth models, the highest determination coefficient (R^2) has been obtained with the Chapman–Richards growth equation, together with the Hegyi competition index (CI) (equation 17.3), where $DBHs_i$ is the *i*-subject diameter, $DBHc_j$ is the diameter of the *j*-competitor and L_{ij} is the subject–competitor distance in metres (Hegyi, 1974). In the competition index, neighbouring trees within 8 m from the subjects have been considered as competitors (Piutti, 1994).

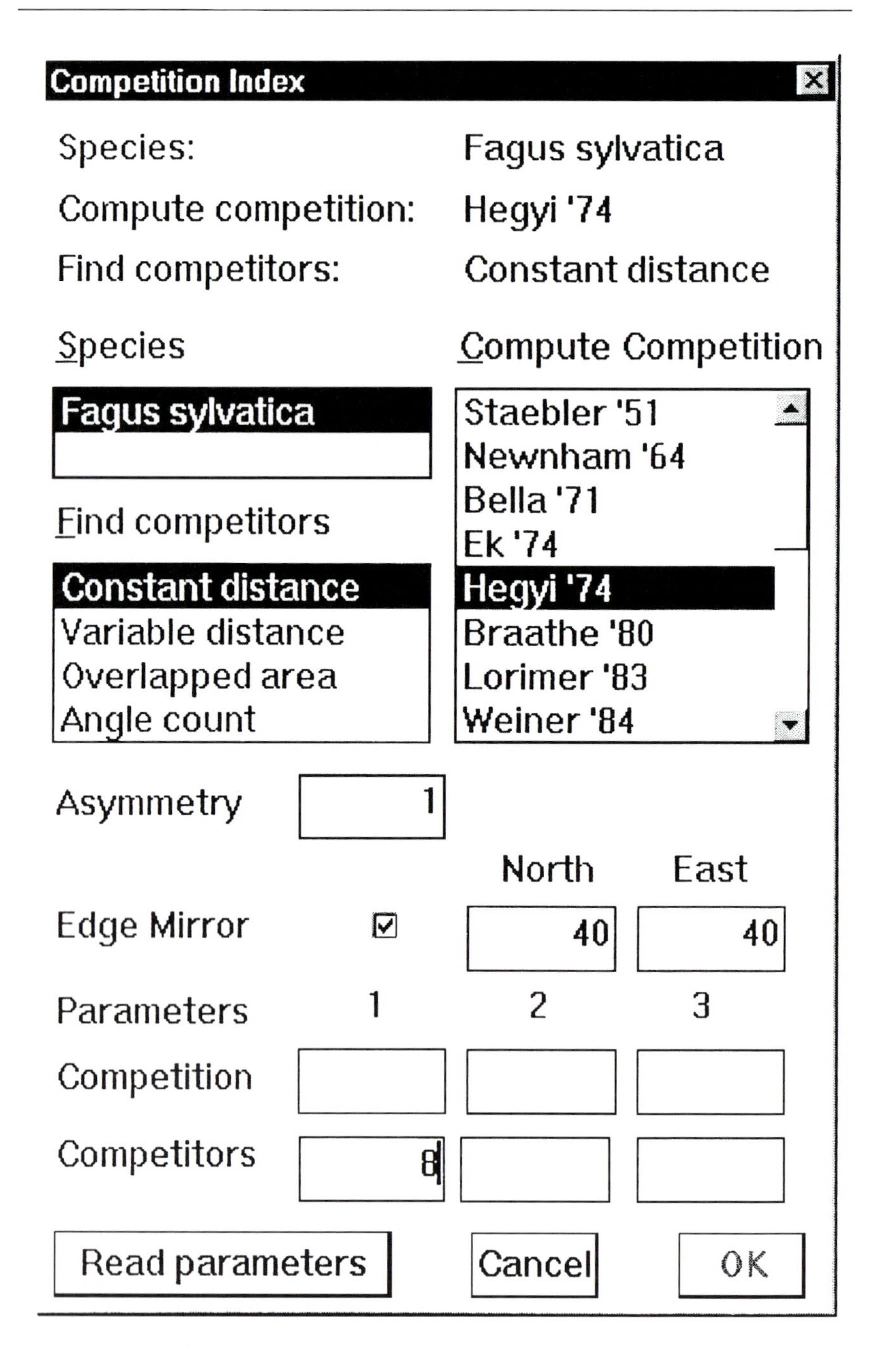

Fig. 17.4. Dialogue windows for the computation of the competition indices as implemented in the FOREST software (Cescatti, 1997).

$$CI_i = \sum_{j=1}^{N} \frac{DBHc_j}{DBHs_i} \frac{1}{L_{ij}} \qquad (17.3)$$

The predicted BAIs of cored trees have been calculated using the growth model with the past values of basal area and competition indices for the 1974–1993 time period. The ratio between observed and predicted BAI has been used as a growth index for the dendroclimatic analysis.

To evaluate the climate–competition interaction, trees have been ranked in four classes on the basis of the competition values estimated for each tree by the single-tree competition index (*CI*). Correlation functions have been computed for trees ranked in competition classes, using the monthly climatic variables from the previous September to the current August and the growth indices obtained from both the detrending procedures.

RESULTS

Climate and Stand Characteristics

In the period 1974–1993 the mean monthly temperatures ranged from −2°C (January) to 14.5°C (July) with a yearly mean value of about 6.1°C; the precipitation regime shows two peaks in June (191 mm) and October (231 mm), and two minimum values in January (105 mm) and July (129 mm) (Fig. 17.5). During the year, precipitation is abundant and hence water deficit rarely occurs during summer. Due to the large amount of precipitation (yearly average 1781 mm) and the limited soil water capacity (60 mm), mean water surplus is high (1278 mm).

Stand statistics computed for each sample plot are summarized in Table 17.1. The variability in stand density, generated by the thinning regime, has produced remarkable differences between the investigated plots. In fact, mean dbh and crown ratio (ratio between crown length and total tree height) decrease while the height/dbh ratio increases moving from area A to area D, as a result of the increased competition. Intraspecific competition is the source of the diameter differentiation, whereas similar values in tree height depend on the environmental homogeneity (altitude, geomorphology, soil, climate) of the experimental plots.

In Table 17.2 the descriptive statistics for the cored trees ranked in competition classes are presented, from low ($CI<0.5$) to high ($CI>3$) competition level. Data shows a strong relation between *CI* values, stem and crown morphology. When competition increases, mean dbh, height and crown ratio decrease whereas height/dbh increases.

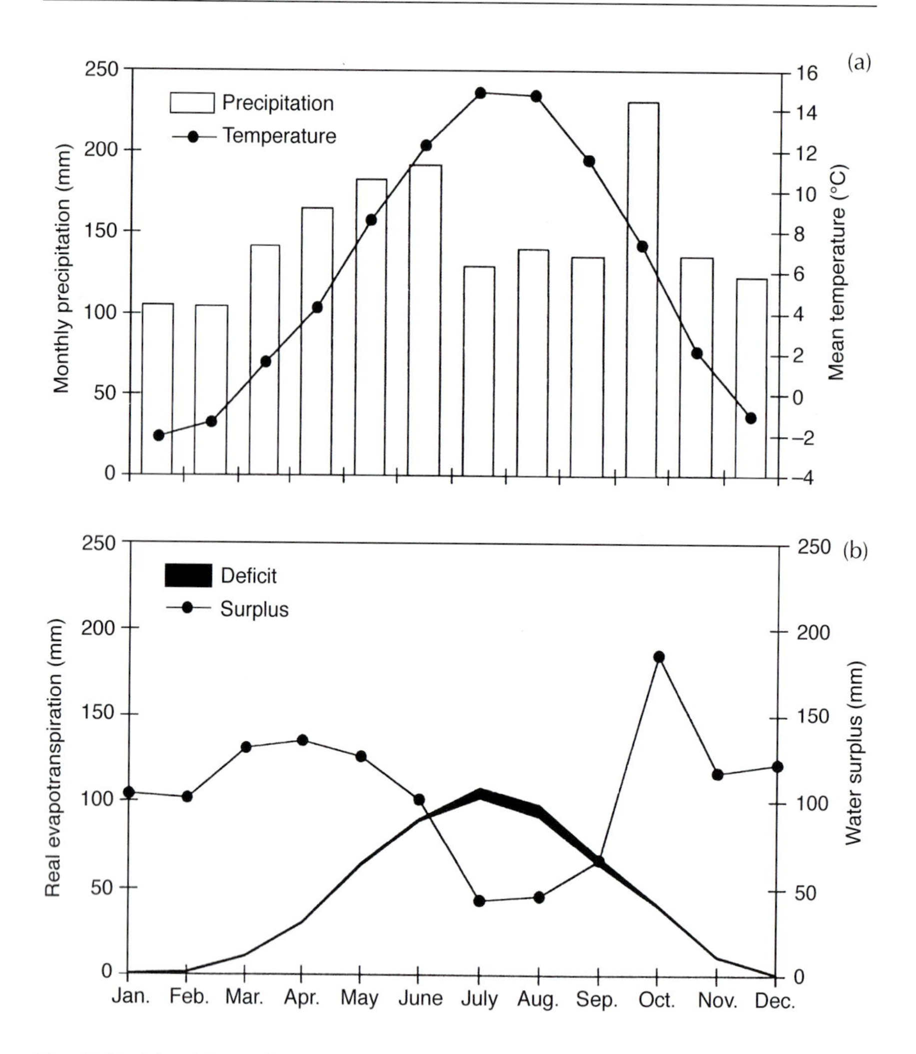

Fig. 17.5. Monthly total precipitation and mean temperature (a) together with the mean water balance (b) for the period 1974–1993.

Tree-Ring Variability Due to Age Trend, Competition and Climate

The growth model is shown in equation 17.4, where Δg is the estimated BAI and g is the basal area at the beginning of the growing season. The first term of equation 17.4 represents the age trend (A_t), which explains 52.1% of the BAI variance; the second term estimates the competition factor ($D1$). With the complete growth model (equation 17.4), a determination coefficient of 0.68 has been obtained.

Table 17.1. Stand statistics and tree biometry for the single experimental plots.

	Area A	Area B	Area C	Area D
Surface (m^2)	7488	2831	4140	1239
Stand density (N ha^{-1})	135	590	570	1267
Basal area (m^2 ha^{-1})	21.9	38.7	39.5	57.9
Mean dbh (cm)	45.4	28.9	29.7	24.1
Ave. height (m)	26.7	22.3	23.5	24.2
Height/dbh ratio	0.60	0.82	0.83	1.11
Crown ratio	0.62	0.46	0.45	0.38
Mean age	119	104	105	106

$$\Delta g = (17.28\, g^{0.9997} - 17.22\, g)\, 1.23^{-CI} \qquad (17.4)$$

The standardization technique with polynomials explains a high percentage of variance (56.1%), because an independent least-square fitting is performed for each tree-ring series instead of estimating a single parameter set for all the cores as in the growth model standardization.

In Fig. 17.6 the response functions computed for the two standardization approaches look completely different, highlighting the needs for the development and application of more mechanistic methods in dendroecology. Correlation coefficients obtained after polynomial standardization are significant only at the lower competition levels but the response to temperature and precipitation is not clear. Large differences occur for trees at the higher (H) and lower (L) competition levels, while similar patterns characterize the two intermediate classes (MH and ML). Results show that spring temperatures have a prevailing negative influence on beeches growing at low and intermediate competition (L, ML and MH). When considering the growth response to precipitation, it may be observed that no specific trend emerges, even if late summer rain is positively correlated with growth in low competition regime, while winter precipitation presents an opposite sign.

A different situation emerges from the growth model based standardization (Fig. 17.6), where precipitation is generally not significantly related with the growth index at intermediate and high competition. On the contrary, at low competition levels, winter and May precipitation negatively affects tree growth, while in March and June the opposite situation occurs. As regards trees living at high competition (H), precipitation during previous September and current August presents the only positive and significant correlation coefficients. On the whole, two opposite trends appear from the comparison of temperature and precipitation diagrams in the four different competition classes. For temperatures, Pearson's *r* changes from a strong positive correlation to an equally strong negative correlation, whereas precipitation switches from negative to non-significant values moving from the lower (L) to the higher (H) competition level.

Table 17.2. Descriptive statistics of cored trees categorized by competition index.

Competition class	Abbrev.*	No.**	Mean *CI*	Mean dbh (m)	Ave. height (m)	Height/dbh ratio	Crown ratio
$CI \leq 0.5$	L	13	0.226	46.2	26.1	0.576	0.629
$0.5 < CI \leq 1.7$	ML	13	1.295	37.1	23.0	0.633	0.503
$1.7 < CI \leq 3$	MH	16	2.187	32.2	23.5	0.761	0.410
$CI > 3$	H	12	5.076	21.3	22.5	1.144	0.341

* Abbreviations of the competition classes as used in the text.
** Number of cored trees in each competition class.

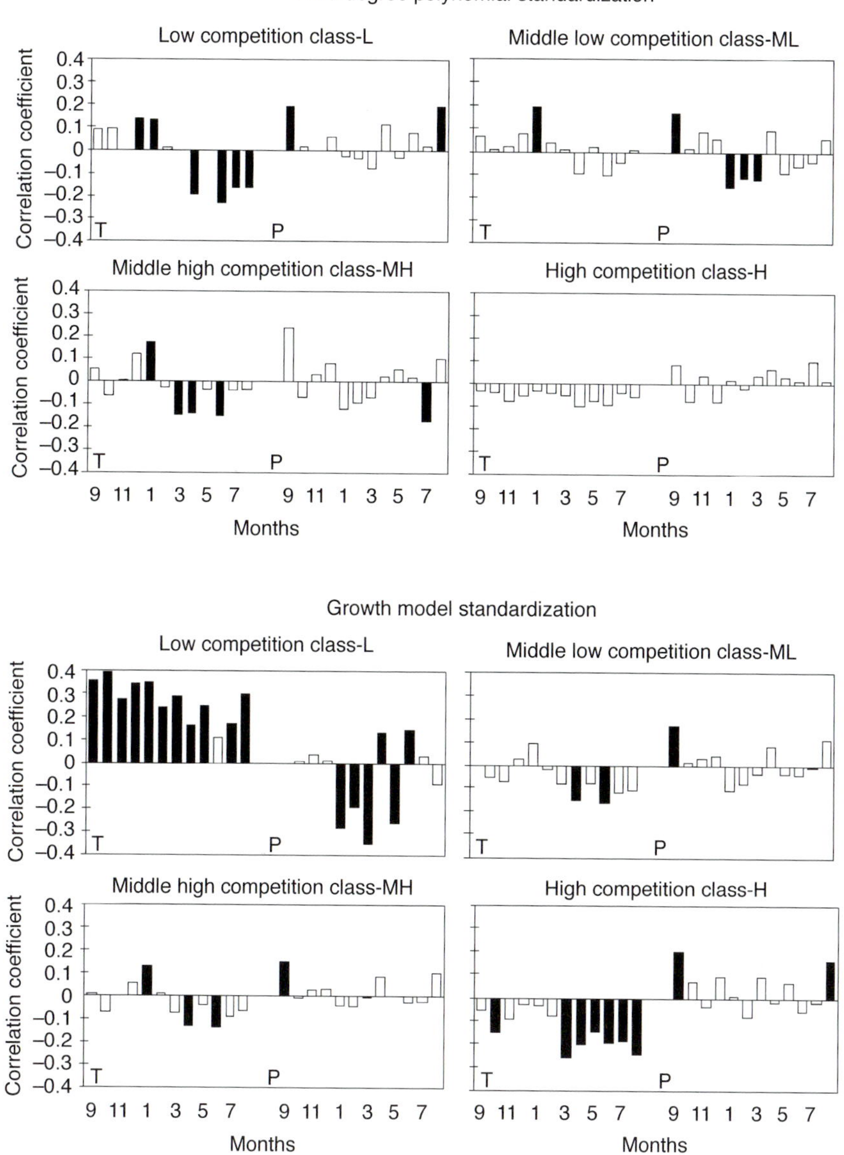

Fig. 17.6. Correlation coefficients between growth index (based on the third-degree polynomial and on the growth model standardization, respectively), temperature (T, °C) and precipitation (P, mm). Black bars mean statistical significance (*P*<0.05).

Further analyses on the seasonal response function to temperature, precipitation and water deficit have been performed exclusively for the growth model detrending method because of its higher biological significance (Fig. 17.7). The seasonal correlation coefficients for temperatures show an inverse relation with the competition level, switching from positive to negative values in all the three examined seasons. On the contrary, precipitation coefficients increase with the competition levels both in winter and spring, while in summer no significant values emerge. These two trends produce the relation between growth index and water deficit, which appears to limit tree growth only during summer time and at the higher competition levels.

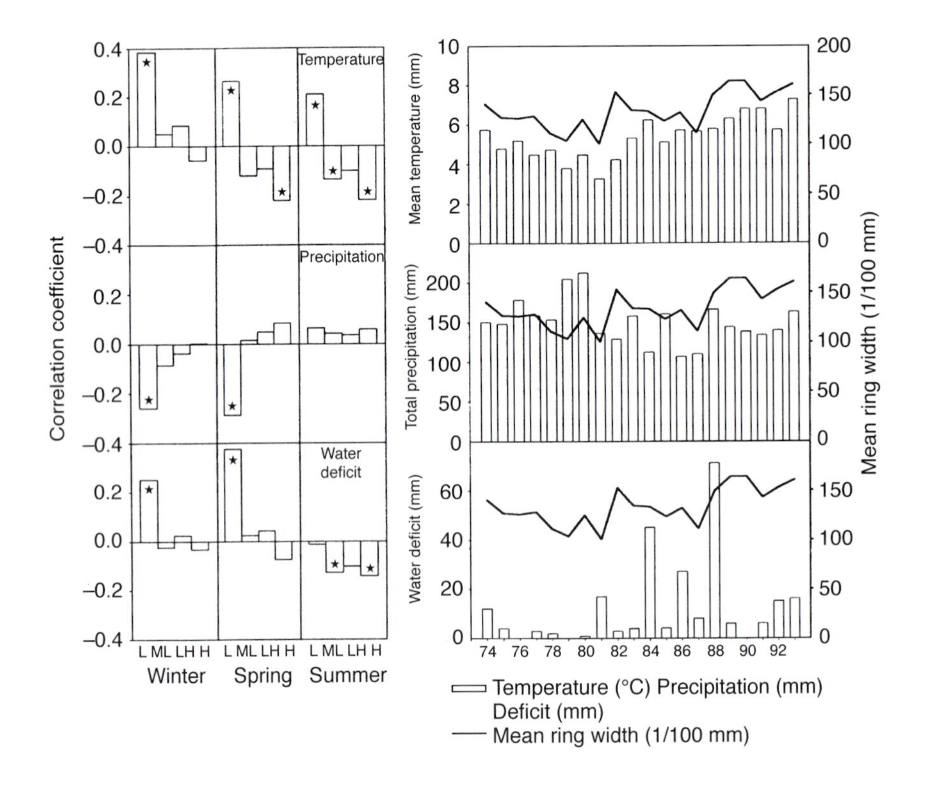

Fig. 17.7. Seasonal response function to temperature, precipitation and water deficit together with the time series of mean yearly temperature (°C), total yearly precipitation (mm), yearly water deficit (mm) and mean ring width (1/100 mm). Statistically significant *r* values ($P<0.05$) are marked with an asterisk (*).

DISCUSSION AND CONCLUSIONS

Methodological Aspects

In dendroclimatological studies, open-grown trees are usually selected to reduce the ring variability due to competition (Schweingruber *et al.*, 1990). However, trees generally compete for resources and, therefore, the analysis of tree response to climate should consider competition phenomena, because they might affect the influence of climate on tree growth.

To quantify competition in tree-ring series a mechanistic approach is required. For this purpose, competition-based growth models seem suitable, because they offer a biologically based interpretation of growth trends, while residuals may be considered as climatic signal and random errors. The application of growth models represents an alternative to the standardization, which simply removes the low-frequency signal induced by age trend and competition, without any mechanistic interpretation of these major sources of variation. Compared with the classical dendrochronological standardization techniques, this methodology requires additional data for the computation of the present and past values of a competition index (*CI*). However, by following this approach, the understanding of tree ecology and of the interplay between forest structure and function may be improved.

Climate, Competition and Tree Growth

Climate and competition are considered crucial factors in the growth process because they affect both the amount and the availability of resources (water, nutrient, light) (Grace and Tilman, 1990; Cannel and Grace, 1993; Nambiar and Sands, 1993). In this work, the hypothesis that climate and competition may affect tree growth in a non-additive way has been tested. This hypothesis has been verified by comparing the response functions to climate of trees living at different competition levels. To quantify the competition, a distance-dependent growth model has been applied. Afterwards, the growth model itself was used to detrend tree-ring series in order to describe in a mechanistic way most of the variance in the data set.

Results show that dominant trees present a higher sensitivity to climate than suppressed trees when data is standardized with the polynomial equation; for this reason, dendroecological studies are generally performed on dominant trees. Using the growth model to detrend series, the response functions for the extreme competition levels appear very different, demonstrating that tree response to climate is competition-dependent and that not only the dominant trees but even the suppressed ones can present a strong climatic signal. In fact, at L and H competition levels, climatic responses to temperature and precipitation are characterized by opposite signs and by the highest *r* values (Fig. 17.7). The variation in both sign and magnitude of the correlation coefficients in the different competition classes clearly demonstrates the non-additive effects of competition and climate on tree growth.

As regards the single environmental variables affecting the growth response, temperature turned out to be extremely important at low and high competition levels. In particular, high temperature seems to limit tree growth of suppressed trees (class H), possibly through a potential increase in the foliage temperature regime or a decreased water availability (Fig. 17.7). In contrast, precipitation shows a negative effect on trees growing at low-competition level while the summer water deficit negatively affects the growth of suppressed trees. This behaviour might be induced by the different water availability for trees living at diverse competition levels. Suppressed trees have to share water with neighbours, while dominant or isolated trees may utilize any available soil water reserve.

As a whole, these results show the complexity of the interplay between climate and competition in affecting the resource availability for plant growth (Le Goff and Ottorini, 1993). In spite of their complexity, these relations open new perspectives for forest management which, through the modification of the stand structure and therefore of the competition regime, might tune the forest sensitivity to climate in a global change scenario.

SUMMARY

Tree-ring width is affected by inner and outer factors, among which climate and competition play a major role. To reduce the competition signal in ring sequences, dominant trees are generally sampled for dendroecological studies. However, open-grown trees have passed through different social positions before reaching the upper canopy and, therefore, their growth trends are influenced by the competition history of the stand. In this work, the topic of filtering the competition signal is faced, and the ring series of trees living at different competition levels in an even-aged beech (*Fagus sylvatica* L.) forest (NE Italian Pre-Alps) are examined. For this purpose, a polynomial detrending method is compared with a new one, based on a single-tree, non-linear, growth model. The proposed method aims to investigate the climate–competition interactions by isolating the competition component from the climate signal. For this purpose the ratio between the observed increment values and the predictions obtained by the growth model is used as a growth index for the computation of the response function to climate. Results based on the growth model show that tree response to climate is highly affected by competition, while the classical dendroecological method presents significant correlation coefficients only for dominant trees. These findings suggest that, in the analysis of tree response to climate, the consequences of intraspecific competition should be taken into consideration using more mechanistic detrending methods, because competition may significantly influence the climate effects on tree growth.

REFERENCES

Alemdag, I.S. (1978) Evaluation of some competition indexes for the prediction of diameter increment in planted white spruce. *Information Report FMR-X-108*, Forest Management Institute, Ottawa, Ontario, 39pp.

Aniol, R.W. (1987) A new device for computer assisted measurement of tree-ring widths. *Dendrochronologia* 5, 135–141.

Biging, G.S. and Dobbertin, M. (1992) A comparison of distance-dependent competition measures for height and basal area growth of individual conifer trees. *Forest Science* 38, 695–720.

Blasing, T.J., Duvick, D.N. and Cook, E.R. (1983) Filtering the effects of competition from ring-width series. *Tree-Ring Bulletin* 43, 19–30.

Box, G.F.P. and Jenkins, G.M. (1970) *Time Series Analysis: Forecasting and Control*. Holden-Day, San Francisco, 575pp.

Cannel, M.G.R. and Grace, J. (1993) Competition for light: detection, measurement, and quantification. *Canadian Journal of Forest Research* 23, 1969–1979.

Cescatti, A. (1992) [A computer program to the Newhall's model application on the water soil regime classification]. *Studies of the Faculty of Agriculture*, University of Padova, Italy 6, 67–85 (in Italian).

Cescatti, A. (1997) Modelling the radiative transfer in discontinuous canopies of asymmetric crowns. I. Model structure and algorithms. *Ecological Modelling* 101, 263–274.

Cole, E.C. and Newton, M. (1987) Fifth-year responses of Douglas-fir to crowding and nonconiferous competition. *Canadian Journal of Forest Research* 17, 181–186.

Cook, E.R. and Kairiukstis, L.A. (eds) (1990) *Methods of Dendrochronology. Applications in the Environmental Sciences*. Kluwer Academic Publishers, Dordrecht, 394pp.

Dale, V.H., Doyle, T.W. and Shugart, H.H. (1985) A comparison of tree growth models. *Ecological Modelling* 29, 145–169.

Daniels, R.F. and Burkhart, H.E. (1988) An integrated system of forest stand models. *Forest Ecology and Management* 23, 159–177.

Daniels, R.F., Burkhart, H.E. and Clason, T.R. (1986) A comparison of competition measures for predicting growth of loblolly pine trees. *Canadian Journal of Forest Research* 16, 1230–1237.

Devall, M.S., Grender, J.M. and Koretz, J. (1991) Dendroecological analysis of a longleaf pine *Pinus palustris* forest in Mississippi. *Vegetatio* 93, 1–8.

Dhôte, J.F. (1994) Hypotheses about competition for light and water in even-aged common beech (*Fagus sylvatica* L.). *Forest Ecology and Management* 69, 219–232.

Fritts, H.C. (1976) *Tree Rings and Climate*. Academic Press, New York, 567pp.

Fritts, H.C. and Swetnam, T.W. (1989) Dendroecology: a tool for evaluating variations in past and present forest environments. *Advances in Ecological Research* 19, 111–188.

Grace, J.B. and Tilman, D. (eds) (1990) *Perspectives on Plant Competition*. Academic Press, San Diego, 484pp.

Graybill, D.A. (1982) Chronology development and analysis. In: Hughes, M.K., Kelly, P.M., Pilcher, J.R. and LaMarche, Jr V.C. (eds) *Climate from Tree Rings*. Cambridge University Press, Cambridge, pp. 21–28.

Guiot, J., Tessier, L. and Serre-Bachet, F. (1982) Applications de la modélisation A.R.M.A. en dendroclimatologie. *Comptes Rendus de l'Académie des Sciences*, Paris, 193, 133–136.

Hegyi, F. (1974) A simulation model for managing Jack-pine stands. In: Fries, J. (ed.) *Growth Model for Tree and Simulation.* Royal College Forestry Research Notes 30, Stockholm, pp. 74–90.

Larsen, D.R. (1994) Adaptable stand dynamics model integrating site-specific growth for innovative silvicultural prescriptions. *Forest Ecology and Management* 69, 245–257.

Le Goff, N. and Ottorini, J.-M. (1993) Thinning and climate effects on growth of beech (*Fagus sylvatica* L.) in experimental stands. *Forest Ecology and Management* 62, 1–14.

Lorimer, C.G. (1983) Tests of age-independent competition indices for individual trees in natural hardwood stands. *Forest Ecology and Management* 6, 343–360.

Marchisio, C., Cescatti, A. and Battisti, A. (1994) Climate, soils and *Cephalcia arvensis* outbreaks on *Picea abies* in the Italians Alps. *Forest Ecology and Management* 68, 375–384.

Mohren, G.M.J., Bartelink, H.H. and Jansen, J.J. (eds) (1994) Contrast between biologically-based process models and management-oriented growth and yield models. *Forest Ecology and Management* 69, 350pp.

Monserud, R. (1986) Time-series analyses of tree-ring chronologies. *Forest Science* 32, 349–372.

Nambiar, E.K.S. and Sands, R. (1993) Competition for water and nutrients in forests. *Canadian Journal of Forest Research* 23, 1955–1968.

Nash III, T.H., Fritts, H.C. and Stokes, M.A. (1975) A technique examining non-climatic variation in widths of annual tree rings with special reference to air pollution. *Tree-Ring Bulletin* 35, 15–24.

Newhall, F. (1972) Calculation of soil moisture regimes from climatic record. Soil Conservation Service, USDA, Washington, DC (mimeo, Rev. 4).

Pienaar, L.V. and Turnbull, K.J. (1973) The Chapman–Richards generalization of Von Bertalanffy's growth model for basal area growth and yield in even-aged stands. *Forest Science* 19, 2–22.

Piutti, E. (1994) [Competition indices as tools to define a single-tree empirical growth model in a beech forest (Cansiglio, Venetian pre-Alps)]. PhD thesis, Faculty of Agriculture, University of Padova, Italy, 156pp. (in Italian).

Schweingruber, F.H., Kairiukstis, L. and Shiyatov, S. (1990) Primary data. Sample selection. In: Cook, E.R. and Kairiukstis, L.A. (eds) *Methods of Dendrochronology. Applications in the Environmental Sciences.* Kluwer Academic Publishers, Dordrecht, pp. 23–35.

Schweingruber, F.H., Briffa, K.R. and Jones, P.D. (1991) Yearly maps of summer temperatures in Western Europe from A.D. 1750 to 1975 and Western America from 1600 to 1982: results of a radiodensitometrical study on tree rings. *Vegetatio* 92, 5–71.

Tessier, L. (1989) Spatio-temporal analysis of climate–tree ring relationships. *New Phytology* 111, 517–529.

Thornthwaite, C.W. and Mather, J.R. (1955). *The Water Balance.* Laboratory of Climatology, Centerton, NY, Publication in Climatology 8.

Tomé, M. and Burkhart, H.E. (1989) Distance-dependent competition measures for predicting growth of individual trees. *Forest Science* 35, 816–831.

Visser, H. and Molenaar, J. (1990) Estimating trends in tree-ring data. *Forest Science* 36, 87–100.

Wensel, L.C., Meerschaert, W.J. and Biging, G.S. (1987) Tree height and diameter growth models for Northern California conifers. *Hilgardia* 55, 1–20.

Zeide, B. (1993) Analysis of growth equations. *Forest Science* 39, 594–616.

18

Dendrochronological Investigations of Climate and Competitive Effects on Longleaf Pine Growth

Ralph S. Meldahl, Neil Pederson, John S. Kush and J. Morgan Varner III

INTRODUCTION

Forests dominated by longleaf pine (*Pinus palustris* Mill.) and maintained by high-frequency, low-intensity surface fires occurred throughout most of the southeastern USA Atlantic and Gulf coastal plains prior to European settlement. Stretching in a broad arc along the Atlantic Ocean to the Gulf of Mexico, longleaf pine and its associated communities covered an estimated 37 million ha, or approximately two-thirds of the southeastern USA (Frost, 1993). These forests were described as open and park-like, with a mono-specific overstorey and the most species-rich understorey in temperate North America (Peet and Allard, 1993). The open canopy was not due to an arid climate or soil infertility, but rather the frequent lightning and aboriginal fires that killed less fire-tolerant vegetation, leaving longleaf pine and its herbaceous understorey to thrive.

Longleaf pine occurs under a variety of environmental conditions. Longleaf forests occupy southerly aspects up to an elevation of 600 m a.s.l., and down to near sea level along the Gulf of Mexico coastline. Though most often associated with deep (often >5 m) sandy soils, or 'sandhills', longleaf pine occurs on all but the most inundated soils in southeastern USA (Wahlenberg, 1946). The constant on all sites that longleaf occupies is fire. Without frequent (once in 2–10 years), low-intensity fire, longleaf sites succeed to a closed-canopy, mixed angiosperm forest, termed the Southern Mixed Hardwood Forest (Ware *et al.*, 1993).

With settlement of the southeastern region of the USA came exploitation and degradation of the longleaf pine forests. Due to tapping of trees for resins utilized as naval stores, the introduction of feral hogs (*Sus scrofa*), timber harvesting, fire exclusion, and the implementation of plantation forestry beginning around 1920, the occurrence of longleaf pine forests became an infrequent sight (Frost, 1993). By 1995, longleaf pine forests occupied an estimated 1.2 million ha, or 3.2% of its former range (Outcalt and Sheffield, 1996). Of this total, only 3900 ha (0.14%) in 14 tracts remain in an old-growth condition (Means, 1996). One of these tracts is a 27-ha stand currently owned by Champion International Corporation. The stand has a rich natural history (with many trees exceeding 200 years in age and more than 50 cm in diameter), and an interesting cultural history. The Alger-Sullivan Lumber Company, one-time owner, preserved this stand through the first half of the 20th century. As part of the preservation effort, the stand was regularly control burned until about 1950. When the company was sold in the early 1950s, the new owners ended the burning regime, although they continued the policy of preservation. Consequently, the stand had remained unburned for more than 40 years. The absence of fire had not only allowed other pine species to grow into the overstorey, but had also permitted a substantial hardwood understorey and midstorey to develop at the expense of longleaf pine regeneration and herbaceous vegetation.

The importance of the stand was recognized by the Society of American Foresters (SAF) in 1963 when they designated the area, then owned by the St Regis Paper Company, as the E.A. Hauss Old Growth Longleaf Natural Area (Walker, 1963). The SAF's definition of a natural area is 'a tract of land set aside to preserve permanently in unmodified condition a representative unit of virgin growth of a major forest type, with the preservation primarily for scientific and educational purposes'.

An agreement has been signed among Champion International Corporation, Auburn University School of Forestry, the Southern Research Station of the US Forest Service, the Nature Conservancy, the Alabama Forestry Commission, and the Alabama Natural Heritage Trust of the Alabama Department of Conservation and Natural Resources for cooperative work on what is now being called the Flomaton Natural Area (FNA) (Meldahl *et al.*, 1994). Efforts are under way to restore, monitor and manage the FNA as an old-growth longleaf pine habitat. As a part of this effort, a subset of longleaf pine are being cored to provide a look into the past. Specific objectives are to study (i) climatic factors influencing growth, and (ii) stand disturbance history.

Life History

Longleaf pine's life history is unique among pine species. Longleaf pine is classified as a very shade-intolerant species but it has none of the characteristics associated with early successional species (Boyer, 1997). It is not a prolific seed producer, the seed is not disseminated great distances, and its early growth is not rapid. Regeneration of longleaf pine occurs erratically. Excellent mast years occur once every 4–7 years, with variations locally (Wahlenberg, 1946; Boyer, 1990). Large canopy gaps (1000–2000 m^2) caused by large storm events, lightning strikes, and high-temperature fires create environments for seedling germination and establishment (Palik and Pederson, 1996). This type of regeneration leads to an all-aged stand with small patches of even-aged trees.

Following seedfall from cones in late autumn (late September–late November), the seeds usually germinate within 14 days. By the following spring, the seedling has shed its cotyledons and begins what is termed the 'grass stage' (Pessin, 1934).

In the grass stage, the seedling reaches a height of a few centimetres and then terminal growth stops. It concentrates all of its growth during the first 3–5 years in its roots, making almost no height growth. In an ecosystem that evolved with a high frequency of low-intensity surface fires, longleaf pine has a few traits to assist in its survival during this grass stage. A sheath of long needles and fire-resistant scales protects its apical bud. Due to food storage in its roots, it has the ability to put out new needles should the existing foliage be killed by fire.

The grass stage lasts for 2 to >14 years (the maximum age to which a seedling can remain in the grass stage has never been found). Important to tree-ring analysis is the fact that longleaf pine produces no annual rings in this stage (Pessin, 1934). This factor underestimates all ages for longleaf pines, perhaps by as much as 14 years or more. When the grass stage seedling attains a root collar diameter of approximately 2.5 cm it begins height growth. Once it does begin, height growth is rapid so that in 2–3 years the seedling will be tall enough that the apical bud will be protected from low-intensity fires.

Diseases and insects rarely cause mortality of longleaf pine. Longleaf is somewhat resistant to the several species of Coleopteran bark beetle (*Dendroctonus* and *Ips* spp.), a severe problem for other southeastern USA pine species (Boyer, 1990). Root decaying fungi (caused by *Heterobasidion annosum*) are minor mortality agents in longleaf pine. Lightning strikes are the primary inciting agent of mortality in longleaf forests (Platt *et al.*, 1988; Palik and Pederson, 1996). Mortality in mature stands of longleaf pine averages 0.5–1 tree ha^{-1} $year^{-1}$ (Chapman, 1923; Boyer, 1979; Palik and Pederson, 1996).

Past Dendrochronological Applications in Longleaf Pine – A Review of Pertinent Literature

Dendrochronological research on longleaf pine began in the late 1920s. It has been erratic and often contradictory. Lodewick (1930) was the first to investigate the impact of climate (both temperature and precipitation) on the growth of longleaf pine. In deep sandy sites in northern Florida, USA, no correlation was found between temperature and growth; however, current year 1 April to 15 October precipitation explained 89% of annual radial growth. Earlywood production was constant and irrespective of precipitation, while mid-June to mid-October precipitation had a significant impact on latewood production. The only year with an inconsistency was a large seed year – 1920. Lodewick found no correlation between crown volume and radial growth.

Five other investigations in tree-ring analysis were undertaken in the 1930s. In the same location as Lodewick, Paul and Marts (1931) investigated timing and cessation of early and latewood production and the impact of nutrient fertilization and irrigation on each. Another study in the same locale in northwest Florida (Pillow, 1931) found compression wood formation in response to crown bending from a 1926 hurricane. Working in southern Georgia, Coile (1936) found that young longleaf pine growth is influenced most by current year February to April precipitation. Current year June to August average temperature had the highest negative correlation on radial growth. Schumacher and Day (1939) again investigated monthly precipitation's impact on radial growth in stands in northeastern Florida. The last note on tree-ring analysis in longleaf pine for nearly 50 years reported on the occurrence of frost cracks in earlywood growth of stands in southern Mississippi (Stone, 1940).

Platt *et al.* (1988) undertook the next effort in longleaf pine dendrochronological research, in an old-growth stand located in southwestern Georgia. This stand, known as the Wade Tract, had undergone timber salvage operations, cattle grazing and annual prescribed burning, leading some to question the value of this stand as a true representative old-growth longleaf pine population. They found strong relationships between age and height and diameters of individual trees within the Wade Tract population. Following up on Platt *et al.* (1988), West *et al.* (1993) looked at the Wade Tract population's larger individuals (>40 cm DBH, >10 m from neighbours or stumps). Of 26 trees cored, the range in ages (ring count at 1.37 m) was 96–396 years. Beginning about 1920, these trees began an increase in growth rates beyond expected rates. Possible explanations listed were increased atmospheric CO_2 concentrations and increased atmospheric SO_x and NO_x concentrations, the latter being less likely (West *et al.*, 1993).

Two other investigations complete contemporary dendrochronological research in longleaf pine. Zahner (1989), examining even-aged stands in southern Alabama, found age of trees, basal area, burning, drought index and individual stand differences were significant factors affecting annual radial growth. Zahner also constructed a tree-ring chronology from 1964 to 1985. Devall *et al.* (1991) constructed a limited chronology (based on cores from 32 trees) from 1921 to 1987 for southern Mississippi longleaf pine. Significant factors for radial growth were current-year August precipitation, September temperatures, and the February Palmer drought-severity index. They found no growth enhancement due to increased atmospheric CO_2 levels.

METHODS

Description of the Study Area

The FNA is located within the city limits of Flomaton, in the south-central part of Escambia County, Alabama, USA, at 31°01′ N mean latitude and 87°15′ W mean longitude (Fig. 18.1). The climate is humid and mild with plentiful rainfall well distributed throughout the year. The warmest months are July and August with average daily maximum and minimum temperatures of 33 and 20°C, respectively. The coldest months are December and January with average daily temperatures of 18 and 3°C, respectively. The growing season averages 250 days. Annual precipitation averages 156 cm with October being the driest month.

The stand is relatively level at 63 m a.s.l. The predominant soil belongs to the Orangeburg series. This soil formed in marine sediments of sandy loams and sandy clay loams. It is low in natural fertility and organic-matter content.

Inventory Procedures

Plot centres have been established throughout the entire tract on a 60 × 80 m grid and located at least 30 m from the edge of the old-growth tract. Each plot is circular, enclosing an area of 0.08 ha (radius = 16.05 m). Data recorded for every tree (>1.25 cm DBH) on each plot include: tree number, azimuth and distance from plot centre, diameter at breast height (DBH) of live trees and standing snags, crown height and total height.

One hundred trees greater than 7.6 cm DBH were cored through the entire tree at 1.22 m. The core was extracted, air-dried, mounted and surfaced. Ring widths were measured to the nearest 0.001 mm. Separation of earlywood and latewood widths was estimated visually. The boundary between earlywood and latewood in longleaf pine is discrete and easily defined by changes in colour

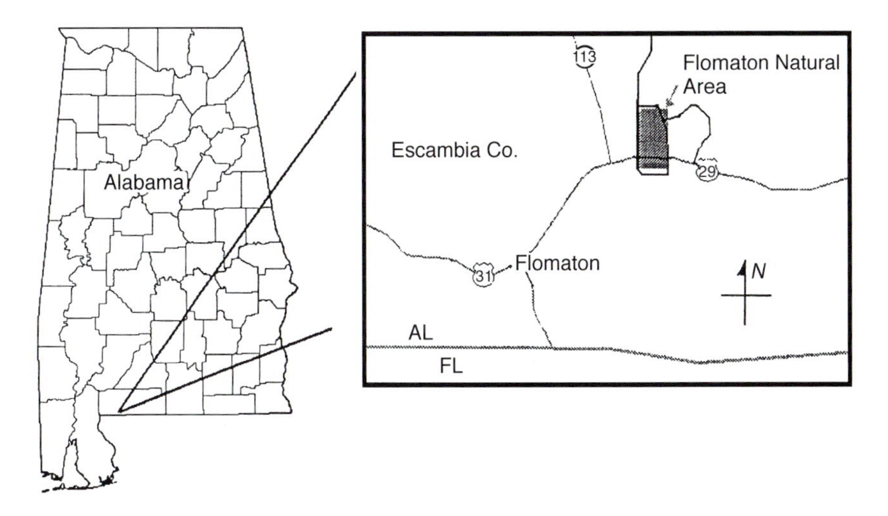

Fig. 18.1. Map of the Flomaton Natural Area and its location within the State of Alabama, USA.

and cell wall thickness. Each core was crossdated visually and re-checked using the computer-based program COFECHA (Holmes, 1983). Tree-ring chronologies were developed for earlywood, latewood and total ring width (Kush *et al.*, 1997). A tree-ring chronology is an index of growth standardized to remove growth trends shaped biologically or by stand dynamics. All chronologies were developed using the ARSTAN program (Cook, 1985). We chose to use the residual chronology for climate comparison. This removes autocorrelation, lowers bias in the indices, and provides a robust signal of climate (Cook, 1985).

Climate

To explore the climatic factors that influence growth in this study, tree-ring chronologies were developed for earlywood, latewood and total ring widths. Coefficients of correlation were calculated for seasonal and monthly precipitation, minimum and maximum temperature, and the Palmer hydrological drought index (PHDI). We define seasonal in this study as a period of 2 months or more. Correlations were calculated back to 2 years before the current year because of longleaf pine's 2-year maintenance of its needles (Wahlenberg, 1946). For example, for 1995 growth, we looked at climate in the years of 1995, 1994 and 1993. The PHDI is an adaptation of the Palmer drought severity index (PDSI). The PDSI utilizes soil type, precipitation and temperature to identify drought conditions (Palmer, 1965). The PHDI approaches real-time and therefore is a hydrological index instead of a meteorological index like the PDSI (Hayes, 1996). The PHDI can range from 8 to -8 with a positive value indicating wet conditions and negative indices representing drought conditions. Indices between 1 and -1 are considered within normal growing conditions. Indices greater than 4 and less than -4 indicate severe states of climate. Because the FNA is near the Alabama–Florida border, we analysed growth for the pertinent climate divisions of each state as defined by the National Climatological Data Center (NCDC).

Disturbance History

During the analysis of the FNA for disturbance history, severe declines in growth were noted directly from the wood and from raw ring-width measurements. We decided to attempt to quantify these dramatic growth trends. We used Lorimer's (1985) tree-ring pattern definition to identify canopy disruption. Lorimer's major interest in disturbance was in stand-initiating events. However, his techniques to identify these events were designed to avoid short-term climate as a potential source of confusion. Therefore, we adapted his definition of canopy releases to determine declines in growth. He identified minor releases as a 15-year, 50% increase in growth over the previous 15 years. Since declines in growth are natural for mature trees, we defined a minor decline as decrease in growth of 125% over 15 years for trees in the dominant and codominant canopy position (Fig. 18.2). We chose trees with crowns in the dominant and codominant position to avoid growth declines caused by overstorey suppression.

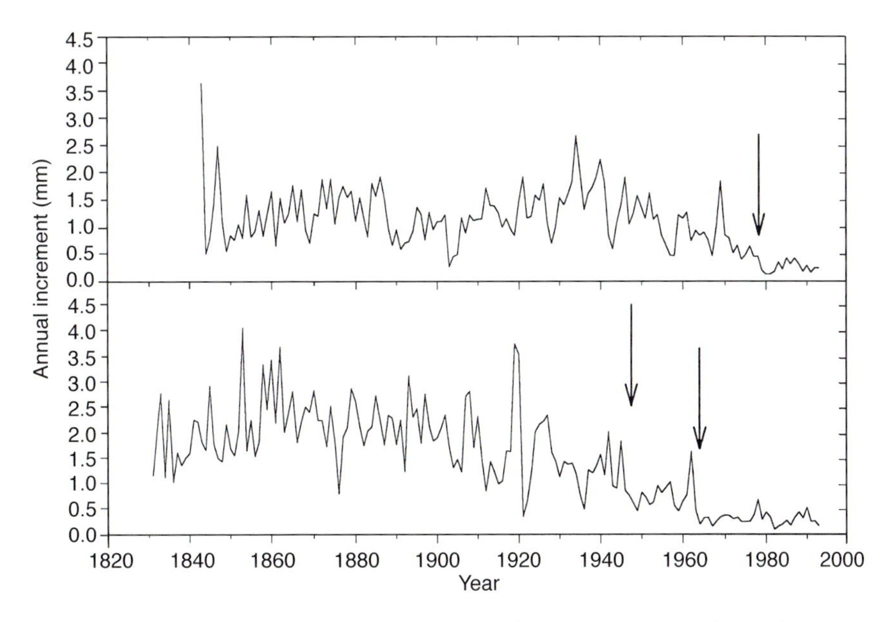

Fig. 18.2. Two chronologies of individual trees showing examples of growth declines of 125% or more over 15 years. Arrows indicate the beginning of each decline.

We assume that trees in these crown classes are mostly being influenced in growth by fluctuations in climate.

RESULTS

Climate

Tree-Ring Chronologies

The total ring chronology has ten cores dating back to 1817 (Fig. 18.3a) and an average tree age of 107 years. The earlywood chronology has ten cores extending back to 1920 (Fig. 18.3b) and latewood chronology has ten cores reaching back to 1900 (Fig. 18.3c). To strengthen each chronology, individual trees and cores with poor COFECHA and ARSTAN correlations were not used. The poor correlations were induced from stand dynamics and overstorey suppressions. We also had great difficulty crossdating earlywood and latewood widths that limited the length of each chronology. Future efforts aim to overcome this shortcoming. All chronologies have at least 45 cores from more than 20 trees. This sampling density produces a chronology with a small standard error (Fritts, 1987). Table 18.1 presents the mean sensitivity and standard deviation for each residual chronology.

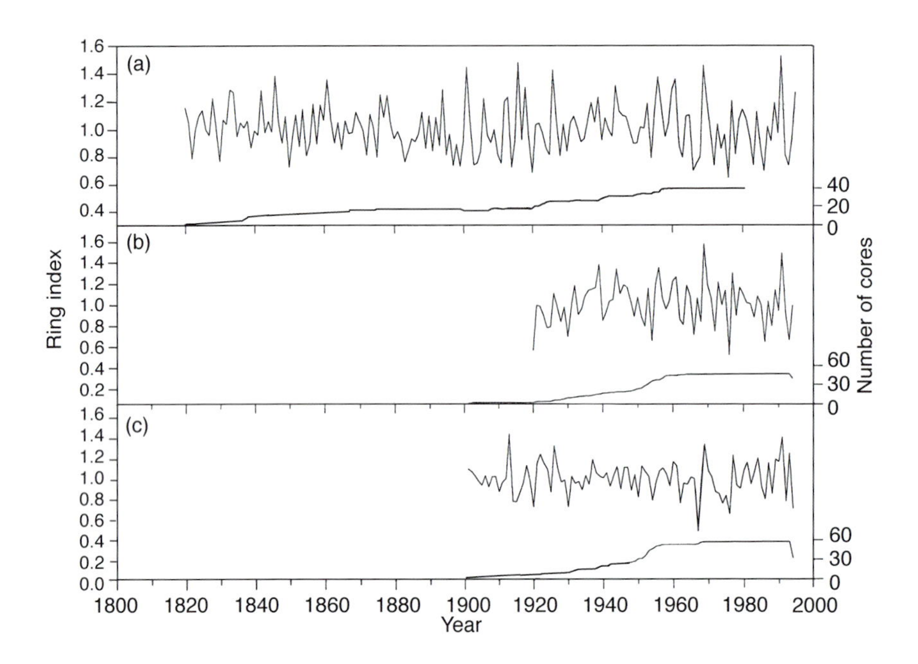

Fig. 18.3. Index of tree growth and sampling depth. a – Whole ring chronology; b – Earlywood chronology; c – Latewood chronology.

Climatic Factors and Growth

Correlation coefficients, between annual chronology growth (total, earlywood and latewood) and (i) average monthly climate values and (ii) seasonal climate values, were used to assess climatic impacts. Graphs of the correlation coefficients versus individual months for maximum and minimum temperature, precipitation and Alabama and Florida PHDI are given in Figs 18.4, 18.5 and 18.6.

Monthly minimum temperature, precipitation and PHDI (Fig. 18.4) significantly influence total ring development (significance was defined at the $P = 0.10$ level). A positive correlation was detected for a low minimum June temperature 2 years prior to the current year's growth (Fig. 18.4b). Precipitation positively influences growth in the current year's March and September (Fig. 18.4c). A strong negative effect was found during February of the previous year. Positive relationships for both Alabama (Fig. 18.4d) and Florida PHDI (Fig. 18.4e) were seen during the later months of the current growing season. Interestingly, the Alabama PHDI extends into November while the Florida PHDI's effect ends in September.

Average seasonal precipitation between March and October was the most important factor driving total ring width. Maximum positive and negative correlations between chronologies and seasonal climatic variables are given in

Table 18.1. Statistics characterizing earlywood, latewood and total ring width residual chronologies.

	Earlywood	Latewood	Total ring
Mean sensitivity	0.1810	0.2441	0.2053
Standard deviation	0.1627	0.2051	0.1781

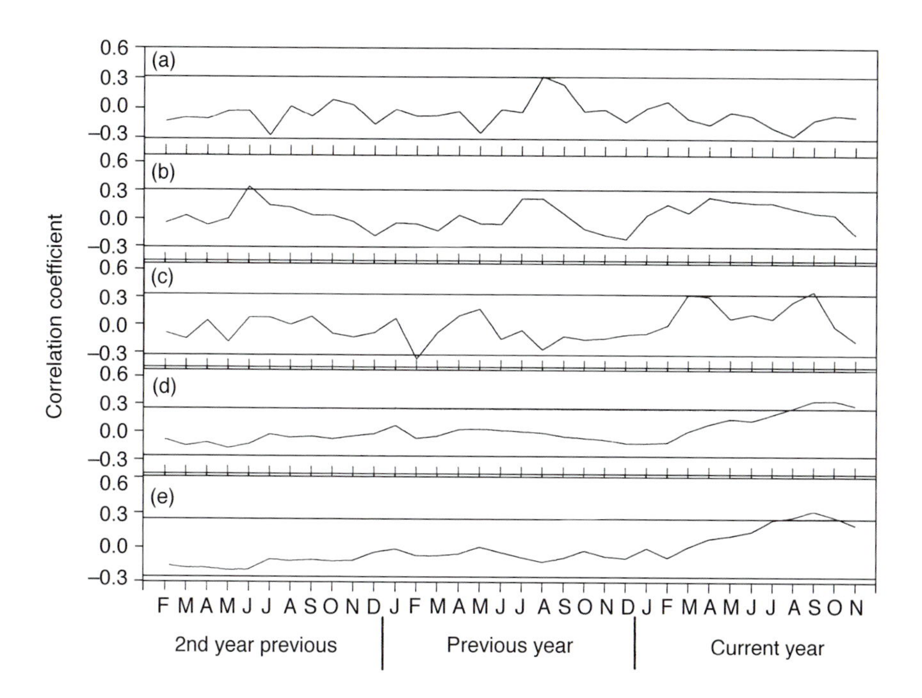

Fig. 18.4. Total ring width correlation coefficients with monthly climatic factors. Reference lines represent a significant relationship at $P = 0.10$. a – max. temp., $N = 66$; b – min. temp., $N = 66$; c – precipitation, $N = 58$; d – Alabama PHDI, $N = 101$; e – Florida PHDI, $N = 101$.

Table 18.2. A significant negative relationship was found with the previous year precipitation from August to December. We also found significant relationships between growth and temperature. Maximum temperature of the previous August and September indicated positive conditions for current year's growth. Late growing-season PHDI values for Alabama and Florida have significant relationships with growth of the current year.

Earlywood formation showed little response to monthly climate factors (Fig. 18.5). The only significant relationship for earlywood growth was a negative

Table 18.2. Maximum positive and negative correlation of ring chronologies and average seasonal climatic variables.

		Season	
Chronology type	Climate variable	Positive (correlation coefficient)	Negative (correlation coefficient)
Total ring width	Maximum temperature	Previous Aug.–Sept. (0.325)*	Current Jul.–Nov. (−0.272)
	Minimum temperature	Current Mar.–Oct. (0.278)	2nd previous Jun.–Aug. (−0.279)
	Precipitation	Current Mar.–Oct. (0.458)*	Previous Aug.–Dec. (−0.374)*
	Alabama PHDI	Current Sept.–Oct. (0.349)*	2nd previous Mar.–Oct. (−0.111)
	Florida PHDI	Current Aug.–Sept. (0.301)*	2nd previous Feb.–Apr. (−0.178)
Earlywood	Maximum temperature	Previous Aug.–Sept. (0.102)	Current Feb.–Apr. (−0.338)*
	Minimum temperature	2nd previous Mar.–Oct.(0.228)	Previous WINTER (−0.172)
	Precipitation	Current Feb.–Apr. (0.252)	Current WINTER (−0.231)
	Alabama PHDI	2nd previous Sept.–Oct. (0.011)	2nd previous Jan.–Mar. (−0.252)
	Florida PHDI	Current Mar.–Oct. (0.077)	Previous Aug.–Sept. (−0.235)
Latewood	Maximum temperature	Previous Aug.–Sept. (0.336)*	Current Jul.–Oct. (−0.429)*
	Minimum temperature	2nd previous Jun.–Aug. (0.241)	Previous Oct.–Dec. (−0.253)
	Precipitation	Current Mar.–Oct. (0.546)*	Previous Aug.–Dec. (−0.444)*
	Alabama PHDI	Current Aug.–Sept. (0.444)*	Current WINTER (−0.282)*
	Florida PHDI	Current Aug.–Sept. (0.373)*	Previous Sept.–Dec. (−0.300)*

* significant at $P = 0.10$. Maximum and minimum temperature $N = 66$. Precipitation $N = 58$. PHDI $N = 101$; WINTER = average of previous November, previous December and current January; current = current year's climate vs. current year's growth; previous = previous year's climate vs. current year's growth. 2nd previous = 2 years previous climate vs. current year's growth. Temperate and precipitation data from Brewton, AL. Alabama PHDI from National Climatological Data Center for Alabama climate division 7. Florida PHDI from National Climatological Data Center for Florida climate division 1.

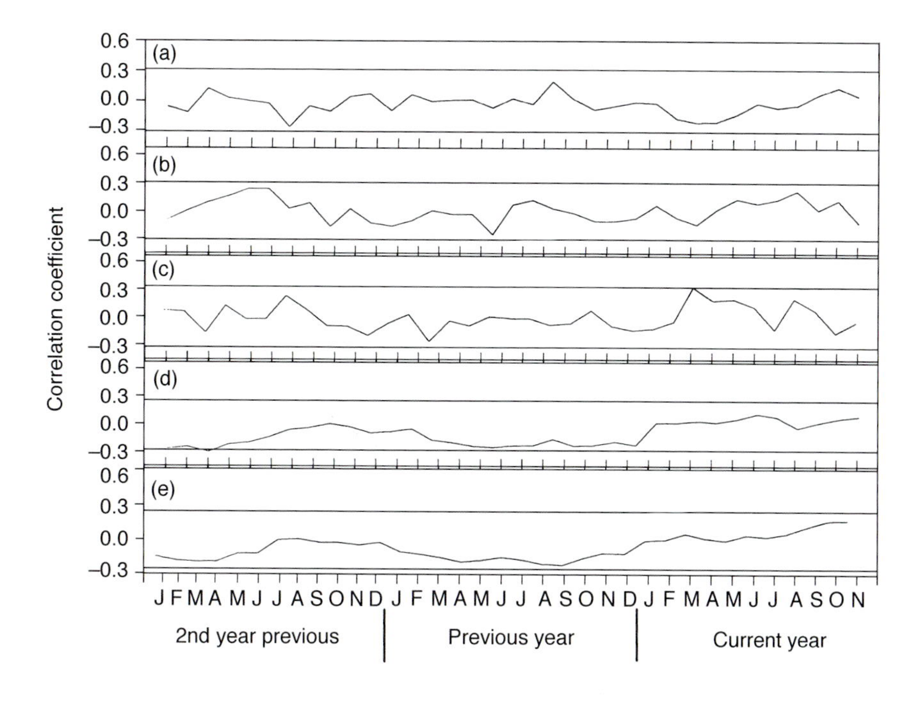

Fig. 18.5. Earlywood width correlation coefficients with monthly climatic factors. Reference lines represent a significant relationship at $P = 0.10$. a – max. temp., $N = 66$; b – min. temp., $N = 66$; c – precipitation, $N = 58$; d – Alabama PHDI, $N = 101$; e – Florida PHDI, $N = 101$.

response to March Alabama PHDI of 2 years prior (Fig. 18.5d). Nearly significant positive influences are minimum temperature for May and June 2 years earlier (Fig. 18.5b), and March precipitation of the current year (Fig. 18.5c). Nearly significant negative correlations were located during the previous year May, September and December Alabama PHDI (Fig. 18.5d), and previous year August and September Florida PHDI (Fig. 18.5e).

Similarly, only one significant relationship between seasonal climate and earlywood growth was found in this study (Table 18.2). Maximum temperature between February and April of the current year was found to have a negative impact on growth. Current year precipitation from February to April and 2 years previous Alabama PHDI from January to March nearly had significant relationships with earlywood growth, positive and negative, respectively. Interestingly, Florida PHDI had a high negative correlation with earlywood growth, but for the previous August and September.

Latewood formation was significantly influenced by each climate factor on a monthly basis (Fig. 18.6). Maximum temperature positively affected latewood during the previous September and negatively during July and August of the

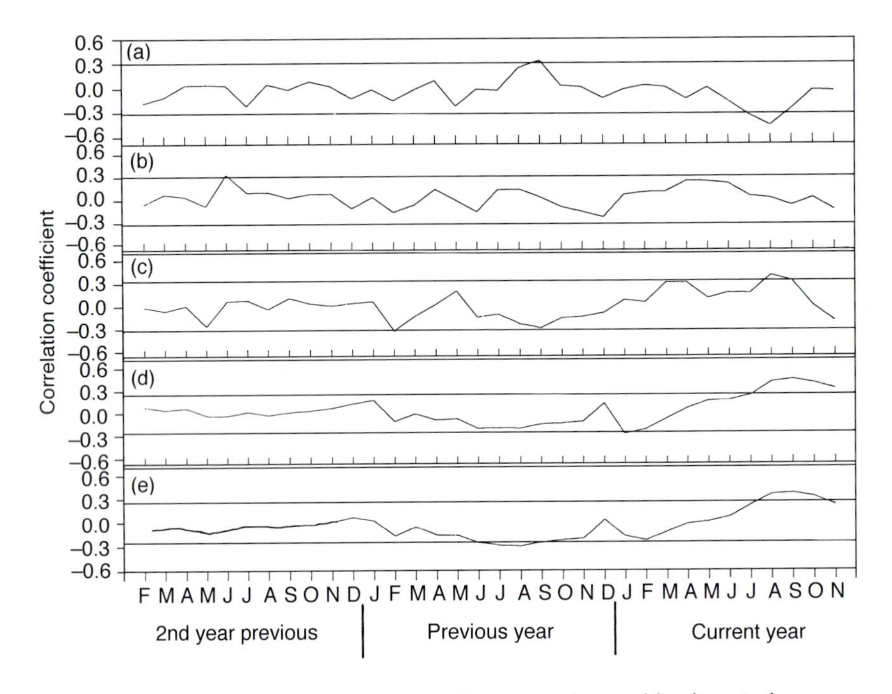

Fig. 18.6. Latewood width correlation coefficients with monthly climatic factors. Reference lines represent a significant relationship at $P = 0.10$. a – max. temp., $N = 66$; b – min. temp., $N = 66$; c – precipitation, $N = 58$; d – Alabama PHDI, $N = 101$; e – Florida PHDI, $N = 101$.

current year (Fig. 18.6a). Minimum monthly temperature influenced latewood during June of 2 years prior (Fig. 18.6b). Precipitation correlated positively during August and September of the current year (Fig. 18.6c). Other strong months of influence were current March and April (positive) and the previous February (negative). Alabama PHDI was more influential on latewood development (Fig. 18.6d) than Florida PHDI (Fig. 18.6e). Significant correlations for Alabama PHDI are present January (negative) and August through November (positive). Significant relationships for Florida PHDI were detected for previous June, July and August (negative) and current year August, September and October (positive).

Growth of stem latewood appears to be significantly influenced by many climatic factors (Table 18.2). Precipitation for the current year March through October was the strongest positive climate variable. It was also negatively correlated with the previous year precipitation between August and December. Significant positive (previous August through September) and negative (current year from July through October) relationships between maximum temperature and latewood formation were detected. Alabama and Florida PHDI during the current year August and September were positively related to growth.

Interestingly, each state's PHDI suggested different negative influences of climate. Alabama's current winter (previous November, December and current January) was strongly negatively correlated while in Florida the most negative period of influence was for the previous September through December.

Disturbance History

The science of detecting natural disturbances that have endogenous (Lorimer, 1985) or exogenous (Brubaker, 1977; Jacoby, 1997) impacts on growth has been well established. Methods to detect more subtle disturbances are still in infancy (Arbatskaya and Vaganov, 1996; Nowacki and Abrams, 1997). Trying to separate natural declines in ring width from declines influenced by other factors, especially anthropogenic, is a difficult and controversial process (Schweingruber, 1987). The many papers and projects stimulated by the red spruce decline in the northeast United States during the last three decades are a testament to the difficulty.

From the trees cored, we constructed a graph portraying the relationship between the FNA longleaf pine diameter (measured at breast height; 1.37 m) and age (ring count) (Fig. 18.7). Several obvious features are worthy of note.

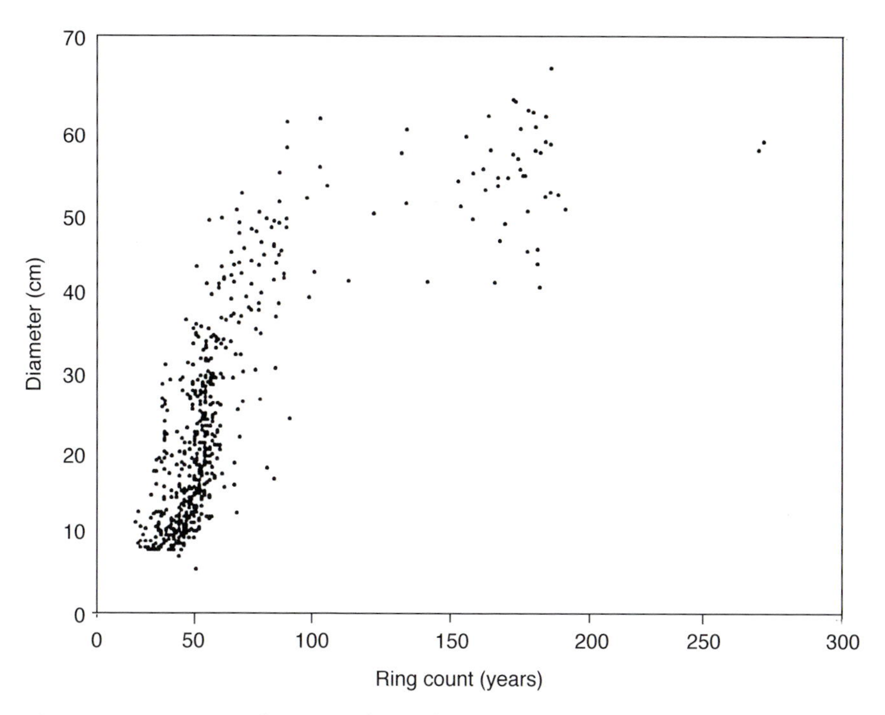

Fig. 18.7. Ring count–diameter relationship of Flomaton Natural Area longleaf pine.

First, the large grouping of trees between 20 and 50 rings is conspicuous. These individuals vary widely in size (<10 cm diameter to nearly 50 cm diameter). This area of the graph illustrates a nearly vertical relationship between age and diameter. These data support the contention that shade-intolerant status of longleaf pine is not as simple as the traditional definition. These individuals are nearly all overtopped and suppressed, having been in this condition all their lives (many >50 years). Furthermore, several individuals have tree-ring patterns showing 75+ years of living in a suppressed state, debunking another long-held myth of longleaf pine (*sensu* Croker, 1990).

In the FNA, periods of growth decline occurred twice in the last 250 years (Fig. 18.8). The first period of decline occurs between 1860 and 1890. The second period of decline starts during the 1930s and extends until the 1980s. The drop in declines during the 1980s is because only the beginning of the decade qualifies for the 15-year requisite to define a growth decline. Growth releases were identified for every decade back to the 1880s.

Beginning in the 1860s a three-decade growth decline was found in the stand. The source of this is not known. The severe declines in growth could have been triggered by below normal growth years in the 1850s and early 1860s (Cook *et al.*, 1997). However, identified suppressions are dramatic declines in growth. Another period of growth decline begins to become a dominant stand feature in the 1940s and continues until the time of this study (Fig. 18.8).

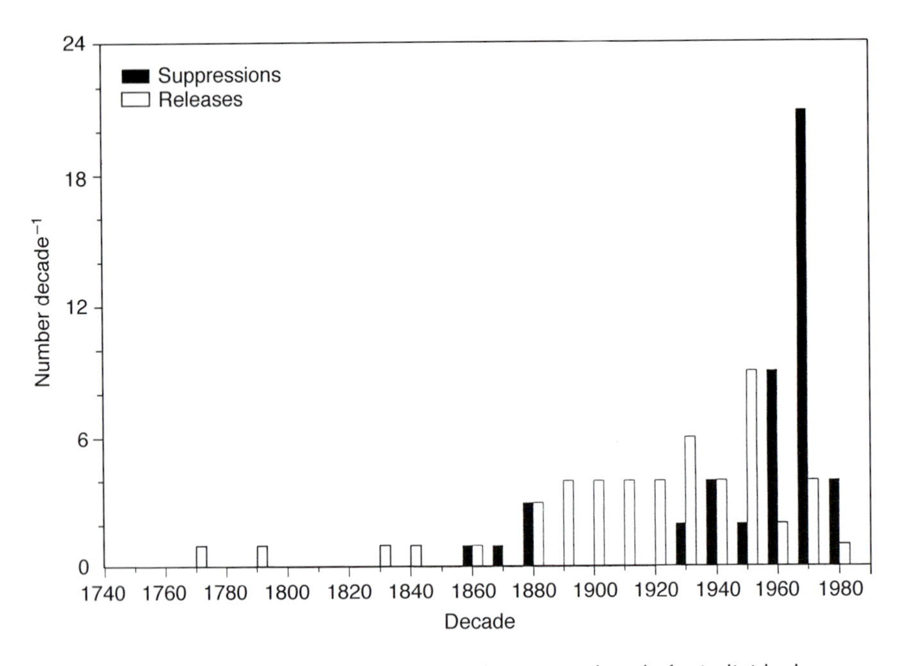

Fig. 18.8. Number of growth declines and releases per decade for individual trees.

Stahle and Cleaveland (1992) noted that the period 1950–1960 had below average growing season rainfall with wet periods between 1937–1949, 1961–1970 and 1972–1982. As stated before, the declines start to dominate in the 1940s and peak in number during the 1970s. The cause of the decreases in growth is not clear.

DISCUSSION

Climate

Our exploratory analysis of factors influencing longleaf pine growth found some trends consistent with previous research and illuminated some novel relationships (Table 18.2). Like Lodewick (1930) and Coile (1936), we found current growing season precipitation (March through October) positively impacted latewood and total ring expansion (Table 18.2; Figs 18.4 and 18.6). High evapotranspiration rates of lower Alabama probably explain the importance of growing season rainfall. Monthly correlations of March and September precipitation were significant for total ring widths (Fig. 18.4). Trends in earlywood development with March precipitation (Fig. 18.5) and latewood growth with August and September precipitation (Fig. 18.6) appear to be the primary factors of the total radial increment–precipitation relationship.

Unexpectedly, for both total and latewood widths, above average precipitation between August and December of the previous year were negatively correlated with growth the following year. One explanation could be an autocorrelation of climate. A second consideration is that an extremely wet late growing season would be associated with more clouds and lower temperatures. This could reduce the amount of photosynthate available for next year's growth.

Unlike the earliest research, significant relationships were found with temperature and earlywood growth. A strong negative correlation was detected with high maximum temperatures for current-year February through April. During that same season, precipitation was positively associated with earlywood formation, though not significantly (Table 18.2). It seems to suggest that an unusually warm late winter–early spring is associated with a lack of clouds and precipitation.

A hot summer and early autumn had negative impact on earlywood formation. This correlation could be related to the costs of high respiration, repair and maintenance of living tissue under extreme growing conditions. Hot summers in the southeast are typically dry, and early needle loss often occurs especially when associated with a dry spring (Heyward, 1934). This would also reduce potential growth by loss of leaf area (reducing photosynthate production) due to premature leaf senescence.

Counterintuitively, a positive correlation was found with high maximum late summer temperatures of the previous year and total ring and latewood development (Table 18.2; Figs 18.4 and 18.6). No good explanation is known for this connection. Again, one link could be an autocorrelation with climate. The strong correlations of late summer temperature and growth in both monthly and seasonal analysis make this unlikely.

A second scenario might involve the taproot of the longleaf pine. A hot late summer appears to make longleaf pine growth slow down. It is possible that this species shuts down growth and induces premature needle loss to reduce its maintenance and repair costs. In this mode the tree could alter its strategy from growth to one of root storage to survive and flourish again another year when conditions are more conducive.

Alabama PHDI was found to be more influential than Florida PHDI on all aspects of radial growth in the FNA. Both had positive associations with current years August through September (Table 18.2), with Alabama PHDI being a little stronger. However, that was where the similarities ended.

Alabama had more statistically significant months during the late current growing season associated with latewood formation than Florida (Fig. 18.6). Furthermore, the strongest negative seasonal effects of Alabama and Florida PHDI did not correspond to tree-ring widths (Table 18.2). Monthly analysis showed similar results (Fig. 18.6). When analysed with total ring widths, the discrepancy became more pronounced. Positive seasonal associations only overlapped for the month of September (Table 18.2). All of these data suggest that both the Gulf of Mexico climate and the inland climate of lower Alabama influence the FNA, but during different seasons.

Relationships were also found going back 2 years before current year's growth (Table 18.2; Figs 18.4, 18.5 and 18.6) in each chronology type. The connections between this lag period are not well understood. As seen above, the previous year's climate can have significant effect on current year's growth. Longleaf pine maintains 2 year's worth of needles, and it seems likely that climatic conditions 2 years previous can still play a role in current tree growth. Jacoby and D'Arrigo (1995) stated that evergreen conifers have a photosynthetic season that extends beyond the radial growth season. Since they retain their needles longer than one year, several years of climate may be integrated in one season's growth ring. Tree memory could have a considerable role in the growth of longleaf pine.

Overall, latewood formation seems to be most affected by climate and is influential on longleaf pine radial growth. Eight out of the ten climate variables tested for seasonal growth played significant roles in growth. Only one of the ten variables was significant for earlywood formation, while half were significant in total ring growth (Table 18.2). It was also found that the strongest climate variables, particularly PHDI values (Fig. 18.5), tended to have a negative impact on earlywood formation. One conclusion that can be drawn from this analysis is that a certain amount of earlywood is 'scheduled' to be produced and extreme climate conditions counteract its development. Taken another way, regardless of normal climatic fluctuations, a nearly regular amount of earlywood is produced annually while latewood production is driven almost solely by an interaction of many climatic factors. Given some unique characteristics of longleaf pine, it seems that a complex relationship exists between climate and growth.

Disturbance History

In the FNA, it is known that an active management programme including prescribed burning ceased during the 1940s. In the half century since fire cessation, we speculate that below-ground competition could have induced the declines. The peak in growth declines for the FNA occurs in the 1970s (Fig. 18.8). The 30-year lag allowed more fire-intolerant tree species like black cherry (*Prunus serotina*), water oak (*Quercus nigra*), and southern magnolia (*Magnolia grandiflora*) to invade and develop extensive root systems (Meldahl *et al.*, 1995). This creates intense below-ground competition that could potentially reduce tree growth. Zahner (1958) reported that a hardwood understorey strongly depleted soil moisture during summer months in mixed pine stands in the mid-South, USA.

Further study is needed to address the issue of the growth declines. Evidence exists that the growth decline in the FNA might be driven more by competition than by climate. First, the large declines are identified in trees with crowns in the dominant and codominant positions. Growth reductions caused by overhead suppression are not possible in this subset of trees. Second, except for the drought conditions of the 1950s, climate fluctuated between normal and wet conditions from 1940 to the 1980s, suggesting that climatic conditions were normal to very good during the period of decline. Third, growth releases occur in every decade for the past 100 years. Growth releases suggest an opening in the canopy (Lorimer, 1985) and an improvement in growth conditions. The numerous growth releases between 1940 and 1980 also suggest overstorey density was normal or decreasing during the period of suppression. Finally, the only change in FNA was the cessation of active fire management that allowed luxuriant woody understorey growth. We propose that decreasing soil resources played a significant role in overstorey growth decreases. The droughts of the 1950s might have intensified the declines, but during better growing conditions of the early 1960s and the especially wet 1970s tree growth failed to recover. More work is needed to clarify causes of growth decline in longleaf pine of the FNA. One answer to the declines in growth might be seen in the next few years. Removal of the understorey trees and the re-introduction of fire might trigger a growth response in the canopy trees.

The new observation of reduced tree growth following fire suppression combined with prior observations of reduced herbaceous species diversity and biomass, higher fuel loads, increased vertical fuel continuity, increased canopy closure, and increased destructive impacts of wildfire all present at FNA agree with observations in western USA ponderosa pine (*Pinus ponderosa*) forests that have undergone fire suppression (Covington and Moore, 1994). The applicability of data and understanding of ponderosa pine dynamics in response to fire suppression (declines and releases) and restoration activities (removal of under- and midstorey invaders and re-introduction of fire) may shed light to less well-understood tree-ring relationships found at FNA.

CONCLUSIONS

Preliminary work in this study and historical research show longleaf pine to be valuable for tree-ring analysis in the southeastern United States. As seen in this study, climate does play an important role in the growth of longleaf pine. As noted in previous work, the production of latewood is most influenced by different climate factors and drives variation in total ring width. Earlywood formation shows a good amount of indifference to climate, but significant relationships have been established with temperature, precipitation and latewood growth. Future analysis of longleaf pine latewood should aid in climate reconstruction of the southeastern United States.

It has also been shown that climate, though important, may not be the only influence on longleaf pine growth. Despite development in open-canopied conditions, competition, especially for below-ground resources, may be a significant component of growth as well. Given longleaf pine's early developmental emphasis on its root system, it seems plausible that soil resources determine a significant amount of growth.

Efforts will continue to examine climatic factors influencing growth, and stand disturbance history as more cores are collected from the FNA. In addition, there are plans to study the effects of fire (the historical burning pre-1940, 1940–1995 cessation, and post-1995 re-introduction) on the growth of old-growth longleaf pine.

SUMMARY

The utilization of longleaf pine (*Pinus palustris* Mill.) as a candidate for dendroecological research is evaluated using preliminary data from an old-growth stand and a synthesis of existing literature. Longleaf pine's range extends throughout the southeastern USA, from eastern Texas to southeastern Virginia, and once covered an estimated 37 million ha. Longleaf pine's wide dominance was largely due to its dependence on relatively high frequency (every 1–12 years), low-intensity surface fires. By 1995, longleaf pine's area had declined to 1.2 million ha, or approximately 3% of its former acreage. Of this total, only 3900 ha (0.14%) remains in an old-growth condition. In this study we use data from the Flomaton Natural Area, a 27-ha remnant virgin stand of longleaf pine located in lower Alabama, USA. The stand has undergone >40 years of fire exclusion. The exclusion of fire enabled less fire-resistant species to enter the stand without a loss of the old longleaf pine trees. Our initial efforts examine (i) climatic factors influencing growth, and (ii) stand disturbance history. Longleaf pine's characteristics of old age (>400 years), wide geographical range, open mono-specific canopies, and disease, drought and insect resistance make it an excellent candidate for ecological research involving past climatic and anthropogenic disturbance histories of southeastern USA forests.

ACKNOWLEDGEMENTS

The authors wish to thank Lorna Pitt, Richard Sampson, Dennis Shaw, Bill Thompson, George Ward, Wandsleigh Williams and Flomo Yanquoi for their assistance in the field. We thank Champion International Corporation and Foster Dickard for allowing us the opportunity to conduct this study and do restoration work in the Flomaton Natural Area. Efforts of Dee Pederson and Ricardo Villalba in manuscript preparation and review are greatly appreciated. Financial support was provided by the US Forest Service Southern Research Station and the Alabama Department of Transportation. H. Chapman, R. Harper, W. Boyer and L. Neel were an inspiration for these efforts. Alabama Agricultural Experiment Station Journal No. 9–985891.

REFERENCES

Arbatskaya, M.K. and Vaganov, E.A. (1996) Dendrochronological analysis of pine response to periodic effects of ground fires. *Lesovedenie* 0(6), 58–61.

Boyer, W.D. (1979) Mortality among seed trees in longleaf pine shelterwood stands. *Southern Journal of Applied Forestry* 3(4), 165–167.

Boyer, W.D. (1990) *Pinus palustris* Mill. In: Burns, R.M. and Honkala, B.H. (tech. eds) *Silvics of North America*. USDA Handbook 654, Washington, DC, pp. 405–412.

Boyer, W.D. (1997) Longleaf pine can catch up. In: Kush, J.S. (comp.) *Proceedings First Longleaf Alliance Conference: Longleaf Pine – A Regional Perspective of Challenges and Opportunities*. Longleaf Alliance Report No. 1, Mobile, AL, pp. 28–29.

Brubaker, L.B. (1977) Effects of defoliation by Douglas-fir tussock moth on ring-width sequences of Douglas-fir and grand fir. *Tree-Ring Bulletin* 38, 49–60.

Chapman, H.H. (1923) The causes and rate of decadence in stands of virgin longleaf pine. *Lumber Trade Journal* 84(6), 11, 16–17.

Coile, T.S. (1936) The effect of rainfall and temperature on the annual radial growth of pine in the southern United States. *Ecological Monographs* 6(4), 533–562.

Cook, E.R. (1985) A time-series analysis approach to tree-ring standardization. PhD dissertation, University of Arizona, Tucson, 104pp.

Cook, E.R., Meko, D., Stahle, D. and Cleaveland. M. (1997) Unpublished PDSI reconstruction. World Data Center-A for Paleoclimatology, Boulder, CO (http://www.ngdc.noaa.gov/paleo/drought.html).

Covington, W. W. and Moore, M. M. (1994) Southwestern ponderosa forest structure: changes since Euro-American settlement. *Journal of Forestry* 92(1), 39–47.

Croker, T.C. Jr (1990) Longleaf pine – myths and facts. In: Farrar, R.M. Jr (ed.) *Proceedings of the Symposium on the Management of Longleaf Pine.* USDA Forest Service Southern Forest Experiment Station General Technical Report SO-75, Long Beach, MS, pp. 2–10.

Devall, M.S., Grender, J.M. and Koretz, J. (1991) Dendroecological analysis of a longleaf pine *Pinus palustris* forest in Mississippi. *Vegetatio* 93, 1–8.

Fritts, H.C. (1987) Principles and practices of dendroecology. In: Jacoby, G.C. Jr and Hornbeck, J.W. (comp.) *Proceedings of the International Symposium on Ecological Aspects of Tree-Ring Analysis.* CONF-8608144, Tarrytown, NY, pp. 6–17.

Frost, C.C. (1993) Four centuries of changing landscape patterns in the longleaf pine ecosystem. In: Hermann, S.H. (ed.) *Proceedings of the 18th Tall Timbers Fire Ecology Conference. The Longleaf Pine Ecosystem: Ecology, Restoration, and Management.* Tall Timbers, Tallahassee, FL, pp. 17–43.

Hayes, M. (1996) Comparison of drought indices. National Drought Mitigation Center (http://enso.unl.edu/ndmc/enigma/indices.htm#pds).

Heyward, F. (1934) Needle browning in longleaf and slash pines during the late summer. *Naval Stores Review* 44(31), 12.

Holmes, R.L. (1983) Computer-assisted quality control in tree-ring dating and measurement. *Tree-Ring Bulletin* 43, 69–78.

Jacoby, G.C. (1997). Application of tree ring analysis to paleoseismology. *Review of Geophysics* 35(2), 109–124.

Jacoby, G.C. and D'Arrigo, R.D. (1995) Tree ring width and density evidence of climatic and potential forest change in Alaska. *Global Biogeochemical Cycles* 9(2), 227–234.

Kush, J., Pederson, N. and Meldahl, R. (1997) Tree-ring data, Alabama Longleaf Pine. *International Tree-Ring Bank. IGBP PAGES/World Data Center-A for Paleoclimatology Data Contribution Series # 97–005.* NOAA/NGDC Paleoclimatology Program, Boulder, CO.

Lodewick, J.E. (1930) Effect of certain climatic factors on the diameter growth of longleaf pine in western Florida. *Journal of Agricultural Research* 41(5), 349–363.

Lorimer, C.G. (1985) Methodological considerations in the analysis of forest disturbance history. *Canadian Journal of Forest Research* 15, 200–213.

Means, D.B. (1996) The longleaf ecosystem going, going ... In: Davis, M.B. (ed.) *Eastern Old-Growth Forests: Prospects for Rediscovery and Recovery.* Island Press. Washington, DC, pp. 210–219.

Meldahl, R.S., Kush, J.S., Shaw, D.J. and Boyer, W.D. (1994) Restoration and dynamics of a virgin, old-growth longleaf pine stand. In: *Proceedings of the 1993 Society of American Foresters National Convention.* Society of American Foresters, Indianapolis, SAF Publication 94–01, pp. 532–533.

Meldahl, R.S., Kush, J.S., Boyer, W.D. and Oberholster, C. (1995) Composition of a virgin, old-growth longleaf pine stand. In: Edwards, M.B. (comp.) *Proceedings of the Eighth Biennial Southern Silvicultural Research Conference.* US Department of Agriculture, Forest Service, Southern Research Station, General Technical Report SRS-1, Auburn, AL, pp. 569–572.

Nowacki, G.J. and Abrams, M.D. (1997) Radial-growth averaging criteria for reconstructing disturbance histories from presettlement-origin oaks. *Ecological Monographs* 67(2), 225–249.

Outcalt, K.W. and Sheffield, R. (1996) The longleaf pine forest: trends and current conditions. *USDA Forest Service Southern Research Station Resource Bulletin SRS-9.* Asheville, NC, 23pp.

Palik, B.J. and Pederson, N. (1996) Overstory mortality and canopy disturbances in longleaf pine ecosystems. *Canadian Journal of Forest Research* 26, 2035–2047.

Palmer, W.C. (1965) *Meteorological Drought*. US Department of Commerce Weather Bureau, Research Paper No. 45, Washington, DC.

Paul, B.H. and Marts, R.O. (1931) Controlling the proportion of summerwood in longleaf pine. *Journal of Forestry* 29, 784–796.

Peet, R.K. and Allard, D.J. (1993) Longleaf pine dominated vegetation of the southern Atlantic and eastern Gulf Coast region, USA. In: Hermann, S.H. (ed.) *Proceedings of the 18th Tall Timbers Fire Ecology Conference. The Longleaf Pine Ecosystem: Ecology, Restoration, and Management*. Tall Timbers, Tallahassee, FL, pp. 45–81.

Pessin, L.J. (1934) Annual ring formation in *Pinus palustris* seedlings. *American Journal of Botany* 21, 599–603.

Pillow, M.Y. (1931) Compression wood records hurricane. *Journal of Forestry* 29, 575–578.

Platt, W.J., Evans, G.W. and Rathbun, S.L. (1988) The population dynamics of a long-lived conifer (*Pinus palustris*). *The American Naturalist* 131(4), 491–525.

Schumacher, F.X. and Day, B.B. (1939) The influence of precipitation upon the width of annual rings of certain timber trees. *Ecological Monographs* 9, 387–429.

Schweingruber, F.H. (1987) Potentials and limitations of dendrochronology in pollution research. In: Jacoby, G.C. Jr and Hornbeck, J.W. (eds) *Proceedings of the International Symposium on Ecological Aspects of Tree-Ring Analysis*. CONF-8608144. Marymount College, Tarrytown, NY, pp. 344–352.

Stahle, D.W. and Cleaveland, M.K. (1992) Reconstruction and analysis of spring rainfall over the Southeastern U.S. for the past 1,000 years. *Bulletin American Meteorological Society* 73(12), 1947–1961.

Stone, E.L. (1940) Frost rings in longleaf pine. *Science* 92, 478.

Wahlenberg, W.G. (1946) *Longleaf Pine: Its Use, Ecology, Regeneration, Protection, and Management*. Charles Lathrop Pack Forestry Foundation, Washington, DC, 429pp.

Walker, L.C. (1963) Natural areas of the Southeast. *Journal of Forestry* 61(9), 670–673.

Ware, S., Frost, C.C. and Doerr, P.D. (1993) Southern mixed hardwood forest: The former longleaf pine forest. In: Martin, W.H., Boyce, S.G. and Echternacht, A.C. (eds) *Biodiversity of the Southeastern United States: Lowland Terrestrial Communities*. Wiley, New York, pp. 447–493.

West, D.C., Doyle, T.W., Tharp, M.L., Beauchamp, J.J., Platt, W.J. and Downing, D.J. (1993) Recent growth increases in old-growth longleaf pine. *Canadian Journal of Forest Research* 23, 846–853.

Zahner, R. (1958) Hardwood understory depletes soil water in pine stands. *Forest Science* 4(3), 178–184.

Zahner, R. (1989) Tree-ring series related to stand and environmental factors in south Alabama longleaf pine stands. In: Miller, J.H. (comp.) *Proceedings – Fifth Biennial Southern Silvicultural Research Conference*. US Department of Agriculture, Forest Service, Southern Forest Experiment Station General Technical Report SO-74, Memphis, TN, pp. 193–197.

19

Influence of Climatic Factors on the Radial Growth of *Pinus densiflora* from Songni Mountains in Central Korea

Won-Kyu Park, Ram R. Yadav and Dmitri Ovtchinnikov

INTRODUCTION

Annual growth rings of many tree species from climate-stressed sites provide coherent quantitative estimates of environmental variables (Fritts, 1976; Hughes *et al.*, 1982; Cook and Kairiukstis, 1990). Some species reported growing for several centuries in subalpine forests in Korea (Kong and Watts, 1993) provide unique opportunities to develop long tree-ring chronologies for climatic reconstruction in this region. Recent dendrochronological reconnaissance (Telewski *et al.*, 1994) also indicated the possibility of finding multi-century-old trees in some of the high mountain areas in Korea. Though the area is very important to our understanding of East Asia monsoon climate, only a few high-resolution tree-ring records are available for the Korean region (Choi *et al.*, 1994; Park, 1994).

Korean red pine (*Pinus densiflora* Sieb. et Zucc.), also known as Japanese red pine, occurs naturally in Korea, Japan and Manchuria, covering a wide ecological spectrum (Yoshioka, 1958; Lee, 1986). This species occupies nearly 40% of the forests of Korea, which cover about 65% of the total land area of Korea (Korea Forestry Administration, 1994). Though the trees attain considerable age and produce distinct datable growth rings, their dendroclimatic potential has not yet been extensively explored. A few studies have been conducted on this species to investigate the effect of pollution on radial growth in Korea and Japan (Yokobori and Ohta, 1983; Ohta, 1987; Kim, 1994; Park *et al.*, 1995). In these studies, the effect of climatic factors on growth is not well documented. Here we describe the influence of climatic factors on the radial growth of Korean red pine from central Korea (Fig. 19.1). The study aimed to evaluate the dendroclimatic potential of this species.

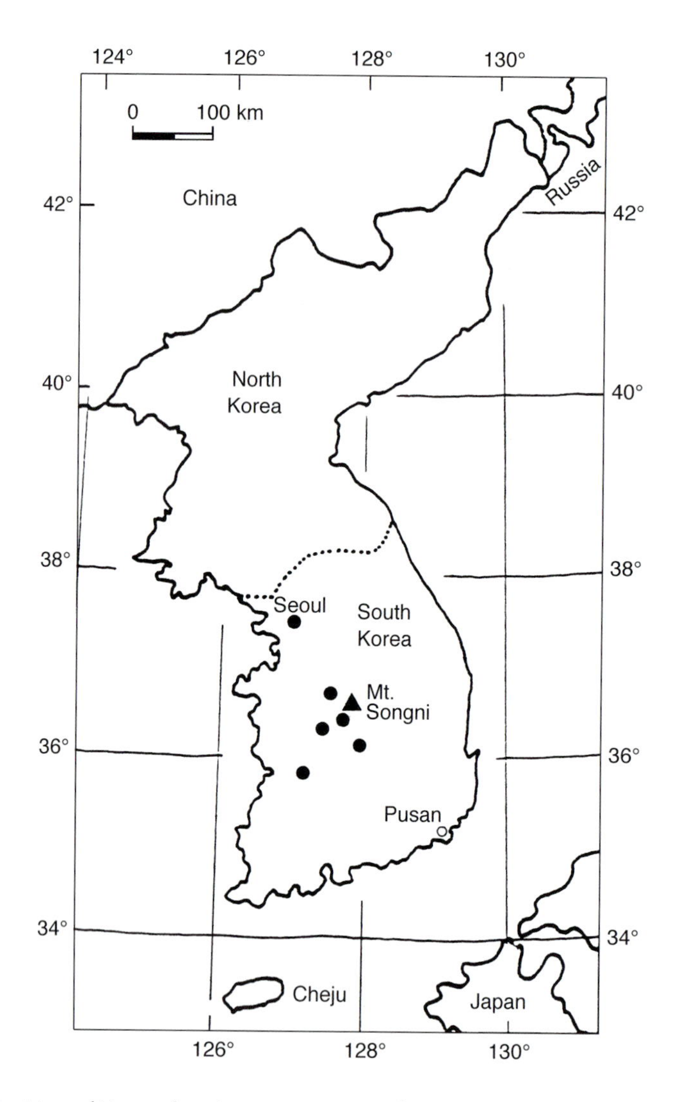

Fig. 19.1. Map of Korea showing tree-ring sampling area (filled triangle) and meteorological stations (filled circles).

STUDY AREA

Five natural open stands of Korean red pine at two sites (one from the Sanggoam site and four from the Bohyunjae site) in the Songni mountain range, cool-temperate zone, at *c.* 850–950 m elevation (near the upper limit of its distribution) were chosen for this study (Fig. 19.1). The two sites are *c.* 4 km apart. Four stands at the Bohyunjae site are separated from each other by 0.5–1 km. In the sample area, *Carpinus laxiflora* Bl. and *Acer* spp. are the main constituents in valleys, while oak species, mainly *Quercus mongolica* Fischer, are occasionally mixed with Korean red pine trees between the valleys and high ridges. Pure stands of Korean red pine are usually found at dry sites along the ridges over rock beds with thin soil cover, indicating that this species is most competitive on xeric sites (Yoshioka, 1958; Lee, 1986). Many Korean red pine trees on the ridges are of stunted stature with prominent crown dieback, both conditions indicative of old age and climatic stress. We selected the trees along the ridges for sampling to maximize the climatic signal.

Climate in general over the Korean Peninsula is under strong polar and tropical influence. During winter continental high-pressure air masses develop over Siberia and bring strong northerly cold dry air. The circulation reverses in summer and southerly winds bring warm moist air masses and monsoon rainfall. The summer monsoon (occurring mainly from July to August) contributes about 45–60% of the total annual precipitation, and winter precipitation about 3–10% (Fig. 19.2). Total annual precipitation in central Korea is 1300 mm. Mean monthly winter (December–February) temperature

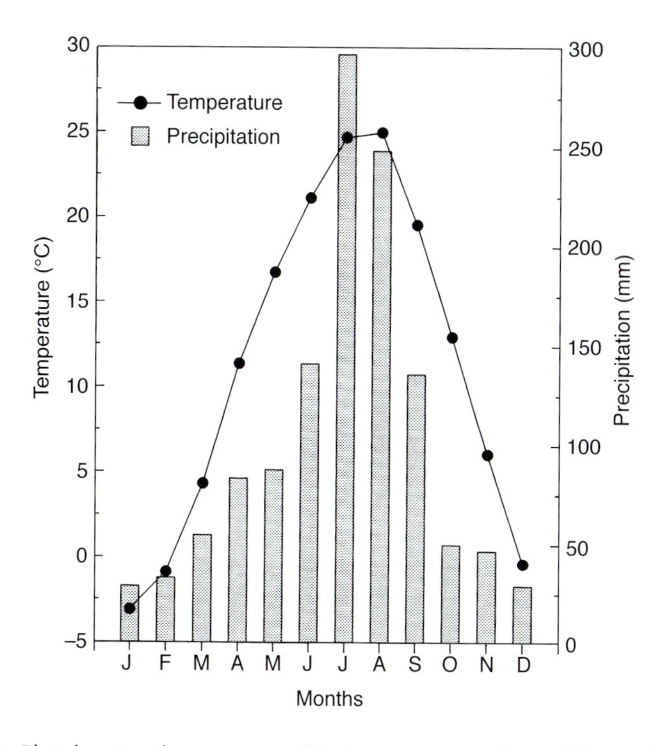

Fig. 19.2. Plot showing the mean monthly temperature and monthly precipitation sums during 1908–1995 in central Korea.

remains below 0°C. January is coldest with a mean temperature of −3°C, while August is hottest with a mean temperature of 25°C.

TREE-RING AND CLIMATE DATA

A total of 88 trees were sampled from five stands during 1996 and 1997. Paired increment cores were taken at breast height using an increment borer, except for a few trees where only one core could be taken due to a steep slope. The tree-ring sequences of the mounted and surfaced core samples were crossdated using the skeleton plot method (Stokes and Smiley, 1968). After fixing the calendar date to each ring, ring widths were measured to the nearest 0.01 mm using a Velmex measuring system. Ring-width plots of each core (log scale) were produced from the ring-width measurements using the TSAP program (Rinn, 1996). These plots were used for visual comparison on a light table to cross-check the dating of sample cores within and between the trees. Another cross-check of dating and measurement accuracy was performed by correlating overlapping 50-year segments of all measured series using the program COFECHA (Holmes, 1983). This program helps identify segments of a core or group of cores where dating or measurement errors might occur. Dating and measurements of a few samples that showed any ambiguity were checked and corrected. Crossdating of Korean red pine samples showed that missing rings are rare; however, intra-annual bands or false rings are quite common especially during the early age of trees.

Ring-width measurements were standardized to ring-width indices using the program ARSTAN (Cook, 1985). Some trees showed non-synchronous patterns of suppression (narrow rings) and release (wide rings) that are probably related to stand dynamics. The detrending methods applied were chosen to remove growth trends related to age and stand dynamics while preserving the maximum common signal. In most of the cases the cubic spline with 50% variance reduction function at 200 years was found to be suitable. However, splines of 120 or 60 years were selected for a few young samples. Each detrended (index) series was prewhitened using an autoregressive model selected on the basis of minimum Akaike criterion, and all residual series were combined using a biweight robust mean (Cook, 1985). The resulting residual chronologies were used for the further analysis.

Basic statistical qualities of each stand chronology, including mean sensitivity and standard deviation, were obtained to evaluate the variability of the chronology. Mean sensitivity is a measure of the relative variance at high frequency from one year to the next. Cross-correlation analysis was conducted to examine the degree to which individual index series agreed with each other; the means of all correlation among different cores (radii), between-tree correlations and within-tree correlations were calculated. The ratio of signal to noise (SNR) and expressed population signal (EPS) were also obtained from the cross-correlation analysis (Wigley *et al.*, 1984; Briffa and Jones, 1990). SNR is an expression of the strength of the observed common signal among trees. EPS is a measure of correlation between the mean chronology derived from the sample of trees and the population from which they are drawn.

Climate records for the meteorological station close to the sampling site are very short, beginning only in the early 1970s. Therefore, we prepared regional temperature and precipitation series (1908–1995) by merging the records from six stations within the homogeneous area (Fig. 19.1). Regional climate data also offer many advantages over single station data as problems associated with record inhomogeneities and differing station microclimates are reduced (Fritts, 1976).

TREE GROWTH–CLIMATE RELATIONSHIP

Since the 1980s, Korean red pine trees in the Songni mountain range have been heavily infested by the defoliating insect, pine leaf gall midges (*Thecodiplosis japonensis* Uchida et Inouye), especially at lower elevations, and this caused severe damage to tree growth (Anon., 1995). It is uncertain whether the trees at higher elevations were infested. Abrupt suppression in ring widths around this period was not noticeable in our samples. However, to avoid any possible noise effect on the tree growth–climate relationship, we excluded the latter part (after 1975) of the chronology for analysis. The chronology length of 68 years (1908–1975) was chosen to assess the influence of climatic variables on tree growth by using simple correlation (Pearson's product-moment). The monthly climate variables over an interval starting from August of the previous growth year and ending in September of the current growth year were used in the correlation analysis.

RESULTS AND DISCUSSION

We developed five tree-ring chronologies (126–265 years) of Korean red pine from central Korea (Fig. 19.3). They include the longest record (AD 1731–1995) developed for this species from Korea. The detailed plots of chronologies (Fig. 19.3b) illustrate that the fluctuation patterns agree well with each other. Particularly, key-year (e.g. 1929, 1936, 1958, 1981) rings are perfectly matched. The average correlation between stand chronologies is 0.844 for the 1900–1995 period.

Chronology statistics and the results of cross-correlation analysis are given in Table 19.1. Mean sensitivities are 0.17–0.26. These data indicate moderate inter-annual variation of the tree-ring series. The correlation between trees (0.31–0.45) and the signal-to-noise ratio (6.29–11.44), which provide measures of the strength of common signal in the samples, are the highest among the ring-width chronologies of any other species so far developed from Korea (Choi *et al.*, 1994; Park, 1994). The EPSs (0.86–0.92) of all chronologies

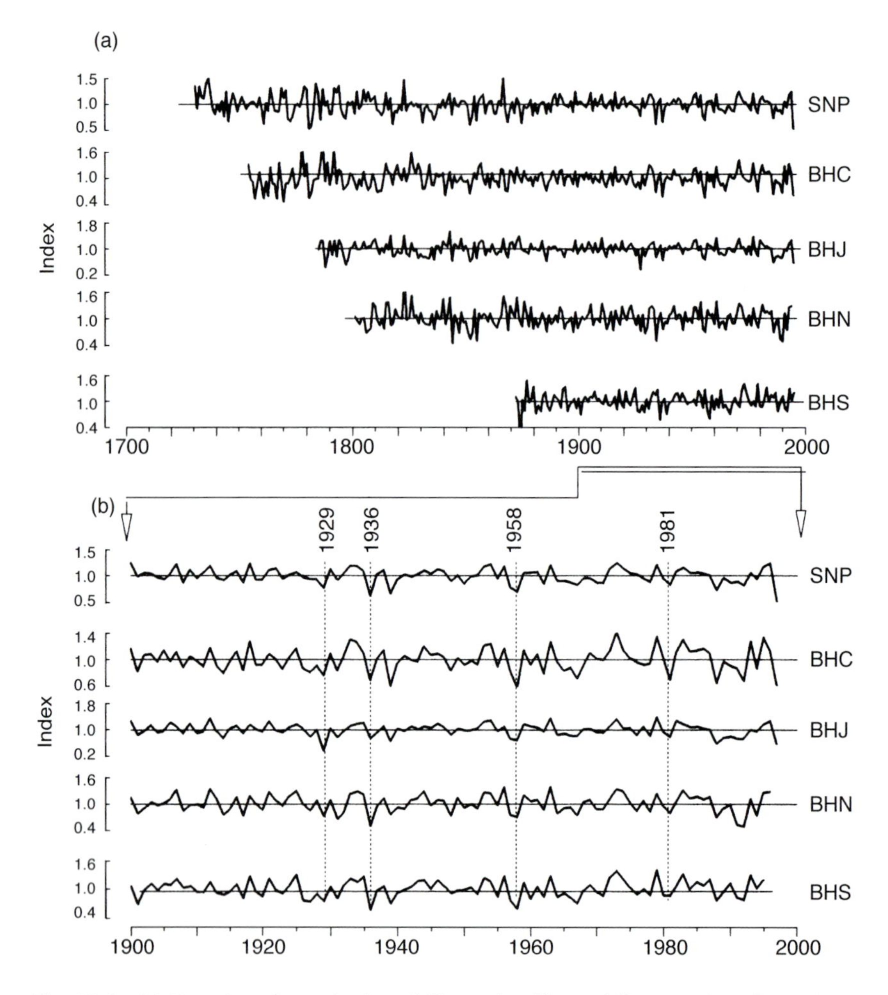

Fig. 19.3. (a) Tree-ring chronologies of *Pinus densiflora* of five stands at Songni Mountains; (b) tree-ring patterns of the chronologies for the 20th century (see Table 19.1 for the chronology abbreviations).

are higher than the EPS limit (0.85) of acceptable statistical quality suggested by Wigley *et al.* (1984). This implies that the chronologies developed in this study possess strong common signals.

Correlations between the tree-ring chronology and monthly climate variables are given in Fig. 19.4. The most significant correlations are positive relationships with May–June precipitation (for the mean chronology; May, $r = 0.45$, $P<0.001$; June, $r = 0.30$, $P<0.008$) and negative relationships with June temperature (for the mean chronology; $r = -0.38$, $P<0.001$). These relationships were consistent for all chronologies.

Table 19.1. Chronology statistics of *Pinus densiflora* (prewhitened series).

Site name:	Sanggoam	Bohyunjae			
Stand location:	North	Low	Centre	North	South
Elevation (m):	900	850	900	950	900
Aspect:	South	South	South	Northeast	Southwest
Chronology ID:	SNP	BHJ	BHC	BHN	BHS
Number of trees (cores)	27 (47)	16 (28)	18 (34)	16 (31)	11 (26)
Period (years)	1731–1995 (265)	1871–1996 (126)	1800–1997 (198)	1754–1997 (244)	1785–1997 (213)
Mean sensitivity	0.17	0.19	0.21	0.26	0.26
Cross-correlation					
Period (years)	1879–1995 (117)	1949–1996 (51)	1914–1997 (84)	1875–1991 (117)	1912–1997 (86)
Among all radii	0.32	0.38	0.40	0.41	0.47
Between trees	0.31	0.37	0.39	0.40	0.45
Within trees	0.65	0.57	0.64	0.64	0.71
SNR	6.29	9.40	11.44	9.36	8.98
EPS	0.86	0.90	0.92	0.90	0.90

SNR, signal-to-noise ratio; EPS, expressed population signal.

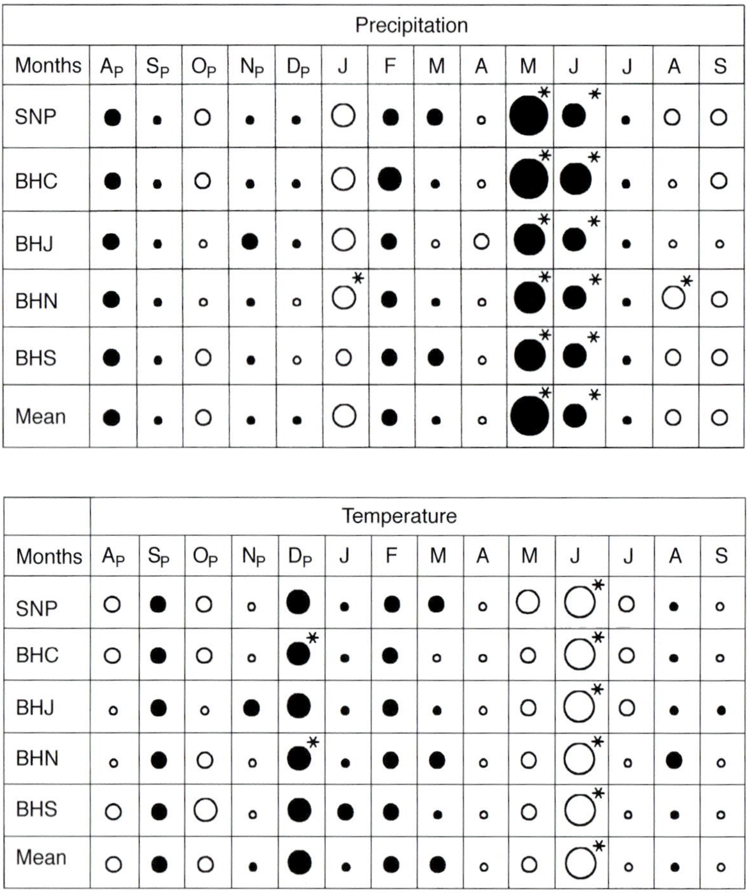

	Precipitation													
Months	A_P	S_P	O_P	N_P	D_P	J	F	M	A	M	J	J	A	S
SNP	●	●	○	●	●	○	●	●	○	●*	●*	●	○	○
BHC	●	●	○	●	●	○	●	●	○	●*	●*	●	○	○
BHJ	●	●	○	●	●	○	●	○	○	●*	●*	●	○	○
BHN	●	●	○	●	○	○*	●	●	○	●*	●*	●	○*	○
BHS	●	●	○	●	○	○	●	●	○	●*	●*	●	○	○
Mean	●	●	○	●	●	○	●	●	○	●*	●*	●	○	○

	Temperature													
Months	A_P	S_P	O_P	N_P	D_P	J	F	M	A	M	J	J	A	S
SNP	○	●	○	○	●	●	●	●	○	○	○*	○	●	○
BHC	○	●	○	○	●*	●	●	○	○	○	○*	○	●	○
BHJ	○	●	○	●	●	●	●	●	○	○	○*	○	●	●
BHN	○	●	○	○	●*	●	●	●	○	○	○*	○	●	○
BHS	○	●	○	○	●	●	●	●	○	○	○*	○	●	○
Mean	○	●	○	●	●	●	●	●	○	○	○*	○	●	○

Fig. 19.4. Correlation coefficients between monthly climate data (mean temperature and precipitation sums) and tree growth (tree-ring indices) for the years 1908 to 1975. Correlations were calculated for 14 months starting in August of the previous year to September of the current year. Filled circles indicate positive correlation coefficients and open circles negative ones. The size of circle represents the magnitude of the coefficients, i.e. 0.0–0.1, 0.1–0.2, 0.2–0.3, 0.3–0.4 and 0.4–0.5, respectively. The coefficients that are significant at the 0.05 level are indicated with an asterisk.

The correlation relationship suggests that Korean red pine trees growing on rocky slopes with thin soil cover in our study area are sensitive to the spring and early summer water regime. High temperature coupled with low precipitation during the early growing season can lead to accentuated conditions of moisture stress, which is detrimental to tree growth. Increased water demand in the early growing season for this species, growing at dry sites in central

Korea, also is indicated by very low leaf water potential during this season (Lee and Kim, 1987). Adequate moisture supply during spring and early summer, therefore, appears to be essential for optimal growth of Korean red pine trees in dry sites. Seasonal soil moisture observations and dendrometer measurements of Korean red pine growth in similar dry sites at Worak Mountain in central Korea also confirms that adequate moisture supply during the early growing season is very critical for the radial growth (W.-K. Park, 1998, unpublished results).

A simple comparison of ring-width indices and May precipitation indicated that droughts in 1917, 1929, 1936, 1958, 1962 and 1967 were associated with very low ring-width indices. Only the low index value for 1939 could not be explained by May precipitation. Correlations with precipitation and temperature were not significant for the monsoon season (July–August). Rather, the summer monsoon has only a weak correlation with tree growth indicating that, with the arrival of monsoon during the summer season, moisture supply no longer remains limiting for the growth of Korean red pine.

CONCLUSIONS

Five tree-ring chronologies of Korean red pine, which were developed in this study, provide the first extensive study of accurately dated chronologies for this species in central Korea. They are highly correlated with each other and produce strong common signals. Comparison of the tree-ring chronology with climate data using correlations indicated that red pine growth is favoured by wet and cool conditions in May and June. This suggests that the soil moisture regime prior to the onset of monsoon rainfall influences most strongly radial growth of Korean red pine in central Korea. The existence of good crossdating and a strong radial growth–climate relationship reinforce the potential for dendroclimatic studies.

The study has shown that the Korean red pine chronologies developed from carefully selected sites could yield valuable proxy data of climatic variables. A preliminary study on the reconstruction for May precipitation (Park and Yadav, 1997) has been successfully conducted using one chronology (SNP) which was initially developed from the present study. We hope that much longer chronologies could be achieved in central Korea with additional intensive sample collecting. Such long chronologies will be useful in reconstructing the climatic variables for this region. A network of climate-sensitive chronologies is possible for this species due to its wide geographical distribution in Korea. Such a network will provide valuable data for global climate change programmes, particularly the Pole–Equator–Pole II (PEP-II) transact of IGBP-PAGES. The long records derived from tree rings can improve our understanding of the natural variability of climate (e.g. Little Ice Age) and anthropogenic impacts on climate change (e.g. greenhouse warming).

SUMMARY

Influence of climate on the radial growth of Korean red pine (*Pinus densiflora* Sieb. et Zucc.) from the Songni Mountains in central Korea was investigated. We prepared five stand chronologies (126–265 years long) using ring widths of 88 trees collected at 850–950 m a.s.l. The rings were well crossdated and the agreements between trees were high. Correlations between the tree-ring index chronologies and monthly temperature and precipitation revealed that Korean red pine growth in this region is favoured by cool and wet conditions in May and June. This relationship was consistent at all stands. The strongest positive correlations were noted with May–June total precipitation. Precipitation during the monsoon season (July and August) is not significantly correlated with tree growth. This indicates that the soil moisture regime prior to the onset of monsoon rainfall may influence most strongly radial growth of Korean red pine in central Korea. The strong association between tree growth and climate demonstrates the potential usefulness of this species in dendroclimatic studies.

ACKNOWLEDGEMENTS

This study was supported by a grant from Korea Science and Engineering Foundation to WKP (KOSEF 951–0608–009–2). We thank Jong-Il Lee, Jae-Myoung Lee, Yong-Ryun Kwon and Min-Hyun Jo for help with fieldwork, Je-Su Kim for assistance in climate data processing, Cheol-Ha Park for providing data on insect damage survey, Jeong-Wook Seo for technical assistance, and Osamu Kobayashi for providing Japanese literature on *Pinus densiflora*. Critical review of the early version of this manuscript by Henri D. Grissino-Mayer, Sang-Joon Kang and Byeong-Jin Cha is gratefully acknowledged.

REFERENCES

Anon. (1995) *Annual Report of the Chungbuk Forest Research Institute*. Cheongju, Korea (in Korean).

Briffa, K.R. and Jones, P.D. (1990) Basic chronology statistics and assessment. In: Cook, E.R. and Kairiukstis, L.A. (eds) *Methods of Dendrochronology: Applications in the Environmental Sciences*. Kluwer Academic Publishers, Dordrecht, pp. 137–152.

Choi, J.N., Park, W.-K. and Yu, K.B. (1994) Central Korea temperature changes reconstructed from tree rings of subalpine conifers: A.D. 1635 to 1990. *Dendrochronologia* 12, 33–43.

Cook, E.R. (1985) A time-series analysis approach to tree-ring standardization. PhD dissertation, University of Arizona, Tucson.

Cook, E.R. and Kairiukstis, L.A. (eds) (1990) *Methods of Dendrochronology: Applications in the Environmental Sciences*. Kluwer Academic Publishers, Dordrecht.

Fritts, H.C. (1976) *Tree Rings and Climate*. Academic Press, New York.

Holmes, R.L. (1983) A computer-assisted quality control in tree-ring dating and measurement. *Tree-Ring Bulletin* 43, 69–78.
Hughes, M.K., Kelly, P.M., Pilcher, J.R. and LaMarche, V.C. (eds) (1982) *Climate from Tree Rings*. Cambridge University Press, Cambridge.
Kim, E.S. (1994) Ecological examinations of the radial growth of pine trees (*Pinus densiflora* S. et Z.) on Mt. Namsan and the potential effects of current level of air pollutants to the growth of the trees in central Seoul, Korea. *Journal of Korea Air Pollution Research Association* 10, 371–386.
Kong, W.S. and Watts, D. (1993) *The Plant Geography of Korea with an Analysis on the Alpine Zones*. Kluwer Academic Publishers, Dordrecht.
Korea Forestry Administration (1994) *Forestry Statistics*. Forestry Administration, Seoul.
Lee, C.S. and Kim, J.H. (1987) Relationships between soil factors and growth of annual rings in *Pinus densiflora* on stony mountain. *Korean Journal of Ecology* 10, 151–159.
Lee, Y.N. (1986) *Korean Coniferae*. Ewha Womans University Press, Seoul (in Korean).
Ohta, S. (1987) The observation of tree-ring structure by soft X-ray densitometry. *Mokuzai Gakkaishi* 24, 429–434 (in Japanese).
Park, W.-K. (1994) Tree-ring networks for detecting global warming in Korea. *Journal of Korean Forestry Energy* 14, 80–87.
Park, W.-K. and Yadav, R.R. (1997) Precipitation reconstruction for late Chosun period using tree rings of Korean red pine. In: Park, W.-K. (ed.) *Proceedings of the East Asia Workshop on Tree-Ring Analysis*. Agricultural Science and Technology Institute of Chungbuk National University, Cheongju, Korea, pp. 37–46.
Park, W.-K., Chong, S.H., Lee, H.Y. and Park, Y.G. (1995) Dendrochronological assessment of forest decline in Korea. In: Ohta, S., Fujii, T., Okada, N., Hughes, M.K. and Eckstein, D. (eds) *Proceedings of the International Workshop on Asian and Pacific Dendrochronology*. Japan Forestry and Forest Products Research Institute, Tsukuba, pp. 102–107.
Rinn, F. (1996) *TSAP Version 2.4 – Reference Manual*. Heidelberg, Germany.
Stokes, M.A. and Smiley, T.L. (1968) *An Introduction to Tree-Ring Dating*. University of Chicago Press, Chicago.
Telewski, F.W., Park, W.-K. and Fuziwara, T. (1994) Dendrochronological reconnaissance of Songni and Sorak mountains, Korea. *Journal of Korean Forestry Energy* 14, 88–90.
Wigley, T.M.L., Briffa, K.R. and Jones, P.D. (1984) On the average value of correlated time series with applications in dendroclimatology and hydrology. *Journal of Climate and Applied Meteorology* 23, 204–213.
Yokobori, M. and Ohta, S. (1983) Combined air pollution and pine ring structure observed xylochronologically. *European Journal of Forest Pathology* 13, 30–55.
Yoshioka, K. (1958) *Ecological Studies on Pine Forests in Japan*. Norin Shupan Co., Tokyo (in Japanese).

Index